Springer Series in Measurement Science and Technology

The Springer Series in Measurement Science and Technology comprehensively covers the science and technology of measurement, addressing all aspects of the subject from the fundamental physical principles through to the state-of-the-art in applied and industrial metrology. Volumes published in the series cover theoretical developments, experimental techniques and measurement best practice, devices and technology, data analysis, uncertainty, and standards, with application to physics, chemistry, materials science, engineering and the life sciences.

The series includes textbooks for advanced students and research monographs for established researchers needing to stay up to date with the latest developments in the field.

More information about this series at http://www.springer.com/series/13337

Songling Huang · Shen Wang
Weibin Li · Qing Wang

Electromagnetic Ultrasonic Guided Waves

清華大学出版社
TSINGHUA UNIVERSITY PRESS

Springer

Songling Huang
Department of Electrical Engineering
Tsinghua University
Beijing
China

Shen Wang
Department of Electrical Engineering
Tsinghua University
Beijing
China

Weibin Li
Xiamen University
Xiamen
China

Qing Wang
Durham University
Durham
UK

ISSN 2198-7807 ISSN 2198-7815 (electronic)
Springer Series in Measurement Science and Technology
ISBN 978-981-10-0562-6 ISBN 978-981-10-0564-0 (eBook)
DOI 10.1007/978-981-10-0564-0

Jointly published with Tsinghua University Press

Library of Congress Control Number: 2016931301

Printed on acid-free paper

This Springer imprint is published by SpringerNature
The registered company is Springer Science+Business Media Singapore Pte Ltd.

Preface

The use of electromagnetic ultrasonic waves has several distinct advantages, including operation without a coupling fluid, non-contact operation, higher temperature operation and flexibility to generate the shear horizontal waves. Since the 1970s, a number of researchers have carried out electromagnetic ultrasonic testing with increasing concern. The use of ultrasonic guided waves has received a great deal attention for nondestructive testing for the advantages of guided wave techniques such as high testing speed and the detection sensitivity on both inner and outer surfaces. Many industrial problems in special conditions can be solved by a combination of the electromagnetic ultrasonic approach and the guided wave technique, such as the online detection of stress corrosion cracks in natural gas pipelines. However, the electromagnetic ultrasonic technique is involved with the coupling and conversion among electromagnetics and mechanics as well as the vibration and propagation of ultrasonic waves, which can make it more difficult to do quantitative research. The multi-modes nature and dispersive characteristics of ultrasonic guided waves, and the mode conversion of the interactions between guided waves and defects, make it the most complicated of uses of guided waves. Consequently, it can be very difficult to combine the electromagnetic ultrasonics with ultrasonic guided-waves for application. In recent years, tremendous progress has been made in the theory and application of electromagnetic ultrasonic guided waves.

This book illustrates the theory and the practical applications of the electromagnetic ultrasonic guided waves. The main parts of the book include: the energy transfer mechanism of electromagnetic ultrasonics as well as the design methods; the calculations and simulations of the electromagnetic ultrasonics based on Lorentz force and magnetostrictive mechanism; the propagation characteristics of ultrasonic guided waves; the mechanism of detection of defects by guided waves; the quantification and location of defects; and the applications of electromagnetic ultrasonic guided wave techniques.

The contents of this book are the summaries of the authors' latest eight years of investigations and practical application in the field of electromagnetic ultrasonic guided waves. Most of them are involved in the dissertations of the graduate and Ph.D. students supervised by the authors, including Shen Wang, Kuansheng Hao, Yongsheng Zhang, Chaofeng Ye, Peng Li, Junjun Xin, Yun Tong et al. In the practical applications of related technology, significant support was provided by colleagues and engineers from related companies and institutes of Petrochina Co. and Sinopec. We express our sincere gratitude to them for helping to improve the technology in real practice.

Chapters 1 and 6 were written by Weibin Li, Chaps. 2 and 3 were written by Qing Wang, Chaps. 4 and 5 were written by Shen Wang, and Chap. 7 was written by Songling Huang.

Considering the rapid growth of online detection and the great advantages of the electromagnetic guided wave technique, we hope that this book will be used as a reference in electromagnetic guided wave nondestructive evaluation and testing by individuals at any level and by graduate students. It is also hoped that this book will expand and promote the use of electromagnetic guided waves at both the national and international levels.

Beijing, China — Songling Huang
Beijing, China — Shen Wang
Xiamen, China — Weibin Li
Durham, UK — Qing Wang

Contents

Chapter 1
Electromagnetic Acoustic Transducer

1.1 Outline

As an important branch of the nondestructive testing field, ultrasonic testing is widely used in the industries of steel, electricity, petroleum, transportation, and medical and others. In the process of ultrasonic testing, the ultrasonic transducer is the core component of the excitation and reception of ultrasonic waves, mainly including the piezoelectric ultrasonic transducer and the electromagnetic acoustic transducer (EMAT). Compared with the piezoelectric ultrasonic transducer, EMAT has many advantages, such as being non-contact, no need for the coupling medium, and easy to produce the shear horizontal wave. In particular, it can be applied under a high-temperature environment, or when an isolation layer exists, and other special applications. It possesses a crucial application value and widespread application prospects.

The limitation of EMAT is the poor efficiency compared with the piezoelectric transducer, which then leads to a low signal–noise ratio of the ultrasonic signal generated from the EMAT. The most important and difficult problems of the research field of EMAT are as follows: improving the energy conversion efficiency and signal–noise ratio, generating and receiving ultrasonic waves with more pure modes, and reducing the interference of other ultrasonic signals with different modes. Early studies of the electromagnetic ultrasonic transducer were based mostly on experimental ways, which is neither time- or cost-efficient. Later, by exploring the establishment of EMAT mode, people tried to carry out further research on the physical mechanism of EMAT by theoretical analysis and numerical simulation. Up to now, the mathematical models of EMAT based on the *Lorentz force* mechanism and the magnetostrictive mechanism have been established. Some related numerical methods and analytical methods have already been put forward.

However, because the theoretical analysis of EMAT is related to the static magnetic field, pulsed electromagnetic field, ultrasonic field, multifield coupling

S. Huang et al., *Electromagnetic Ultrasonic Guided Waves*, Springer Series
in Measurement Science and Technology, DOI 10.1007/978-981-10-0564-0_1

problem, and so on, as well as material characteristics of the tested sample, its numerical simulation can be very complex, especially for the problem of EMAT based on the magnetostrictive mechanism. When it is used to test the ferromagnetic specimen, its complete energy conversion process is related to the magnet's mechanical properties, magnetostrictive properties, inverse magnetostrictive properties, and so on, and it can be more difficult to do numerical simulation analysis. In order to have a deeper understanding of the energy conversion mechanism of EMAT, grasp the physical properties of EMAT comprehensively, and then make an optimal design for it, it has great theoretical and practical significance in solving the problems of theory existing in the numerical simulations. And, furthermore, with the application of new materials and the improvement of the EMAT design, the coil has a more complex structure when used in EMAT, and thus, it is very important to introduce some new analyzing methods for EMAT with new structures.

1.2 Research Status of EMAT

An experiment on the generation of acoustic elastic vibration produced by *Lorentz* force was conducted by Randall R.H in 1939. Aksneov and Vkin et al. reported their findings of the disturbed resonance effects generated by *magnetic resonance coils* in the magnetic field in 1955, which was recognized as the earliest report about EMAT. In the late 1970s, the study on EMAT began to develop rapidly, and the theoretical and experimental research of EMAT were carried out successively by the researchers of UK, USA, Russia, Germany, Japan, and so on, which greatly expanded the innovation and application range of the EMAT technology. After 50 years of improvements, EMAT technology has gradually entered the stage of industrial application. Its application field, from the original testing of metallic plate and train wheel testing and the thickness measurement of the metal components with high temperature, gradually developed into the testing of metal weld, steel bar, steel pipeline, railway track, composite material, and many other fields.

1.2.1 Structure of EMAT

EMAT consists of three parts—the magnet that provides a bias magnetic field; the coils that produce a pulsed magnetic field; and the test sample of the material in which the ultrasonic wave is excited and propagated. A variety of types and patterns of ultrasonic waves can be generated by the different combinations of coils and magnet. For example, *Lamb waves*, *Rayleigh waves*, *SH waves*, and body waves can be emitted in the plate-like specimen, and *L*-mode and *T*-mode guided waves with pipe axial direction can be emitted in the pipe-like specimen. EMAT not only can work in pulse-echo mode, in the state of which only one probe is needed, but also can be used in through-transmission mode, in the state of which two probes are needed to play the roles of excitation and reception, respectively. Generally, the structures of the transmitting probe and the receiving probe are similar to each other.

1.2.1.1 The EMAT Applied to Plate and Pipe Circumferential Detection [1]

In the plate and pipe circumferential testing, where the dimension of plate specimen or diameter of pipe-shaped specimen is considerably larger than the dimension of EMAT itself, the coils of EMAT are all of a flat type. On the other hand, the magnet that provides the bias magnetic field is usually a permanent magnet. The widely used structures of EMAT include meander coil, spiral coil, and racetrack coil; the structures of the magnet include square type, horseshoe type, and periodic permanent magnet. These are shown in Figs. 1.1 and 1.2.

Different types of ultrasonic waves or modes that have different applications can be generated inside the specimen by the various combinations of coils, magnet, and specimen. With the combination of meander coil and square magnet, *Lamb waves* and *Rayleigh* waves can be generated in the non-ferromagnetic specimen. The specific type of ultrasonic waves is related to the coils' dimension and the frequency of the excitation signal. It is also possible to generate guide waves of *SH* mode in the magnetic sample. The spiral coil and the racetrack coil are mainly used to generate body waves in the ferromagnetic specimen and thickness measurement of the sample under test. They can generate transverse waves when combined with the square magnet, or combined with the horseshoe permanent magnet in the circumstance for the generation of longitudinal waves. With the combination of

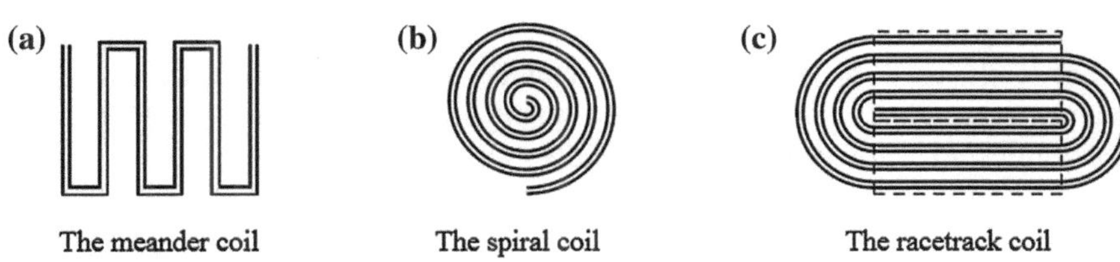

Fig. 1.1 The coils' structure of EMAT

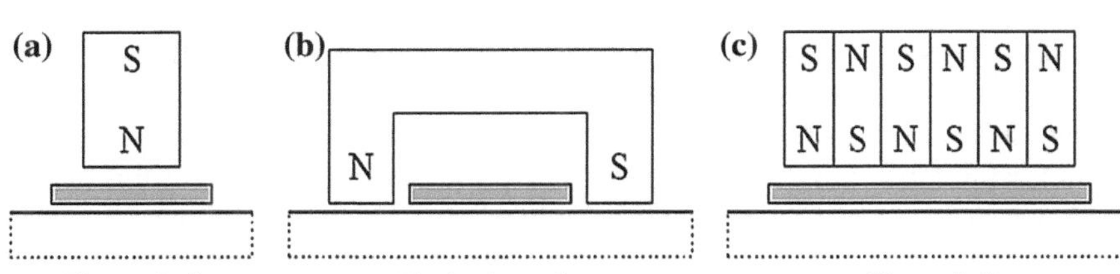

Fig. 1.2 The magnet structure of EMAT

racetrack-shaped coil and periodic permanent magnet, the *SH* waves can be generated and received in the ferromagnetic and non-ferromagnetic materials.

The types, strength, frequency, propagation direction, and some other parameters of the ultrasonic waves can be controlled by changing the strength and orientation of the bias magnetic field, or the strength and frequency of a high-frequency current, as well as the shape and size of the coils of the EMAT with different structures. In the meantime, the bandwidth of the received ultrasonic waves and some other parameters can also be changed in the same way.

For the formation of a strong bias magnetic field in the surface or near the surface, the magnet should be made by using magnetic materials with high field intensity; specifically, the materials can be a permanent magnet or electromagnet. The advantage of using electromagnets is that the strength of the magnetic field can be adjusted easily. There is a certain level of skin effect in the magnetic field when the electromagnets are used in *AC* magnetization, in the circumstance of which the intensity of the bias magnetic field can be enhanced. The advantage of using a permanent magnet is the small size of the magnet, which can make the structure more compact. But it is important to note that although the structure of EMAT can become more compact using the permanent magnet, it can be restricted by the temperature and the magnetic field intensity. In order to generate stronger magnetic field and get the right direction of the magnetic field, the magnet needed by EMAT should be well designed. The analysis of the magnetic field and structure optimization of ferromagnetic material or non-ferromagnetic material with different structures is made by Maxfield et al. The purpose is to reduce the magnetic leakage and magnetic loss and then get the largest effective magnetic field; thus, the energy conversion efficiency of EMAT is improved. The analysis and calculation of the magnetic field distribution generated by the horseshoe permanent magnet and its influence on energy conversion efficiency were done by Mirkhani K using *ANSYS* software. He pointed out that when the width of the permanent magnet is equal to the coils' width, the efficiency of energy conversion will increase by 10 % compared with that when the width of the permanent magnet is 20 % bigger than the coils' width.

The coils used by EMAT mainly include spiral type, racetrack type, and meander type. The manufacturing mechanisms of the EMAT coils include the flat cable, thin film, wire, printed circuit, and so on.

1.2.1.2 The EMAT Applied to the Testing Along the Axial Direction of Pipe [2]

An EMAT based on a magnetostrictive mechanism generating waves that propagate along the axial direction of the pipe was first proposed by Doctor Kwun from the Southwest Research Institute, USA. It was applied to the crack detection along the circumferential direction of the pipeline. The principle and structure of this EMAT are shown in Fig. 1.3. The working principle of the EMAT is as follows: A certain width of coil is wrapped around the pipeline; the bias magnetic field along the axial direction of the pipeline is applied on both sides of the coil using a permanent

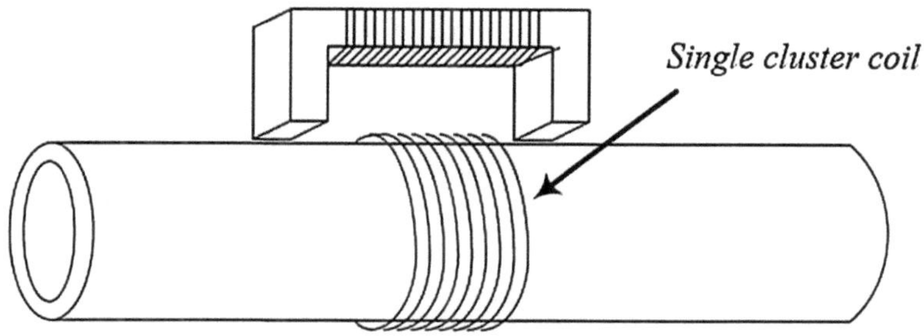

Fig. 1.3 The EMAT with *L*-mode guided waves of axial direction

magnet or solenoid electromagnet; when the excitation of periodic pulses is applied on the coils, ultrasonic waves of *L* mode can be generated along the axial direction of the pipeline. Because of the symmetrical structure of the pipeline, the ultrasonic waves can be generated continuously in it, and the diffusion components along the circumferential direction can be offset against each other, which reduce energy diffusion and attenuation of guided waves propagating along the axial direction of the pipe. Ultrasonic guided waves can propagate longer, and experimental investigation shows that this type of ultrasonic waves can travel hundreds of meters along the pipeline. In addition, this EMAT coil structure can be used in the transmission and reception of the ultrasonic waves with a broader frequency range, but the specific relationship between coils and frequency is still not clear.

Based on the study of electromagnetic ultrasonic guided waves of *L* mode, an EMAT structure used to produce ultrasonic guided waves of *T* mode was developed by Doctor Kwun. This EMAT includes coils, a nickel strap, and the pipeline under test. In practice, the nickel strap should be magnetized first and the direction of magnetization should be along the circumferential direction of the pipe. Because some residual magnetism may be retained in the nickel strap, it is necessary to provide some kind of bias magnetic field in the circumferential direction of the pipeline. When the periodic pulsed magnetic field is applied to the coils, the superposition of static magnetic field and dynamic magnetic field may cause the periodical deformation of the nickel strap; thus, a guided wave of *T* mode will be generated in the nickel strap. The guided waves are coupled to the pipeline and then generated on the pipeline along its axial direction. Because the speed of a T_0 guided wave mode does not change with the variation of the pipeline's thickness and excitation frequency, there is no frequency dispersion phenomenon, which gives it better applicability (Fig. 1.4).

The advantages of the EMAT proposed by Doctor Kwun are its simple structure and convenience of its use. However, the coils' structure is too simple to achieve the best frequency and the optimum selection of ultrasonic mode. Hence, Professor Kim Y and the researchers in his group tried to give the EMAT a frequency selection characteristic by changing the usage mode of the nickel strap, specifically matching the excitation frequency of EMAT by adjusting the space between many narrow nickel belts. In order to improve the efficiency of energy conversion, they also made

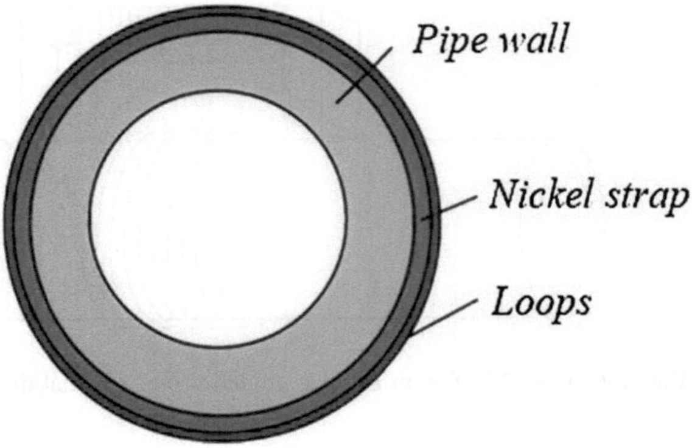

Fig. 1.4 The cross section of EMAT with *T*-mode guided wave of axial direction

the nickel belt group tip at a certain angle, or used the z-structure nickel belt. The experiment results indicated that better results were obtained by the modification. But compared with the EMAT structure proposed by Doctor Kwun, the EMAT structure proposed by Professor Kim Y is inconvenient for manufacture and use.

1.2.2 Energy Conversion Mechanism and Analytical Method of EMAT

According to the different mechanisms of ultrasonic wave generation, EMAT can be divided into two kinds: the EMAT based on the *Lorentz force* mechanism and the EMAT based on the magnetostrictive mechanism. The former is suitable for the testing of non-ferromagnetic specimens and the latter for the testing of ferromagnetic specimens.

The mechanism of energy conversion based on *Lorentz force* is somewhat simpler. Relevant reports of the research achievements of EMAT based on magnetostrictive mechanisms already existed a long time ago. In the 1960s, Voltmer F.W et al. published an article on *Applied Physics Letters*, introducing magnetostrictive mechanism-based equipment for generating ultrasonic waves. Thompson R.B from Rockwell International Science Center, California, USA, carried out further study on this research. The principle structure of EMAT based on a magnetostrictive mechanism and *Lorentz force* was also introduced by him in 1973 and concerned with the generation of body waves. When he proceeded with the analysis of the generation mechanism of the *Lamb waves* in the ferromagnetic material, Thompson R.B discussed the working principle and energy conversion efficiency of EMAT when the skin depth is smaller than the wavelength of the elastic wave. It turned out that magnetostriction plays a leading role in the lower field region, while in the higher field region, due to the saturation of the magnetization, the hindered movement of the magnetic domain, and the weak magnetostrictive effect, *Lorentz force* will play a dominant role. Later on, from 1976 to 1979, the following issues were analyzed by Thompson R.B successively: the dependency problem of the force generated by the ultrasonic surface wave in ferromagnetic metal materials; the generation and reception mechanism of the electromagnetic ultrasonic *Lamb waves* in iron–nickel polycrystalline materials; and the model of the generation efficiency of electromagnetic ultrasonic *Rayleigh* and *Lamb waves* in ferromagnetic metal crystals. The structure of EMAT based on magnetostrictive mechanism was also developed, which can be used to generate *SH* guided waves. When the directions of the static magnetic field and dynamic magnetic field are parallel to the propagation direction of the ultrasonic waves, there are two different kinds of mechanism: the magnetostriction force and the *Lorentz force* of the inductive coil. According to the given magnetic field and material elasticity and magnetic property, the use of the given equations in this book can accurately predict some important parameters, such as the dependency of the ultrasonic wave generation efficiency on the field; the importance of magnetic permeability, magnetostriction coefficient, and electrical conductivity; and the effect of liftoff to the generation of ultrasonic waves. Comparisons between those predicted values and the experiment results were performed. Results showed that the maximum efficiency appeared on the rotating area of the magnetic domain and indicated that the magnetic permeability and the variety of stress magnetostrictive coefficients are very important to the identification of efficiency. This is different to the mechanism of *Lorentz force*, where the magnetostriction coefficient is very sensitive to the electrical resistivity and the value of magnetic permeability because there is an exponential attenuation relationship between the amplitude that is used to generate the signal of the ultrasonic waveform and the liftoff value of EMAT. During the same period, Il'in and Kharitonov used the meander coil to generate *Rayleigh surface waves* and performed some theoretical analysis and experiments for the corresponding phenomenon.

The mathematical model presented by Thompson R.B and I1'in et al. in theoretical analysis of EMAT is the one-dimensional harmonic analysis model. But for the energy conversion problem, it is essentially a coupled problem among the electromagnetic field, ultrasonic field, and force field generated by the interaction between electromagnetic field and material. Apparently, it is difficult to pinpoint such a complex field coupling problem using only the one-dimensional analytical model. The theoretical model of EMAT proposed by Wilbrand first summarized the integral solving method of the differential equation that is used to characterize the characteristics of EMAT with three different structures in three-dimensional space. The magnetization mechanism and magnetostrictive mechanism were discussed in this report. However, the conclusions about generation and propagation of elastic waves summarized by Wilbrand were all based on the assumptions of an infinite half space with isotropic characteristic; the given integral results were also based on the far-field assumption. Consequently, the ideal models show an obvious limitation in the actual design of EMAT. In real industrial demands, a more reasonable EMAT mode and a more effective method are required to solve the mathematical model. From the late 1980s to the early 1990s, the two-dimensional finite element model was established by German scholar, Ludwig R. Based on the electromagnetic field theory, Ludwing R made an analysis of the eddy current distribution generated in the pulsed current sample in EMAT coils and calculated the Lorentz force in the sample and the particle displacement under the action of force. By using the momentum conservation law, Dai X.W, a PhD student of Ludwig R, divided the EMAT into three coupled subsystems to analyze machinery, electromagnetism, and materials. He derived the system control equation of EMAT and the boundary condition based on *Lorentz force*, magnetic force, and magnetostriction force. In addition, he also focused on the analysis and finite element simulation of the receiving mechanism of EMAT on the inspection of non-ferromagnetic materials. The model developed by Dai X.W consists of five parts: the simulation of the static magnetic field, the distribution of the pulsed eddy current, the distribution of *Lorentz force*, the generation of ultrasonic waves under the effect of *Lorentz force,* and the electromagnetic ultrasonic wave detection using the EMAT receiver coil. Based on the mathematical model and calculation of Ludwig R, a two-dimensional calculation method was developed by Kaltenbacher, which was used for the simulation of *Lamb waves* generated by EMAT. The experimental measurements were taken, and the numerical results were compared with the measurement results. A non-holonomic equation is used to calculate the density of the adopted source current by Ludwig R, Sugiura. Maruyama also developed a finite element method for ultrasonic detection. All of their approaches are based on the space electric field in the space between coil conductors to calculate the induced electromotive force received in the coils. However, a fact was ignored in all of the above-mentioned methods using the finite element method to do the calculation of mathematical modeling and simulation to the EMAT systems, which is the influence of skin effect and proximity effect on the coil conductor. Because of the high operation frequency of EMAT and the wires of the coil being close to each other, this leads to obvious skin effect and proximity effect. That is why the accuracy of theoretical analysis and numerical calculation can be reduced if the skin effect and proximity effect are ignored. Besides that, although Dai X.W had already derived the governing equations and boundary conditions of an EMAT system based on the magnetostrictive mechanism, the simulation and analysis of the EMAT were done; consequently, it is difficult to confirm the correctness of the model proposed by him, and the FE model is invalid to an EMAT based on a magnetostrictive mechanism.

By using the complete equation of the source current density, Shapoorabadi R.J calculated the two-dimensional finite element simulation of EMAT in ferromagnetic metal materials and non-ferromagnetic metal materials. The skin effect and proximity effect were taken into consideration. In his study, by using the modified finite element equation, he calculated the internal force of the tested specimen in the excitation mode and the induced electromotive force of the receiving mode in the receiving coil. The results showed that the modified equation had better stability and convergence. For the ferromagnetic metal material, the general control equation which is used to characterize the EMAT was given by him based on the magnetomotive force vector and particle displacement vector. The force inside the specimen was also calculated using the third-order finite element analysis. However, his research work was still at a preliminary stage since he did not calculated the particle displacement vector of the ultrasonic waves. The main work he completed is the calculation of the electromagnetic field and the force field, which has no relation to the coupling problem between force field and sound field. The whole stimulation and receiving model of EMAT was not established. Consequently, the overall energy conversion process of EMAT cannot be described effectively, and the ultrasonic excitation efficiency of EMAT also cannot be estimated accurately. In addition to that, his calculation of induced electromotive force inside the receiving coil was conducted using the conditions of the time and space analytical functions of ultrasonic waves already given, but the signal of the ultrasonic wave received by EMAT is difficult to describe by analytical expression, so it is difficult to guarantee the practicability of the approach proposed by him.

Additionally, in the paper reported by Ogi, the coupling excitation of the ultrasonic body waves and electromagnetic field was discussed in detail. The dependence of the magnetostrictive coefficient matrix on the overall magnetic field was emphasized. The deviation between theoretical calculation and experimental result was analyzed, and it was concluded that the magnetostrictive mechanism always plays a dominant role for generating and receiving body waves in ferromagnetic specimens, irrespective of the direction of the bias magnetic field.

Some scholars studied the energy conversion mechanism and numerical method of EMAT in China. For instance, the mechanism of EMAT was studied by Zhu Hongxiu using the experimental method and the theoretical calculation model of EMAT was established by Lei Huaming, and his research result is similar with that of Dai X-W.

Using the FEM analysis software *COMSOL Multiphysics* combined with Abaqus, the simulation of the electromagnetic field of *Lorentz force*-based EMAT and the ultrasonic wave propagation in the tested specimen were achieved by R. Dhayalan and Krishnan Balasubramaniam. Based on the research achievements of Ogi, intensive studies were made by Remo Ribichini from Imperial College London, for EMAT of *SH* guided wave based on a magnetostrictive mechanism.

For EMAT with guided wave propagation along the pipeline axial direction, Kwun and Kim have done some theoretical analysis and experimental research work aimed at the basic working mechanism and characteristics of EMAT in the process of its design and application. Sablik M.J actually built the theoretical calculation of EMAT with the guided waves along the pipeline axial direction and also built the analytical calculation model of EMAT with the guided waves along the pipeline axial direction. The analytical calculation of the electromagnetic field as well as the mechanical field was included in this model. However, many simplifications in his calculations caused a large deviation between calculations and experimental results. In China, the analytical model for the *L*-mode guided waves of pipeline axial direction was also established by Wang Yueming, Naval University of Engineering, and some related experimental research was also conducted by him. The research group of He Cunfu, from the Beijing University of Technology, has also done some basic experimental research related to EMAT with guided waves of pipeline axial direction. The numerical simulation method of the guided wave propagation in the pipeline was developed.

1.3 Optimal Design of EMAT and Its New Configuration

1.3.1 Optimal Design of Meander Coil

The driving force that generates ultrasonic waves in the specimen is derived from *Lorentz force* and magnetostrictive force. For the EMAT that is based on *Lorentz force*, the *Lorentz force* applied on the specimen is proportional to the bias magnetic field and eddy current in the specimen. For the EMAT that is based on the magnetostriction mechanism, the magnetostrictive force applied on the specimen is directly proportional to the piezomagnetic coefficient of the specimen's working point and the derivative of the dynamic magnetic field with respect to the coordinates. The magnitude and direction of those two forces are related to those of the static magnetic field and dynamic magnetic field. For the meander coil EMAT, the ultrasonic mode generated in the specimen is determined by the direction of the force and the magnitude of the ultrasonic wave is determined by the magnitude of the force. The character of ultrasonic generation and propagation in the static magnetic field will be similar to that in the dynamic magnetic field if not considering the mechanism of force generated and just taking the specimen as the analysis object. Therefore, we just take the *Lorentz force*-based EMAT as an example to analyze the effect of meander coil configuration and the coil conductor's size on the performance of EMAT to optimize the scheme of the meander coil. The results we get are also suitable to the EMAT configuration based on the magnetostrictive mechanism.

The simulation model is shown in Fig. 1.5. The length of the specimen is 500 mm, and its thickness is 2 mm. The transmitting and receiving probes are placed on the specimen, with the distance between the two probes being 200 mm. The excitation coil and receiving coil are both meander coils of one layer; the coil size and the material parameter are listed in Tables 1.1 and 1.2. The excitation current signal is a tone burst signal with a frequency of 310 kHz and three cycles; its waveform is shown in Fig. 1.6. For the simulation of the pulsed signal with a frequency of 310 kHz, two kinds of *Lamb waves* A_0 and S_0 will be excited at the same time; their group velocities are 3007 and 5302 m/s, respectively.

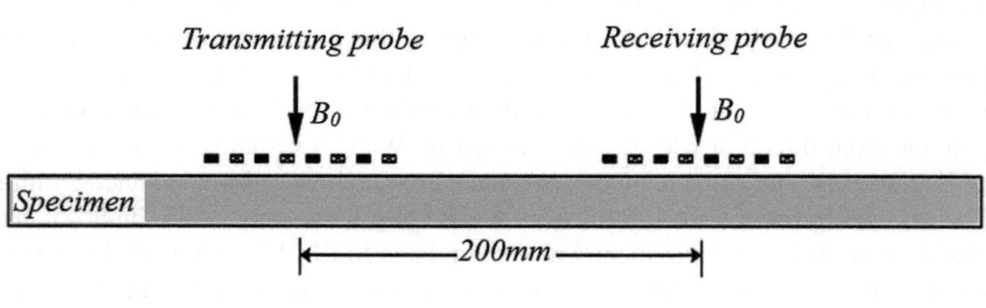

Fig. 1.5 Simulation model

Table 1.1 Dimensions and parameters of experimental aluminum plate

Item	Value
Length	500 mm
Width	350 mm
Thickness	3 mm
Conductivity	$3.5 \times e^{-7}$ S/m
Elasticity modulus	70 Gpa
Poisson's ratio	0.33

Table 1.2 Coil's size and material parameter

Item	Value
Plate thickness	0.500 mm
Copper layer width	0.720 mm
Copper layer thickness	0.035 mm
Line spacing	0.905 mm
Baffle spacing	3.25 mm
Copper layer permeability	$4\pi \times 10^{-7}$ H/m
Copper layer conductivity	2.667×10^{7} S/m

1.3.1.1 Effect of Coil Wire Spacing on the Performance of EMAT

Generally, the matching condition should be satisfied between the wavelength of the guided wave mode and the coil's configuration for the design of the meander coil as (1.1).

$$D = (2n + 1)\lambda, \quad n = 0, 1, 2, \ldots \tag{1.1}$$

In the above equation, D is the spatial period of the meander coil and λ is the wavelength of the guided wave. Generally, take $n = 0$, and thus, $D = \lambda$. The purpose is to enhance the energy of the excited ultrasonic wave mode signal and weaken the ultrasonic signal of other modes through the superposition or offset of the ultrasonic waves based on the periodic characteristics of the meander coil.

The relationship between the wavelength and frequency of the guided wave mode can be expressed as (1.2).

$$D = \lambda = c/f \tag{1.2}$$

In the above equation, c is the phase velocity of the ultrasonic wave for the chosen wave mode.

Next we investigate the effect of line spacing on the performance of EMAT by changing the line spacing. To generate an A_0 *Lamb wave* mode at the frequency of 310 kHz, the corresponding spatial period of the meander coil is $D = 6.5$ mm. In this case, observation of the variation of the received signals by the receiving coil through changing the spatial period of the exciting and receiving coils separately is presented in this work.

Figure 1.7 shows the normalized amplitude of the simulated signal under the condition of the spatial period of the exciting and receiving coils being 6.5 mm. The number of folds is 8 and the height of the coil conductor is 0.5 mm, while the width is 1.625 mm. The first wave packet in the figure is the S_0 *Lamb wave* mode, while the second bigger wave packet corresponds to the A_0 wave mode. The peak value of the second wave packet is taken as the baseline for normalized processing. It can be seen from Fig. 1.7 that the design of the coil span is based on the wavelength of the A_0 *Lamb wave* mode. So, the A_0 mode has greater signal amplitude compared with that of the S_0 mode.

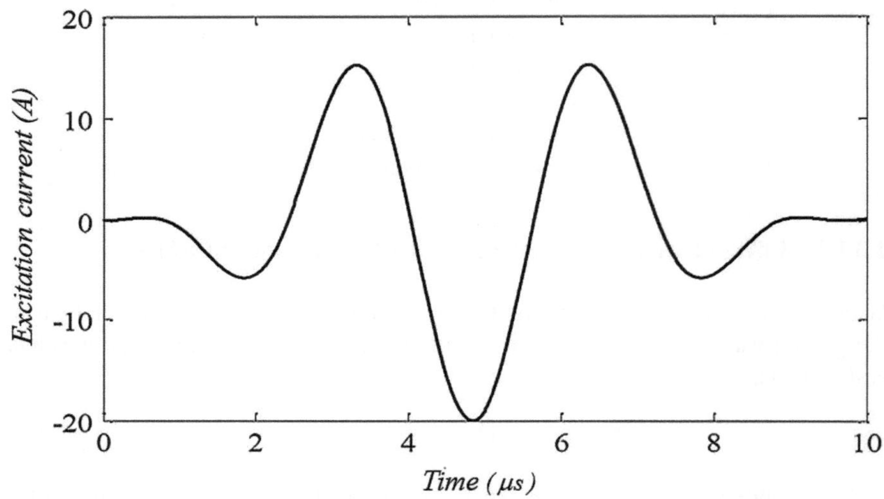

Fig. 1.6 Signal waveform

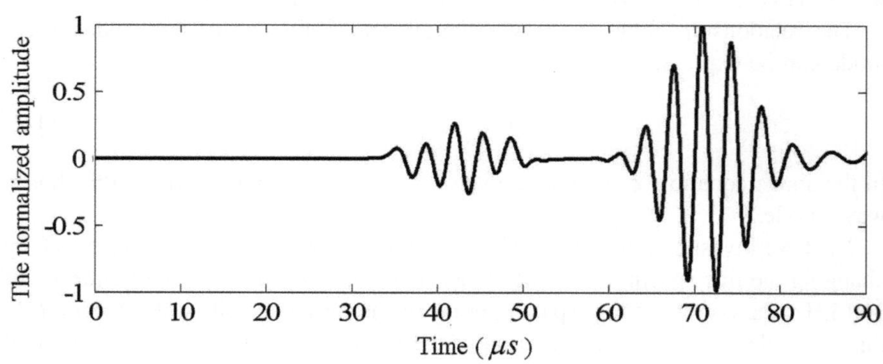

Fig. 1.7 Received signal under the condition of D = 6.5 mm

Figures 1.8 and 1.9 show the normalized amplitude of the received signal when the spatial period of the receiving coil is kept constant, while the spatial period of the exciting coil is changed to 7.875 and 8.125 mm, respectively. Compared with the results shown in Fig. 1.7, it is found that the amplitude of the received signal decreases, and the waveform of the signal also changes when the spatial period of the excitation coil is increased or decreased.

Figures 1.10 and 1.11 show the normalized amplitude of the received signal when the spatial period of the exciting coil is kept constant, while the spatial period of the receiving coil is changed to 4.875 and 8.125 mm, respectively. Compared with the results shown in Fig. 1.7, it is found that when the spatial period of the receiving coil is increased or decreased, the amplitude of the received signal will decrease, and the signal's waveform also changes. It shows that a change in the spatial periods of the exciting coil and receiving coil has the same effect on the received signal.

Therefore, the maximum efficiency of exciting and receiving coils can be achieved by ensuring the exciting and receiving coils have the same spatial period,

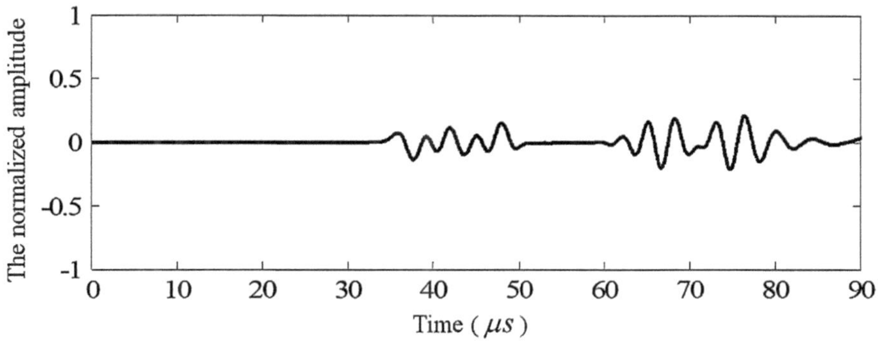

Fig. 1.8 Received signal under the condition of D = 7.875 mm

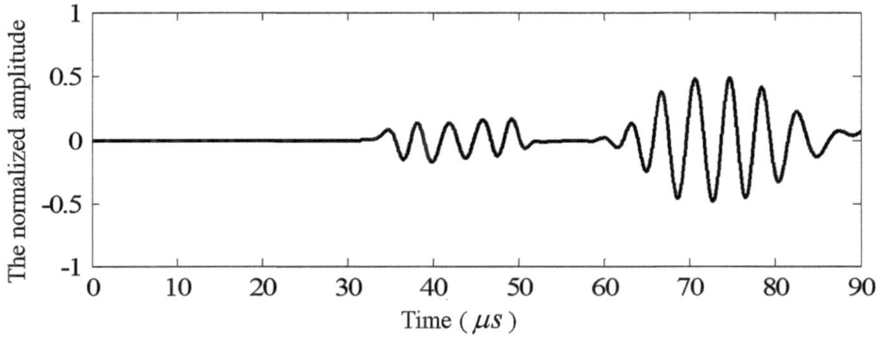

Fig. 1.9 Received signal under the condition of D = 8.125 mm

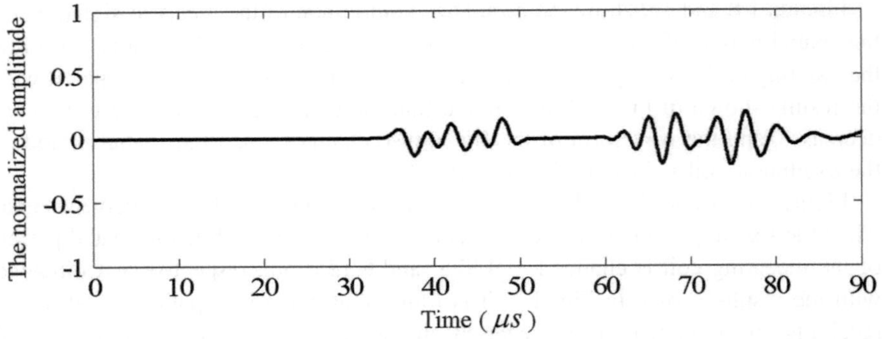

Fig. 1.10 Received signal under the condition of D = 4.875 mm

as well as the periods matching each other with the wavelength of the generated ultrasonic wave signal for the design of the meander coil.

1.3.1.2 Effect of Coil Fold Number on the Performance of EMAT

As mentioned above, the multifolded structure of the meander coil can cause the superposition and counteraction of ultrasonic waves, thus enhancing the signal of the desired ultrasonic wave and weakening the ultrasonic wave signals of other modes. In this book, the effect of the variation of the coil fold number on the performance of EMAT was presented. Figure 1.7 shows the observation of the change of the receiving signal under the conditions of changing the fold number of the exciting and receiving coils, while the excitation current remains constant.

Figures 1.12 and 1.13 show the normalized amplitude of the received signal when the exciting coil is onefold and fourfold, respectively, while the receiving coil is eightfold. As we can see from the figure, when the fold number of the excitation coil is 1, the amplitude value of the S_0 *Lamb wave* mode is bigger than that of the A_0 wave mode, even if the spatial period of the receiving coil is designed based on the A_0 *Lamb wave* mode. When the fold number of the excitation coil is 4, the amplitude of the A_0 *Lamb wave* mode is bigger than that of the S_0 *Lamb wave* mode, because of the regulating effect on the bending structure of the coil. Moreover, the amplitude value of the receiving signal will increase with the increase in fold number and the width of the wave packet will also increase.

Figures 1.14 and 1.15 show the normalized amplitudes of the received signal when the exciting coil fold number is *8*, while the receiving coil fold number is onefold and fourfold, respectively. It is found that a change in the fold numbers of the exciting coil and receiving coil has the same effect on the performance of EMAT.

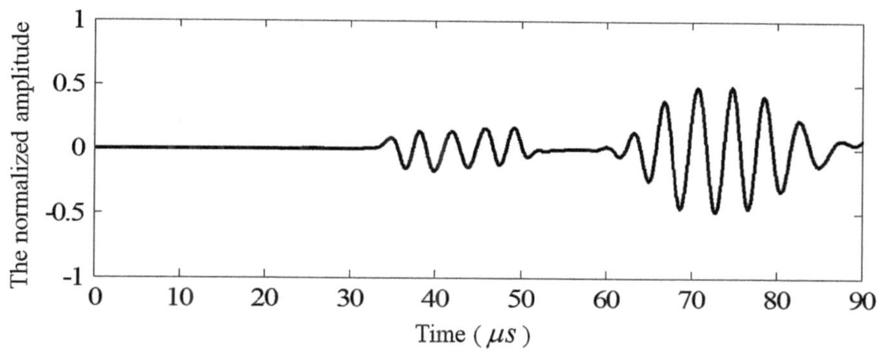

Fig. 1.11 Received signal under the condition of D = 8.125 mm

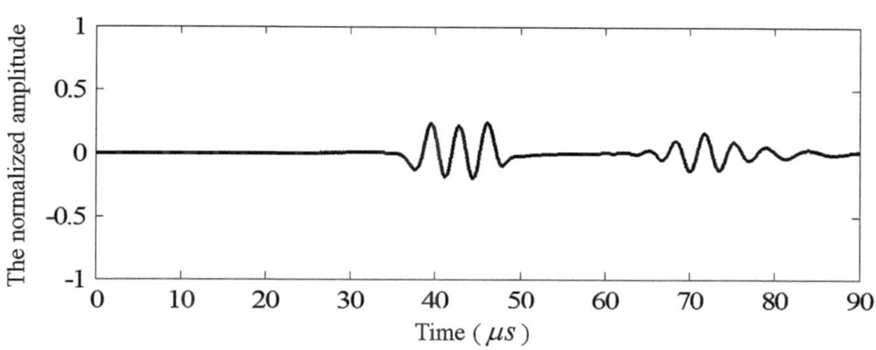

Fig. 1.12 Received signal where the exciting coil fold number is 1

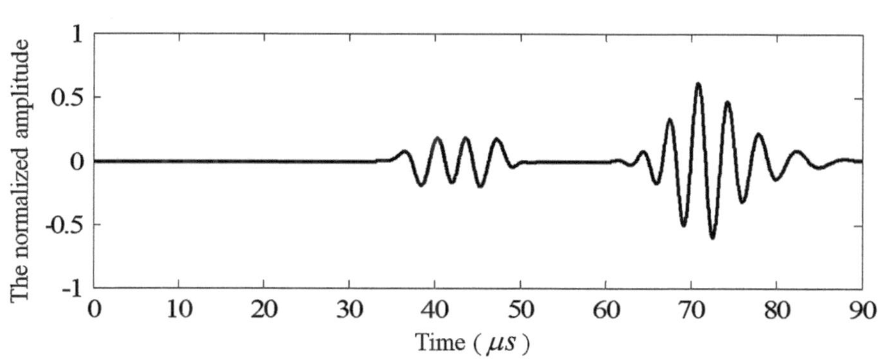

Fig. 1.13 Received signal where the exciting coil fold number is 4

Figures 1.16 and 1.17 show the normalized amplitude of the received signal when the exciting coil fold number is 1 and the receiving coil fold number is 4. The comparison between those two figures and Fig. 1.7 shows that the amplitude value of the A_0 *Lamb wave* mode grows quickly with the increasing fold number, while the amplitude value of the S_0 *Lamb wave* mode gradually decreases. This finding shows us that the structure of the meander coil can effectively strengthen the amplitude value of the required ultrasonic wave mode, while weakening the wave amplitude value of other modes. Also, the time duration of the signal wave packet will increase with an increase in the fold number.

Therefore, when we choose the fold number of the meander coil, the effect of the fold number of the coils should be considered comprehensively. To improve the testing sensitivity, an increase in the fold number of the meander coil could work to strengthen the signal amplitude and reduce the interfering effect of other signal modes, without affecting the width of the signal wave packet. When we want to improve the testing resolution by reducing the width of the signal wave packet, a

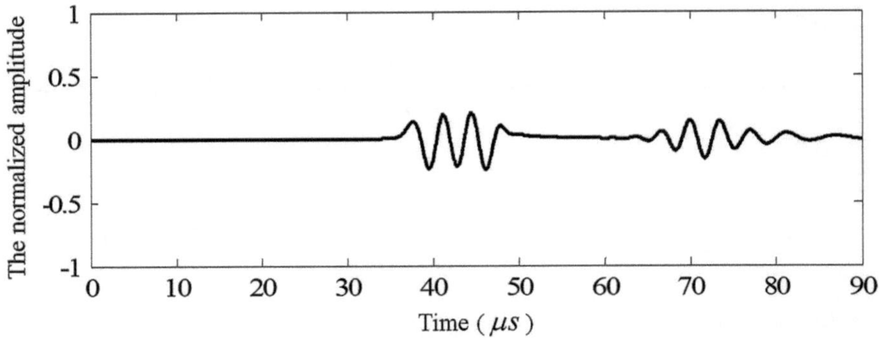

Fig. 1.14 Received signal where the receiving coil fold number is 1

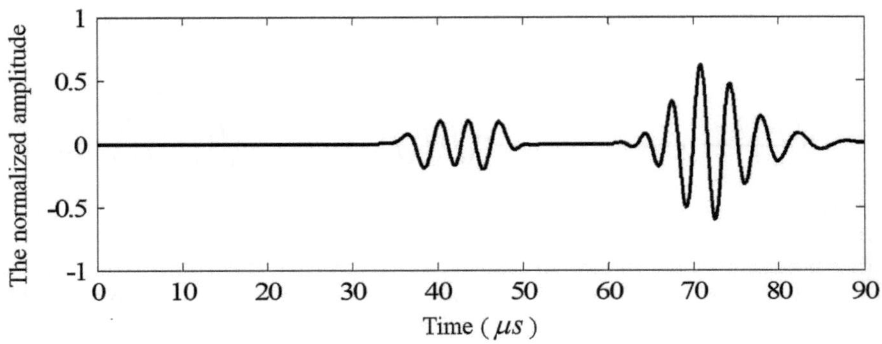

Fig. 1.15 Received signal where the receiving coil fold number is 4

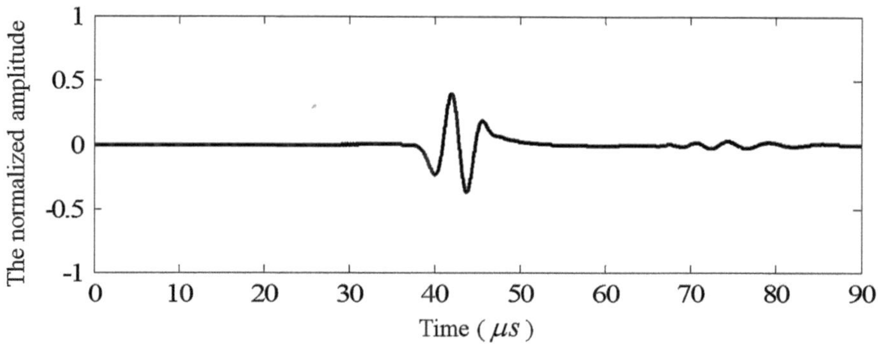

Fig. 1.16 Received signal of the exciting and receiving coils where fold number is 1

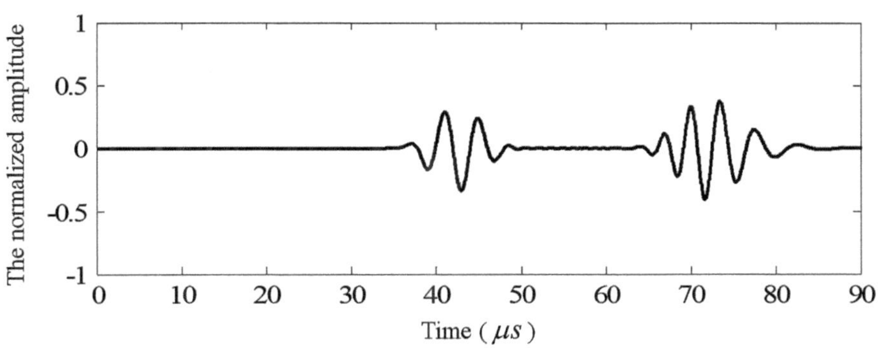

Fig. 1.17 Received signal of the exciting and receiving coils where fold number is 4

reduction in the fold number of the meander coil can make this occur. Generally, we should take both factors into consideration for the design of the fold number of the meander coil. The exact number of coil folds is dependent on the practical situation.

1.3.1.3 Effect of Coil Conductor on the Performance of EMAT

Besides the fold spacing and fold number, the performance of EMAT can also be affected by the dimension of the coil conductor. When the fold spacing and fold number and current density of the coil are kept constant, observing the effect of the coil conductor on the performance of EMAT is performed by only changing the width and height of the coil conductor.

Figures 1.18 and 1.19 show the normalized amplitudes of the received signal when changing the excitation coil conductor's height to 1 and 0.1 mm, respectively, while keeping the dimension of the receiving coil the same as in Fig. 1.7.

Comparing the results shown in those two figures with those of Fig. 1.7, it is found that the amplitude value of the received signal also grows gradually with the increase in the coil conductor's height.

Figures 1.20 and 1.21 show the normalized amplitudes of the received signal when the dimension of the excitation coil remains the same as in Fig. 1.7, while changing the excitation coil conductor's height to 1 and 0.1 mm. Comparing the results shown in these two figures with those of Fig. 1.7, it is found that the amplitude value of the received signal decreases with the increase in the coil conductor's height. The value of signal amplitude is approximately inversely proportional to the conductor's height.

Figures 1.22 and 1.23 show the normalized amplitudes of the received signal when the dimension of the receiving coil remains the same as in Fig. 1.7, while changing the excitation coil conductor's width to 0.8125 and 2.4375 mm. Comparing the results shown in the two figures with those of Fig. 1.7, it is found that the amplitude value of the received signal will increase from the point of view of the increased amplitude with the increase in the excitation coil conductor's

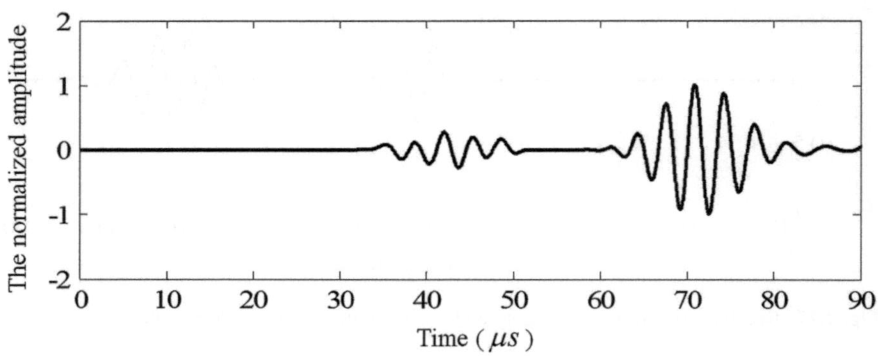

Fig. 1.18 Received signal of the exciting coil conductor with a height of 1 mm

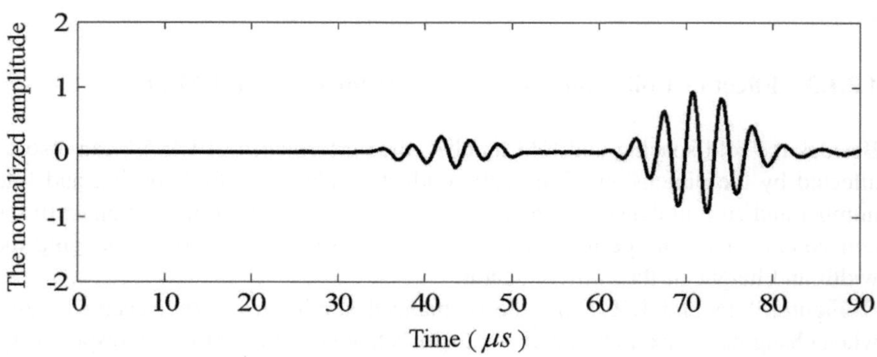

Fig. 1.19 Received signal of the exciting coil conductor with a height of 0.1 mm

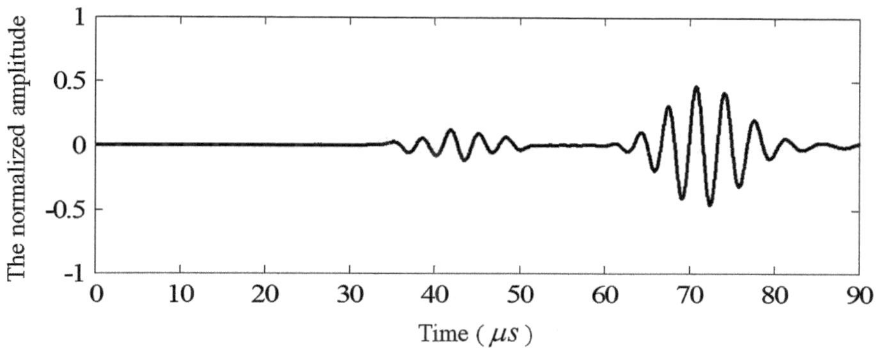

Fig. 1.20 Received signal of the receiving coil conductor with a height of 1 mm

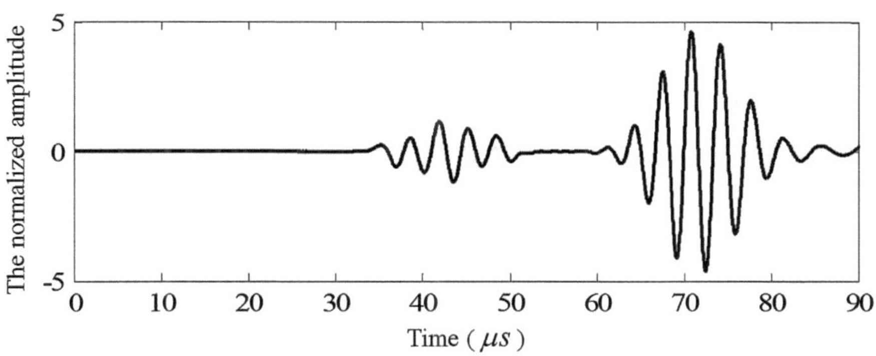

Fig. 1.21 Received signal of the receiving coil conductor with a height of 0.1 mm

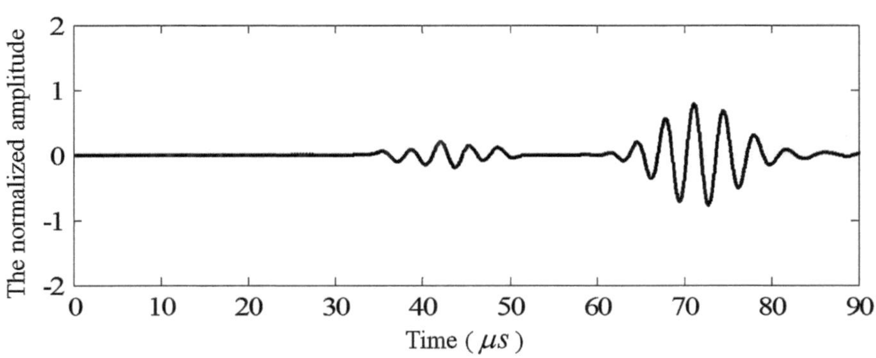

Fig. 1.22 Received signal of the excitation coil conductor with a width of 0.8125 mm

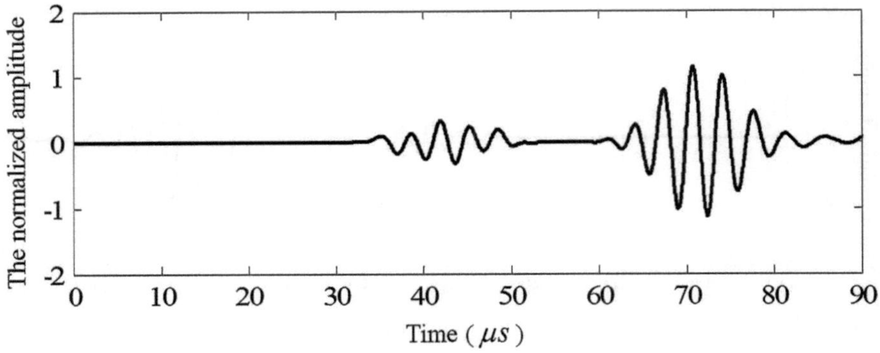

Fig. 1.23 Received signal of the excitation coil conductor with a width of 2.4375 mm

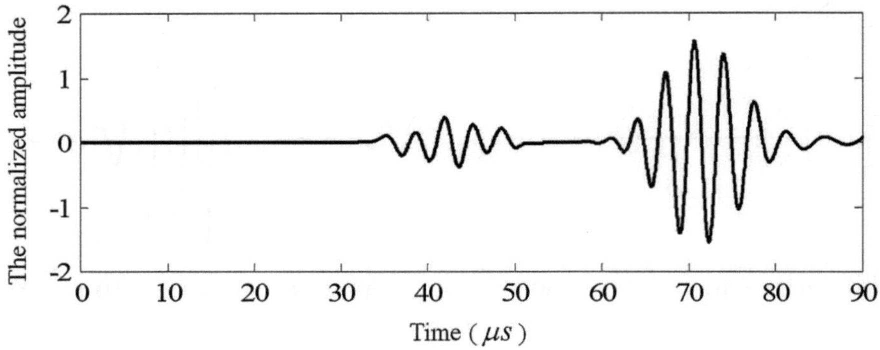

Fig. 1.24 Received signal of the receiving coil conductor with a width of 0.8125 mm

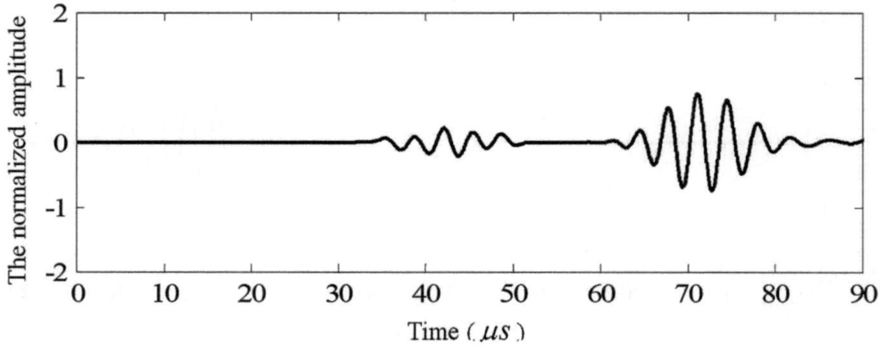

Fig. 1.25 Received signal of the receiving coil conductor with a width of 2.4375 mm

width. The influence of the width increase on signal amplitude is greater than that of the height.

Figures 1.24 and 1.25 show the normalized amplitudes of the received signal when the dimension of the excitation coil remains the same as in Fig. 1.7, while changing the exciting coil conductor's width to 0.8125 and 2.4375 mm. Comparing the results shown in the two figures with those of Fig. 1.7, it is found that the amplitude value of the received signal will decrease with the increase in the excitation coil conductor's width. The value of the signal amplitude is approximately inversely proportional to the conductor's height.

Therefore, for the design of the dimensions of the meander coil, we should increase the width and height of the excitation conductor appropriately and minimize the size of the receiving coil conductor to increase the amplitude and signal-to-noise ratio of the received signal.

1.3.1.4 Effect of Liftoff on the Performance of EMAT

The coil's liftoff is one of the key factors that affect the performance of EMAT. We changed the liftoff distance of the transmitter and receiver, respectively, to investigate the effect of liftoff by changing the distance between coil and specimen and calculating the amplitude value of the received signal. The results are shown in Figs. 1.26, 1.27, and 1.28.

It can be seen from Figs. 1.26,1.27, and 1.28 that the amplitude exponentially decreases versus the liftoff distance for both transmitter and receiver. So, it is critical to minimize the coil's liftoff distance for the design of EMAT.

Through the above analysis, the effect of the coil fold's number on the performance of EMAT can be explained by the investigation of the mechanical field. The way to achieve the optimal design is reinforcing the particle vibration of the simulated ultrasonic signal. This purpose can be achieved by the appropriate design of

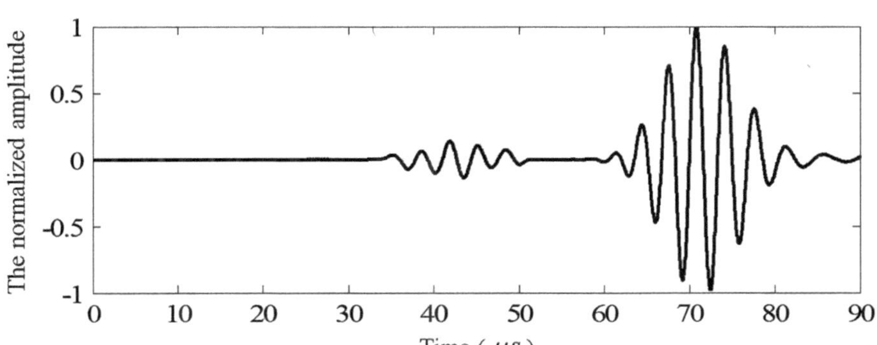

Fig. 1.26 Received signal when the height of coil's lift is 0.5 mm

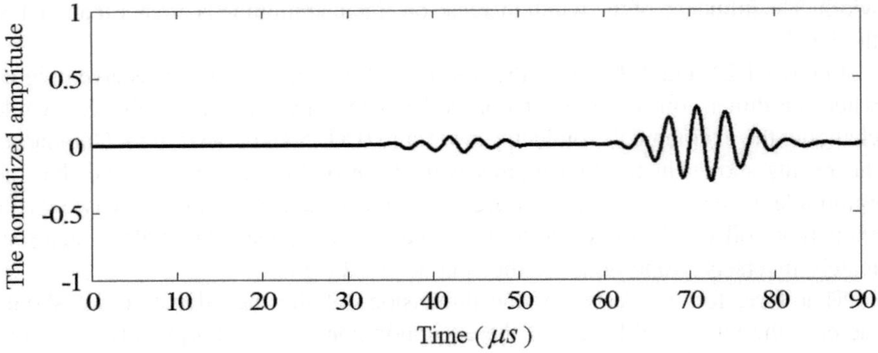

Fig. 1.27 Received signal when the height of coil's liftoff is 1 mm

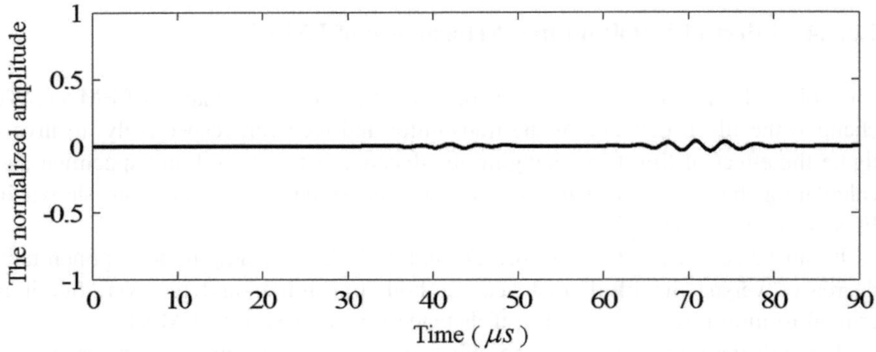

Fig. 1.28 Received signal when the height of coil's liftoff is 2 mm

the fold spacing and increasing the number of coil folds. The effect of the change of the coil conductor's dimension on the performance of EMAT can be explained by the analysis of the electromagnetic field that is generated by the excitation coil and particle motion inside the specimen. Under the condition that the electric current density of the excitation coil reminds unchanged, increasing the area of the coil conductor will lead to an increase in the field intensity of the dynamic magnetic field inside the specimen, consequently intensifying the amplitude value of the ultrasonic wave. At the receiving part, the closer range between receiving coil and specimen, as well as the smaller size of the receiving coil conductor, will lead to larger amplitude of the received signal.

In order to increase the amplitude and signal-to-noise ratio of the received signal, the meander coil can be designed with multiple layers and multiple turns. Increases in ampere-turns and the number of sensing conductors were introduced into the new configuration to improve the signal amplitude.

1.3.2 *Multibelt Coil Axisymmetric Guided Wave EMAT*

By controlling the matching of coil span of meander coil and excitation frequency, the dimensional design is made simple, and the types and modes of the excited ultrasonic wave can be better controlled; while at the same time, the amplitude value and width of ultrasonic signal can be adjusted through the tuning of the coil fold number and the conductor's dimensions. By drawing lessons from such coil structures and applying them to the defect detection of pipelines made by ferromagnetic material, the drawbacks of the single-belt coil structure can be effectively overcome.

1.3.2.1 Longitudinal Guided Wave Mode and Frequency Dispersion

As far as pipes are concerned, propagation of the longitudinal guided wave is dispersive. The general phase velocity and group velocity dispersion curves of longitudinal waves in pipes are shown in Figs. 1.29 and 1.30. As shown in the figures, there are four different modes: *L*(0,1), *L*(0,2), *L*(0,3), and *L*(0,4). Since the magnetostrictive transducer typically operates up to a few hundred kHz frequency, *L*(0,1) and *L*(0,2) wave modes are generally generated. However, *L*(0,1) and *L*(0,2) coexist in a large frequency range; both *L*(0,1) and *L*(0,2) wave modes could be generated by the same excitation. Which mode should be generated is determined by the dimensions of the coils.

1.3.2.2 Transducer Configuration and Transduction Principles

A schematic diagram of the new transducer configuration is shown in Fig. 1.31a. To keep the bias magnetic field homogeneously distributed around the circumference of the pipe, a solenoid electromagnet is used in place of the U-shaped magnet in the single-belt transducer, which can improve the convergence performance of the longitudinal guided wave. The multibelt coil, which can be made by winding wires

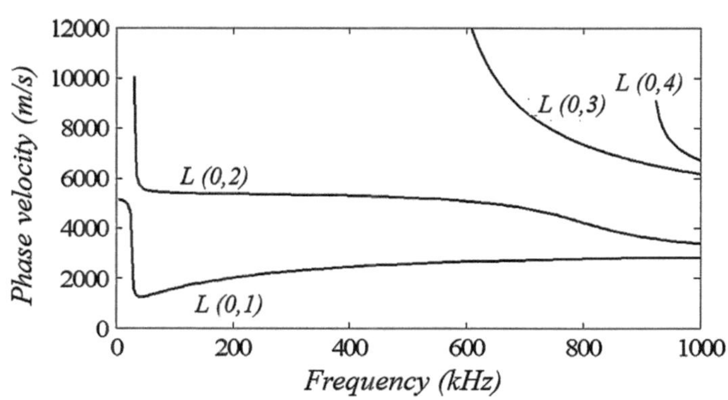

Fig. 1.29 Phase velocity dispersion curves of pipe

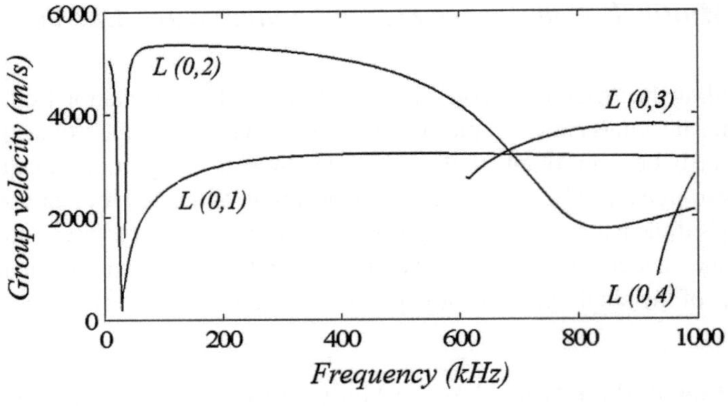

Fig. 1.30 Group velocity dispersion curves of pipe

or by flexible printed circuit (FPC) technology, is used for providing a periodic dynamic field. Both the transmitting coil and receiving coil have the same configurations.

Figure 1.31b is a detailed description of the multibelt coil, the coil extending around the circumference and neglecting the end connection between two adjacent belts for inducing opposite dynamic magnetic field. According to the principle of EMAT, the distance l of the adjacent belts should be half a wavelength of the

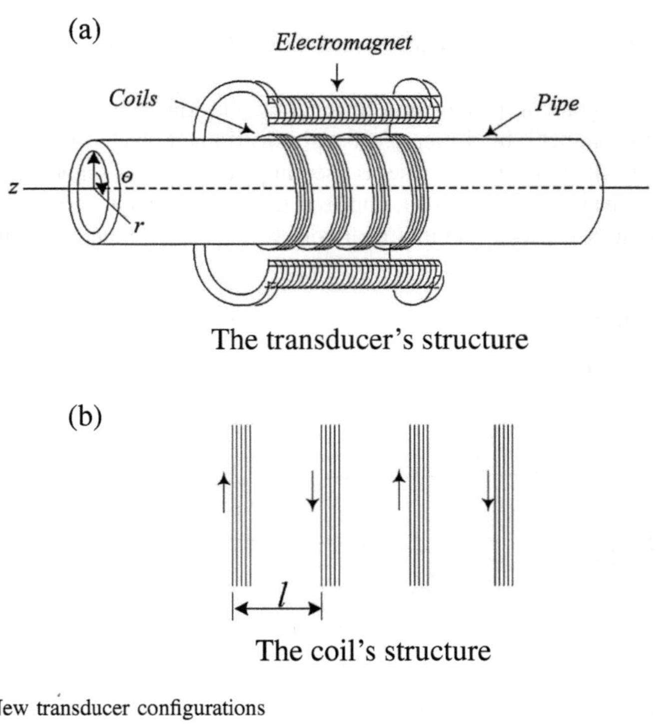

Fig. 1.31 New transducer configurations

generated longitudinal guided waves. Therefore, the wave frequency could be expressed as (1.3).

$$f_{\mathrm{L}} = \frac{c_{\mathrm{L}}}{\lambda} = \frac{2c_{\mathrm{L}}}{l} \tag{1.3}$$

In the above equation, f_{L} is the frequency of the longitudinal guided wave, c_{L} is the phase velocity, λ is the wavelength, and l is half the wavelength, which equals the distance between adjacent belts. If c_{L} is fixed, f_{L} is solely determined by l.

When the transducer works, the electromagnet will induce the static bias magnetic field H_{s} and the multibelt coil will induce dynamic magnetic field H_{d}, both directions of which are along the axial direction of the pipe. The superposition of H_{s} and H_{d} causes the periodically changed magnetic field H_{c}, as shown in Fig. 1.32. Under the action of H_{c}, the magnetostrictive force is brought about, which will cause periodic deformation of the pipe, and thus, an ultrasonic wave is generated. The magnetostrictive force under one belt of the multibelt coil can be described as (1.4).

$$F_{ms} = -\frac{1}{2}(3\tau + 2\mu)(1 - 2v)\frac{\partial \zeta}{\partial M_0}\frac{\partial m_z}{\partial_z} \tag{1.4}$$

In the above equation, τ and μ are *Lame constants*; υ is the *Poisson's ratio*; ζ is the line magnetostriction; M_0 is the static bias magnetic field magnetization intensity; and m_z is the z component of the dynamic magnetic field magnetization intensity. The receiving principle of the ultrasonic wave is opposite to its generating principle. The principle of wave detection is the inversed process of wave generation. When waves propagate to the position of the receiving coil, the deformation of the pipe will causes a change in the magnetic field. Thus, the voltage is induced in the receiving coil.

1.3.2.3 Experimental Study of L-mode EMAT [3]

To verify the performance of the proposed EMAT, an experiment needs to be performed. The schematic diagram of the experimental setup is shown in Fig. 1.33. There are two probes: One is the transmitter for generating the ultrasonic signal, and the other is the receiver for detecting the directly arriving waves and reflecting waves from the crack and the end. A DC power supply (*DF1730SL40A*) was used as the power supply providing the bias magnetic field to the electromagnets. The transmitter was supplied by a RF power amplifier (*AG1024*), which could provide a

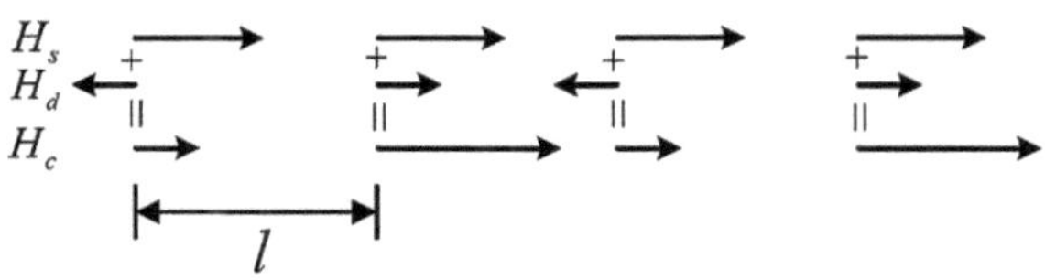

Fig. 1.32 Superposition of dynamic magnetic field and static magnetic field

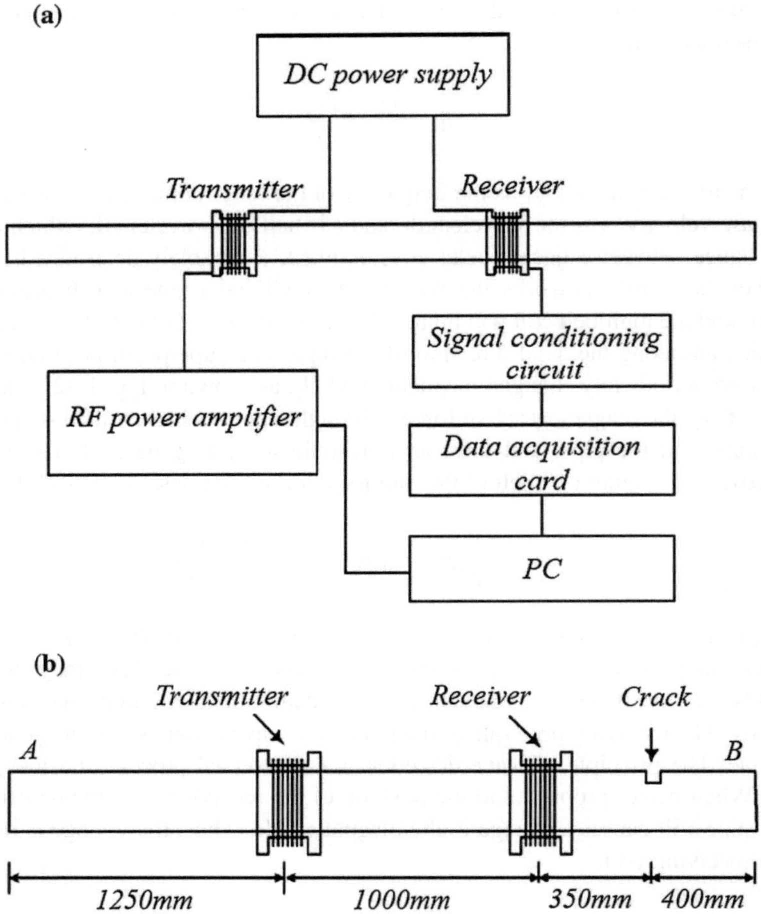

Fig. 1.33 Schematic diagrams of the experimental setup

tone burst signal for the EMAT coil to generate the dynamic magnetic field. The amplitude, frequency, and periodicity of the tone burst signal are all adjustable. The detected signal of the receiver was filtered and amplified by a signal-conditioning circuit and then collected by a data acquisition card in a PC.

According to the phase velocity dispersion curves shown in Fig. 1.29, $L(0,2)$ mode would be generated and the frequency was chosen as 100 kHz. Thus, the corresponding phase velocity under this frequency is 5400 m/s. Calculated through Eq. (1.3), l was obtained as 27 mm. Four-belt coils were used as the transmitter and receiver coils with 48 turns and 160 turns, respectively.

Figure 1.33a shows the detailed positions of the transmitter and receiver on a chosen bare steel pipe, which has dimensions of length 3000 mm, inside diameter (ID) 53 mm, and outside diameter (OD) 59 mm. The transmitter was 1250 mm from one end of the pipe, and the distance between the transmitter and receiver was

1000 mm. To verify the crack detection ability of the new transducer, an artificial circumferential crack was made 400 mm from the *B* end of the pipe, whose dimensions were width 4 mm, depth 2 mm, and circumferential length 45 mm.

In order to verify the practicability of the EMAT structure, a series of experiments were carried out. In the experiments, direct current (8 A) was provided to the bias electromagnet to generate a bias magnetic field. A tone burst of 100 kHz was applied to the multibelt coil, and the number of pulse periods was 6. Its wave form is shown in Fig. 1.34. The signals generated and received by EMAT and the reflected signal from the end of the pipeline are shown in Fig. 1.35.

In Fig. 1.35, signal a shows the initial pulse, representing the space-induced signal of the receiving coil; b shows the signal of the direct wave; d, e, and f show

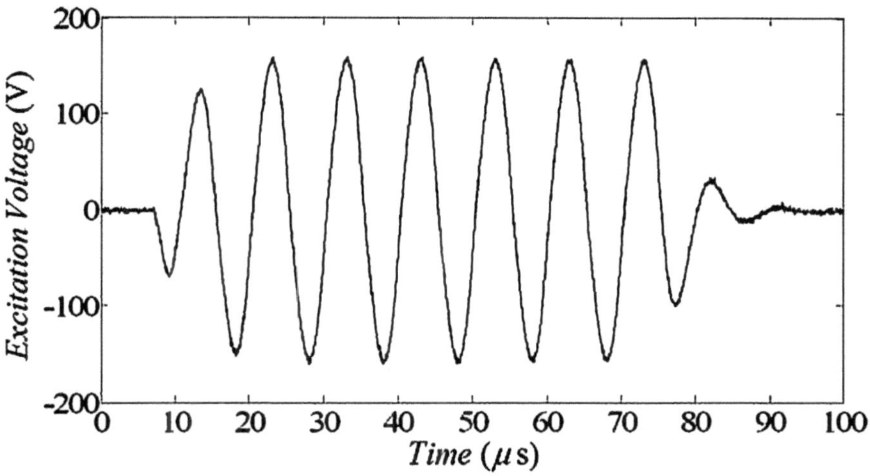

Fig. 1.34 The excitation voltage signal generated by *RF* power amplifier

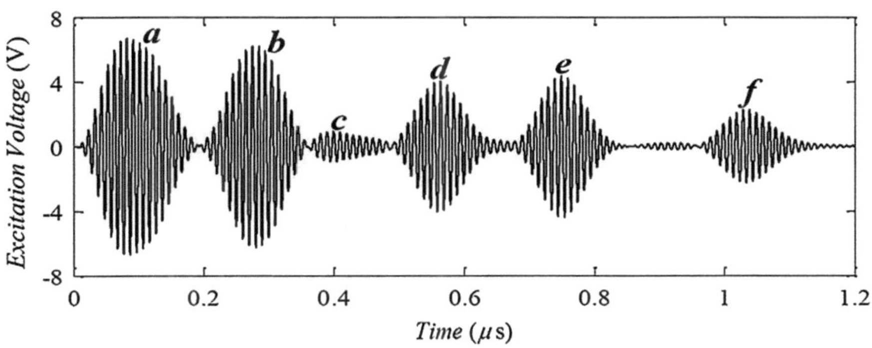

Fig. 1.35 The measured signal of EMAT

the reflected signals from the end; and c is the reflected signal from the defect. We can see from Fig. 1.35 that the ultrasonic waves generated and received by EMAT have a big amplitude value and signal-to-noise ratio, as do the reflected signals from the defect and the far end.

Through the multiplication of the time difference between each reflected signal and the initialized signal, and the group velocity of the *L*(0,2) mode ultrasonic waves, the propagation distance of every pulse signal with respect to the transmitting probe can be estimated. The correctness of the defect's position and the ultrasonic mode excited by EMAT can be verified by checking the estimated value and actual distance. The calculation of the distance and its actual value is shown in Table 1.3, where Δt represents the time difference; Δd_{es} represents the difference in the distance obtained from the experimental measurement; and Δd_{ex} represents the accurate difference of the distance. As shown in Fig. 1.3, the distance that is calculated by the experimental measurement is consistent with the actual distance (the error is less than 3 %). It shows that the EMAT can be used to detect the defects along the pipeline in the axial direction.

1. ***The frequency tuning characteristic of the multibelt coil***

In the experiment, the amplitude values of the bias magnetic field and pulse signal remain consistent while changing the excitation frequency of the multibelt

Table 1.3 The corresponding relationship between the signal packet and ultrasonic propagation distance

Pulse signal	b	c	d	e	f
Δt (ms)	0.1823	0.3070	0.4700	0.6579	0.9456
Δd_{es} (mm)	993	1672	2561	3584	5152
Δd_{ex} (mm)	1000	1700	2500	3500	5000
Error (%)	0.7	1.6	2.4	2.4	3

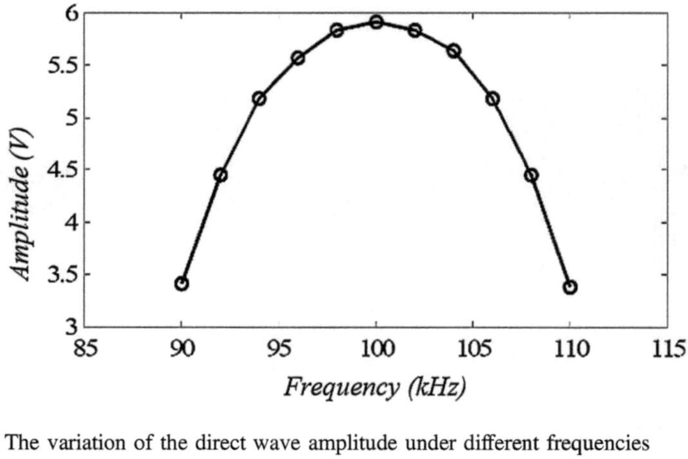

Fig. 1.36 The variation of the direct wave amplitude under different frequencies

coil, thereby inspecting the selectivity and adjustability of the multibelt coil to frequency. Figure 1.36 shows the relationship between the direct wave amplitude value and frequency variations. From the figure, the direct wave amplitude values rise to maximum at the frequency of 100 kHz. The amplitude value will always reduce with the frequency increase or decrease from this point. The result shows that the structure of the multibelt coil has a frequency-dependent characteristic, and the ultrasonic wave can reach the maximum amplitude value at the designed frequency.

2. ***The influence of static magnetic field and dynamic magnetic field***

Equation (1.2) shows that the static magnetic field and dynamic magnetic field are the key factors which can affect the EMAT characteristics. An experimental approach is used to investigate the effect of variations of the static magnetic field and dynamic magnetic field on the characters of EMAT.

The amplitude value of the excitation current was kept constant at 8, 14, and 20 A, while the signal amplitude values of every DC excitation value were recorded by increasing the DC excitation value of the electromagnet from 7 to 35 A. Results are shown in Fig. 1.37. At the beginning, the amplitude value of the direct wave almost increases linearly, while the increasing slope of the amplitude value gradually slows down until there is no increase any more, when the electric current reaches a certain value.

The value of the excitation current was kept constant at 8 and 22 A, while the amplitude value of every forward wave signal under every amplitude value was recorded by increasing the amplitude value of the coil's dynamic excitation current. The result is shown in Fig. 1.38. The amplitude value of the forward signal has an approximate linear relationship with the pulsed current.

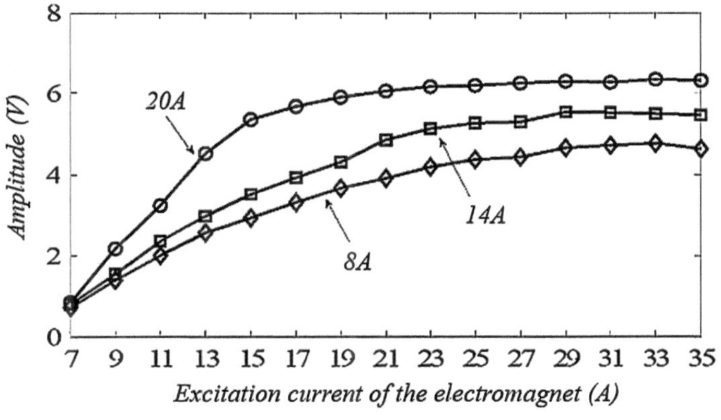

Fig. 1.37 The direct wave amplitude under the exciting current of different electromagnets

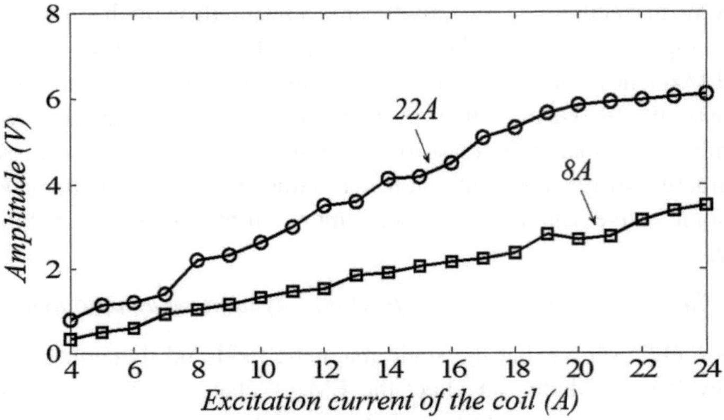

Fig. 1.38 The direct wave amplitude under different coil excitation currents

The results shown in Figs. 1.37 and 1.38 indicate that the static bias magnetic field and the coil dynamic magnetic field have significant influence on the performance of EMAT. The amplitude value of the excited ultrasonic wave will be changed by changing either one of the values, but when the bias magnetic field reaches a certain value, continual increases of the bias magnetic field have no significant effect on the ultrasonic amplitude value. However, if the dynamic magnetic field keeps increasing, the amplitude value of the ultrasonic wave will continue to increase.

3. ***The influence of the coil's liftoff*** [4]

In many testing fields, the surface of pipes under testing will be packaged with a coating or insulating layer, and thus, *EMAT* works under a certain liftoff value. Therefore, it is necessary to study the liftoff effect on EMAT.

Setting the coil dynamic excitation current at 20 A and changing the liftoff values of the excitation coil and receiving coil, respectively, the direct wave amplitude values were measured under different liftoff states. As shown in Fig. 1.39, it was found that the amplitude values of the ultrasonic wave decay exponentially with increasing liftoff values of the excitation and receiving coils, and the excitation coil is more sensitive to the change in the liftoff value.

4. ***The influence of the excitation signals' cycles***

In the above experiment, the number of periods of the pulsed current is 7. The periodicity of the pulsed current can also affect the property of EMAT. Under a specific amplitude and frequency, the periodicity determines how much energy of the current source can transfer into the coil. Part of the energy is converted into mechanical energy to produce ultrasonic waves.

We wish to evaluate the effect on the amplitude value and waveform of generated ultrasonic waves by changing the number of periods of the driving signal from 1 to 10. In the experiment, the exciting current of the electromagnet is 13.5 A

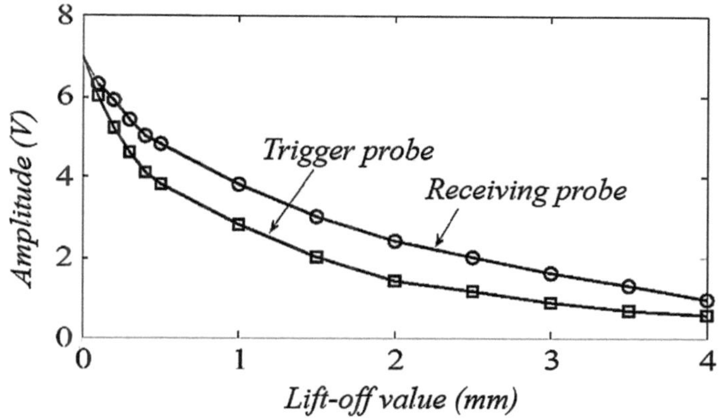

Fig. 1.39 The influence of liftoff on the excitation and receiving probes

and the amplitude value of the exciting current is 16 A. The amplitude values of the direct signal are measured when the number of periods of the pulsed current are 4, 6, 8, and 10. As shown in Fig. 1.40, the result illustrates that with the increase of the number of periods, the wave packets of the direct wave will grow monotonically and the amplitude value of the signal also increases accordingly.

1.3.2.4 The Experimental Research of T-Mode EMAT

A similar coil structure with that of the *L*-mode guided wave EMAT is employed with the *T*-mode EMAT in the axial direction of the pipe. The generation and receiving of the *T*-mode guided waves in the axial direction of the pipe can be realized in the pipeline, when the bias magnetic field is made using soft magnetic materials. The setup of the coil and nickel strap on the pipeline is shown in Fig. 1.41.

The premagnetized nickel strap is fixed along the axial direction in the pipe, and the coupling method or chucking appliance can be used as the fixation technique. After the fixation, the magnetization direction is in the circumferential direction of the pipe. The multibelt coil is wrapped around the thin nickel strap, or placed on the flexible plate, multibelt coil is made with the FPC technique. The spaces between the multibelt coils meet the matching relationship with the wavelength of the guided wave.

The principle of EMAT is the bias magnetic field H_0 in the pipeline circumferential direction is provided by the residual magnetic flux density of the nickel strap, when the pulsed excitation is applied on the coil. The dynamic magnetic field H_d is generated from each cluster of the multibelt coil. The total magnetic field H_t will be generated by the combination of the dynamic and static magnetic fields. Under the action of the total magnetic field, the magnetic unit in the nickel strap

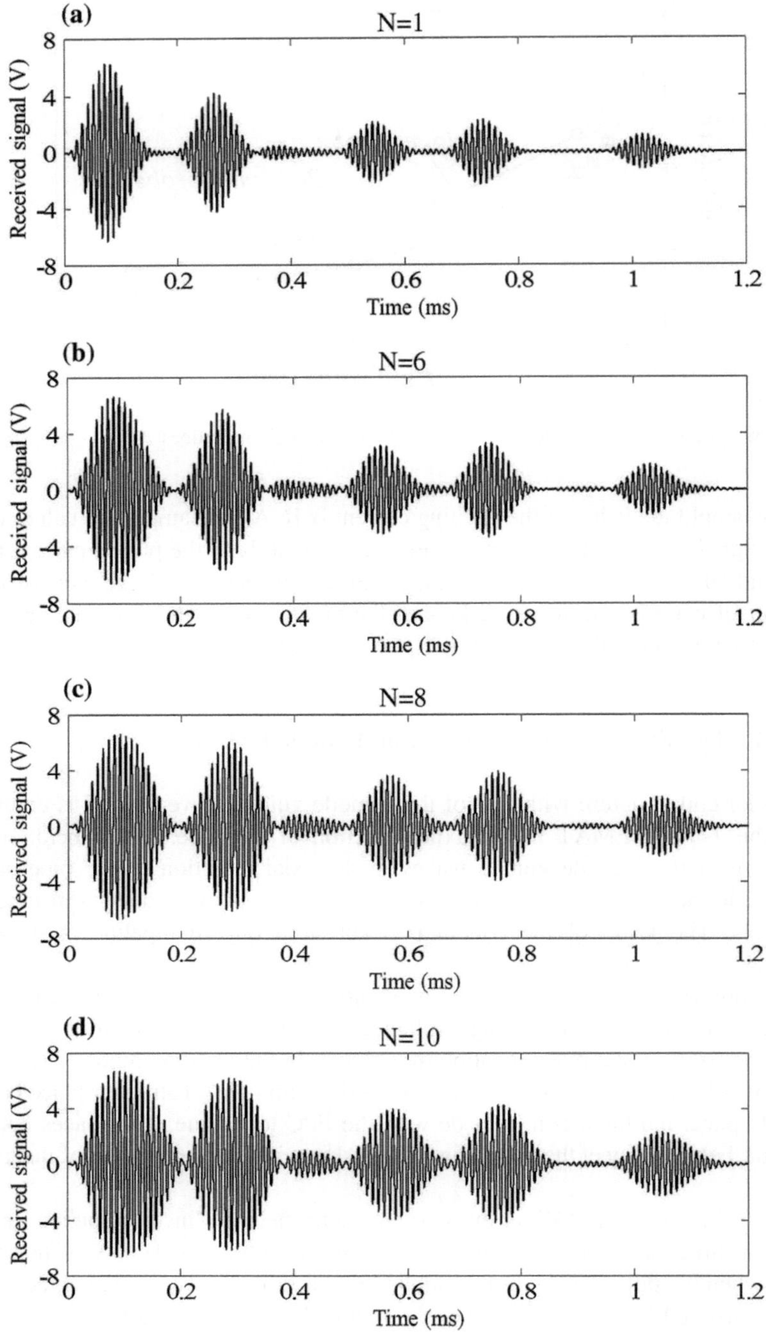

Fig. 1.40 The receiving signals under different exciting cycles

will be deformed, which causes the expansion of the nickel strap material and then generates ultrasonic waves.

In order to study the practicability of generating and receiving guided waves in the axial direction of the pipe using the EMAT, a flexible multibelt coil was made using the FPC technique. The shapes of the coil and nickel strap used are shown in Fig. 1.42. In the experiment, the inside diameter of the tested steel pipeline is 53 mm, its outside diameter is 59 mm, and the length is 1250 mm. The nickel strap and FPC coil were fixed on one end of the pipeline. The pulse echo method was used in this work.

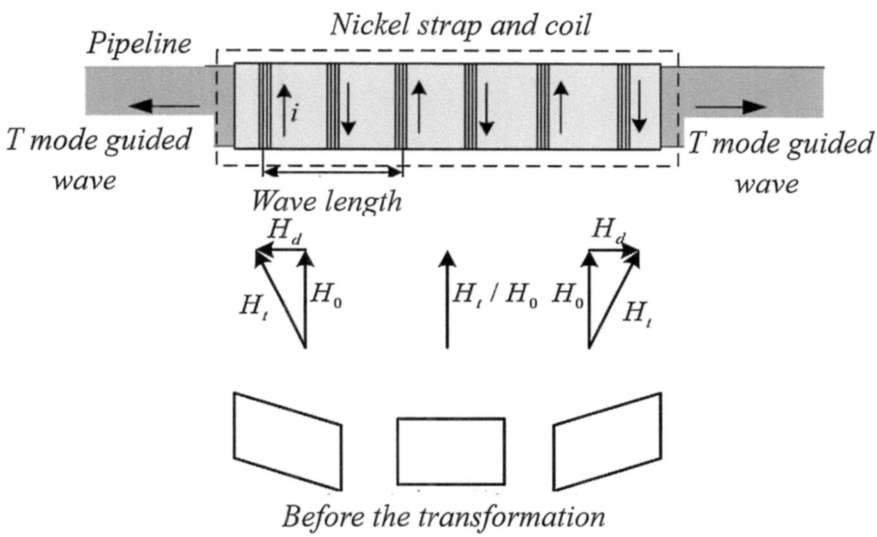

Fig. 1.41 The structure and principle of T-mode EMAT with the pipeline axial direction guided wave

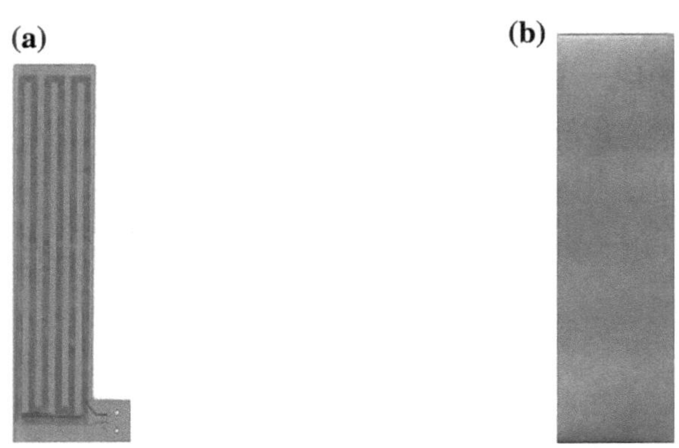

Fig. 1.42 The FPC coil and nickel strap of EMAT used in the experiment. **a** Meander coil and **b** Nickel strap

For the tested steel pipe in the experiment, the dispersion curves of its phase velocity and group velocity were calculated as shown in Fig. 1.43.

The *T*(0,1) mode guided wave along the pipeline axial direction was selected to do the test. According to the frequency dispersion curves shown in Fig. 1.43, the phase and group velocities of the *T*(0,1) mode guided wave are 3000 and 3200 m/s, respectively. The excitation frequency was chosen as 180 kHz, and 6 cycles of impulse excitation were applied to the coils. The received echo signal by EMAT from the end of steel pipe is shown in Fig. 1.44.

In order to verify the excited guided wave mode experimentally, the time difference between the two reflected signals from the end was calculated, and the propagation distance corresponding to the time difference is double that of the pipeline length (2500 mm). The group velocity of the guided wave can be checked

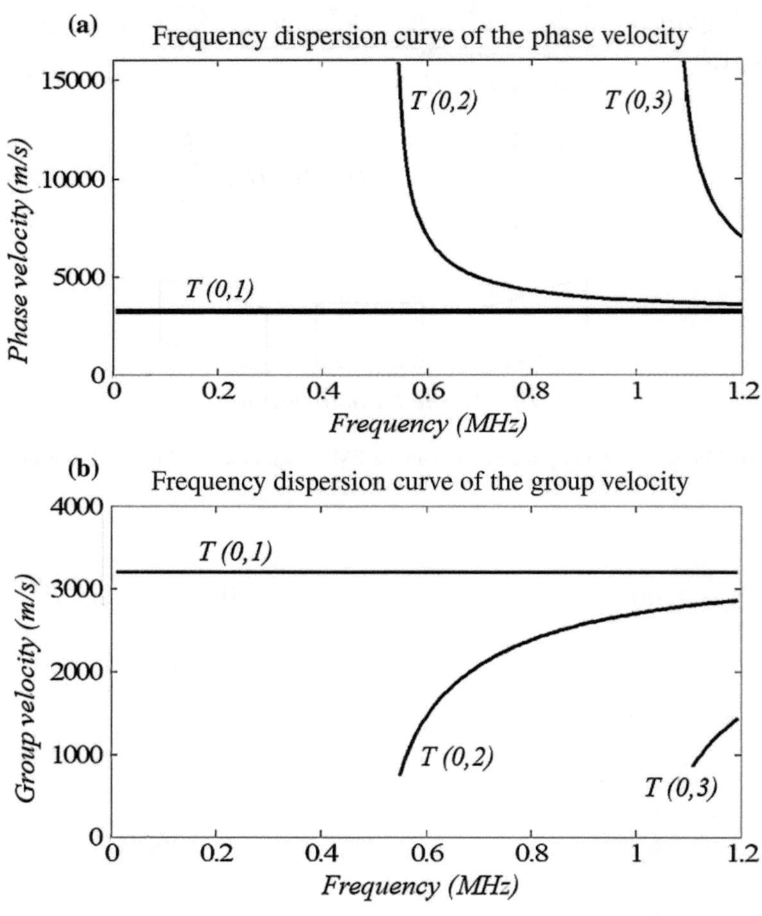

Fig. 1.43 The dispersion curves of the phase and group velocities of the *T*-mode guided wave in the tested steel pipeline

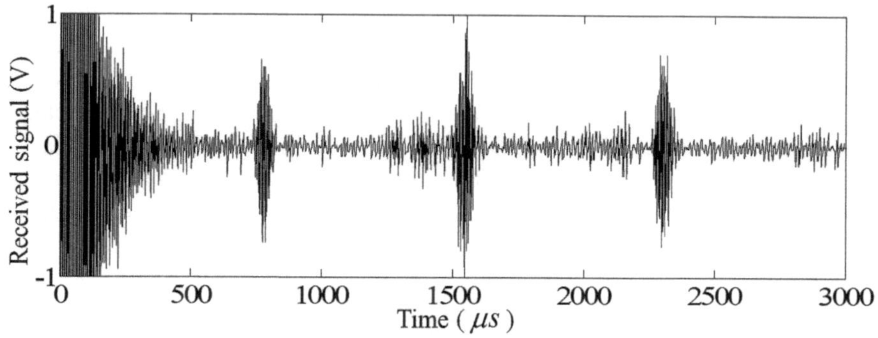

Fig. 1.44 The reflected signal from the end of the steel pipeline in the experiment

using the propagation distance divided by the propagation time, which is 3255 m/s. There is quite a comparable wave velocity between the experimental one with that shown in the dispersion curve, and thus, the correctness of the excited ultrasonic wave was verified. It also verified the effectiveness of the proposed method to generate and receive a *T*-mode guided wave along the axial direction of the pipe using EMAT.

1.3.3 SH Guided Wave EMAT Used in Non-ferromagnetic Material

Based on a magnetostrictive mechanism, ferromagnetic material can be used to excite ultrasonic waves with a greater amplitude value than that in non-ferromagnetic material. Conveniently, a *SH* guided wave can also be excited based on a magnetostrictive mechanism. From this view point, the excitation and receiving of a *SH0* guided wave can be achieved in a platelike non-ferromagnetic specimen.

The specific approach is as follows: A thin nickel strap is used as the medium to excite and receive the *SH* guided wave. The *SH* guided wave is first excited in the thin nickel strap by using the meander coil and then by using the direct coupling method or coupling medium, the ultrasonic waves are coupled into the non-ferromagnetic specimen. When receiving the ultrasonic wave, the propagation of the *SH* guided wave in the non-ferromagnetic specimen will cause vibration in the specimen, which will propagate into the thin nickel strap that is used to receive the ultrasonic wave. The *SH* guided wave can be achieved through the inverse magnetostrictive effect.

Because of the tiny thickness of the nickel strap, a strong ultrasonic signal can be excited into it. Although there are energy losses in the coupling process of the thin nickel strip and tested specimen, a *SH* guided wave with higher amplitude and signal-to-noise ratio can also be excited into the specimen. There are two methods to realize the measurement of *SH* guided waves by designing the structures of EMAT: contact and non-contact methods. For the contact method, the thin nickel strap, meander coil, and biasing magnet are taken as a whole body. The operating mode of EMAT is similar to that of the piezoelectric ultrasonic transducer. For the non-contact method, the thin nickel strip is fixed on the surface of the specimen. In the process of measurement, the meander coil and biasing magnet should be placed above the nickel strip.

In practical applications, the EMAT can be assembled into the same form as the piezoelectric transducer because this EMAT just needs the biasing magnet, coil, and

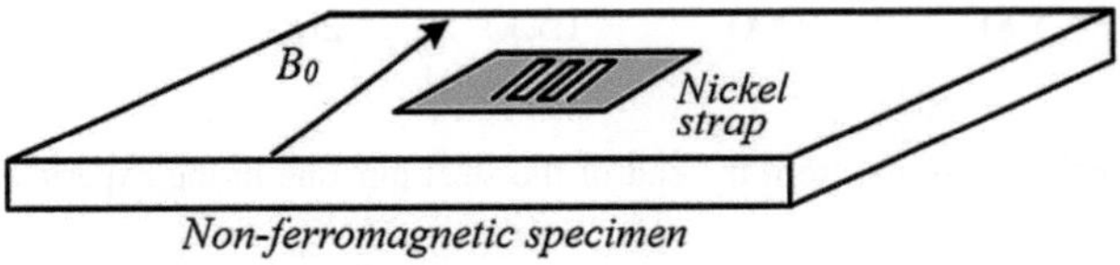

Fig. 1.45 The principle structure of EMAT

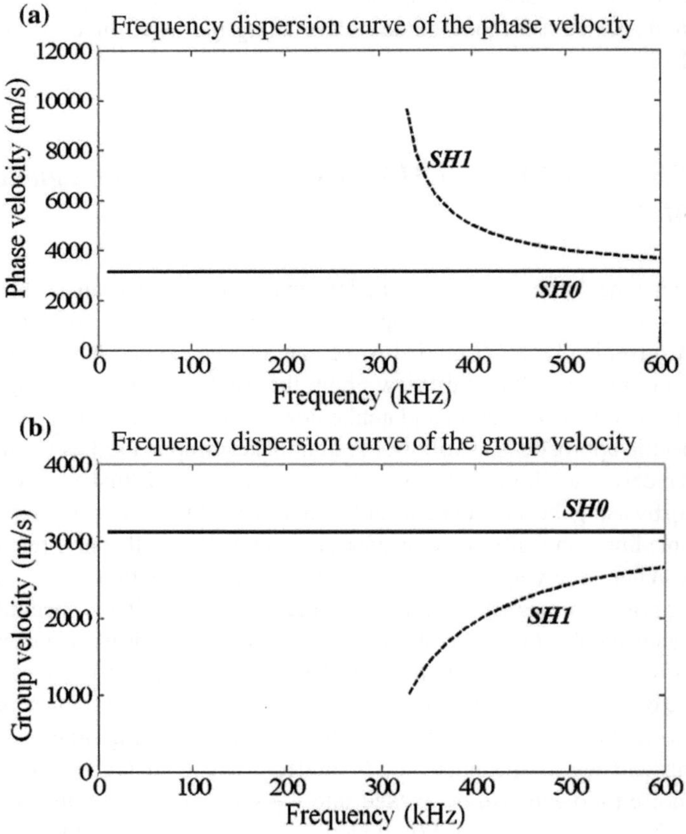

Fig. 1.46 The dispersion curves of the phase and group velocities in the aluminum specimen

thin nickel strap. The cost of this transducer can be lower than that of piezoelectric transducers.

The principle structure of the EMAT is shown in Fig. 1.45.

In order to verify the effectiveness of this EMAT structure, an experimental platform was set up. A thin nickel strap (thickness 0.2 mm, length 50 mm, width 40 mm) was fixed on the aluminum plate with 5 mm thickness using epoxy resin. The epoxy resin plays the function of fixation as well as the coupling medium of the ultrasonic guided wave between the nickel strap and the aluminum specimen. The dispersion curves of the *SH* guided wave modes in the aluminum specimen under test are shown in Fig. 1.46.

As shown in Fig. 1.46, the phase velocity and group velocity do not change with the product of frequency and thickness of the specimen, which means the *SH0* guided wave mode generated by EMAT has a wide application for non-dispersive characteristics.

In the experimental study, the excitation frequency was chosen as 350 kHz considering the character of the coil structure. The incident and receiving signals excited by the coil are all in contact with the signal excitation and receiving device (*RITEC EPR-4000*). The final waveform signal will be collected by a data acquisition card and processed and shown on the computer.

The test setup is shown in Fig. 1.47.

The waveform of the incident signal used in the experiment is shown in Fig. 1.48.

The reflected signal received by the coil from the end of the aluminum specimen is shown in Fig. 1.49. It is found that the signal has a higher signal–noise ratio and can propagate a long distance. The wave velocity of the signal is calculated by

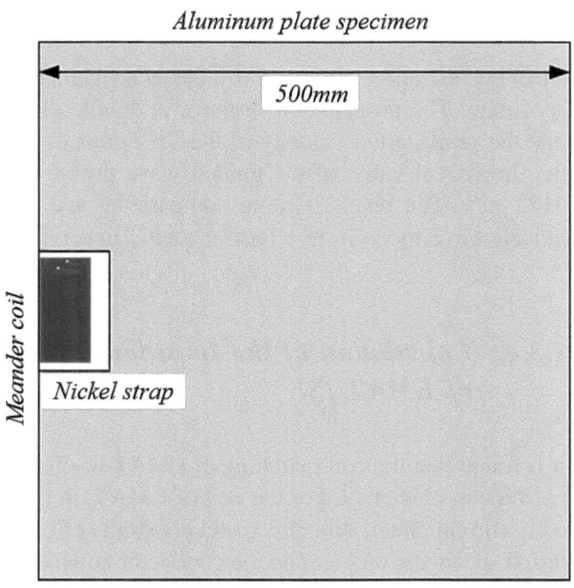

Fig. 1.47 The sketch of the experimental arrangement of the *SH* guided wave excitation and receiving in the aluminum plate

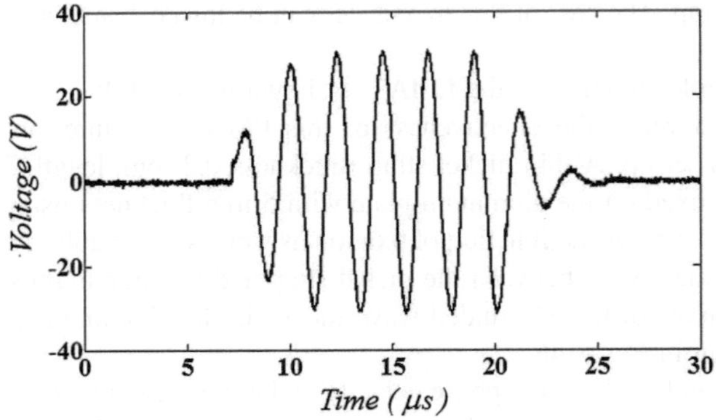

Fig. 1.48 The waveform of the excitation voltage applied by the coil

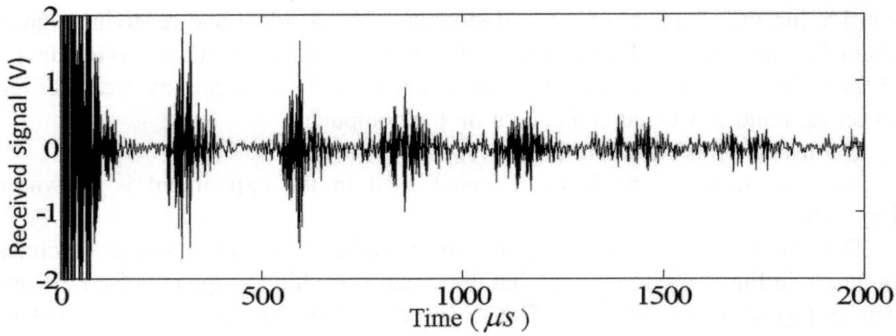

Fig. 1.49 The reflected signal received by the coil from the end of the aluminum specimen

checking the time difference of the echo signal from the end of the aluminum specimen. The propagation distance is double the length of specimen. It is found that the propagation velocity of the *SH0* guided wave is 3225 m/s, compared with the theoretical value of the guided wave group velocity provided in Fig. 1.46 of 3122 m/s. The result is quite comparable, and thus, the generation of the *SH0* guided wave mode in non-ferromagnetic materials by EMAT is verified.

1.3.4 Calculation of the Impedance Matching Capacitance of EMAT [5]

It is found that the coil matching of EMAT is effective and necessary to improve its conversion efficiency. For the coil of EMAT, in the energy conversion process, the eddy current effect, skin effect, and proximity effect between the EMAT coil and the tested specimen, and among the conductor coils, can all lead to impedance changes.

In addition, the scale of the impedance can also be influenced by the material property and the coil's liftoff. Moreover, the coil's distribution under high frequency cannot be ignored. All the factors mentioned above can make the matching of the coil's impedance difficult. At present, according to specific inspection situations, several experimental attempts are generally introduced for the matching of the coil. In this way, it is not only time-consuming, but also involves great effort. Besides, it is difficult to achieve the best matching result. The best method of matching the coil is to calculate the impedance value of EMAT's coil accurately and then calculate the value of the shunt capacitance needed by the matching. The optimum matching of the coil can be achieved conveniently using this approach.

In Fig. 1.50, L_{eq} is the equivalent inductance, R_{eq} is the equivalent resistance, and C_d is the equivalent distributed capacitance.

When the EMAT is in its working status, a tone burst signal generated by high-frequency power source is applied to the coil with capacitance matching through the wire, to excite ultrasonic waves in the specimen. It is important to note that because the wire that connects the coil and signal source is relatively short, its capacitance value is too small to be considered. In Fig. 1.50, R_{in} indicates the internal resistance of the signal source, C_{in} indicates the port capacitance of the signal source, and C_m indicates the matching capacitance of the signal source. The shunt capacitor model shown in Fig. 1.50 is an effective matching model of the narrow band. The coil can obtain maximum energy from the signal source after the matching.

The purpose of matching the coil is to make it obtain maximum energy from the excitation power supply. In other words, it is to make the coil current reach its maximum under the given excitation. The matching method is making a parallel connection for the capacitor at the two ends of the actual coil, specific to the coil's equivalent circuit. As shown in Fig. 1.50, the actual coil's AC resistance under the AC steady state can be considered to have a fixed value. The reaction at the two

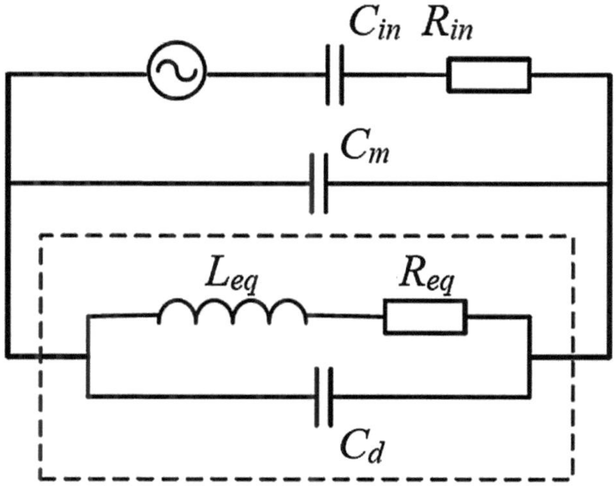

Fig. 1.50 The coil's equivalent circuit and driving circuit

ends of the coil can be changed by shunt capacitance. Thereby, the admittance after the parallel connection of the capacitor can be expressed as (1.5).

$$Y_{\rm eq}=\frac{R_{\rm eq}}{R_{\rm eq}^2+(\omega_0 L_{\rm eq})^2}+j(\omega_0(C_{\rm d}+C_{\rm m})-\frac{\omega_0 L_{\rm eq}}{R_{\rm eq}^2+(\omega_0 L_{\rm eq})^2}) \quad (1.5)$$

In the above equation, ω_0 is the angular frequency of the excitation signal.

When there is resonance between the actual coil and the matching capacitor, the impedance of this parallel link reaches its maximum value and the coil can obtain the maximum energy from the excitation power supply. It can be expressed as (1.6).

$$\omega_0(C_{\rm d}+C_{\rm m})-\frac{\omega_0 L_{\rm eq}}{R_{\rm eq}^2+(\omega_0 L_{\rm eq})^2}=0 \quad (1.6)$$

Therefore, (1.7) can be obtained.

$$C_{\rm m}=\frac{L_{\rm eq}}{R_{\rm eq}^2+(\omega_0 L_{\rm eq})^2}-C_{\rm d} \quad (1.7)$$

Taking the double-layer and three-split meander coil as an example, it explains the principle structure of the double-layer and multisplit coil. It is illustrated in Fig. 1.51, where Fig. 1.51a is the top view of the coil and Fig. 1.51b is the cross section drawing of each coil fold. The length of the coil conductor is l and the center distance between two adjacent fold lines is w, which corresponds to one half of the generated ultrasonic wave. Each fold of the coil is made up of six conductors which are independent of each other. Those six conductors are divided into two layers arranged parallel to each other, the current direction of each conductor being consistent with the rest. In Fig. 1.51b, w_d and h_d represent the width and thickness of each conductor, respectively, w_j represents the conductor's interval of the same layer, and h_j represents the interval between the two conductor layers.

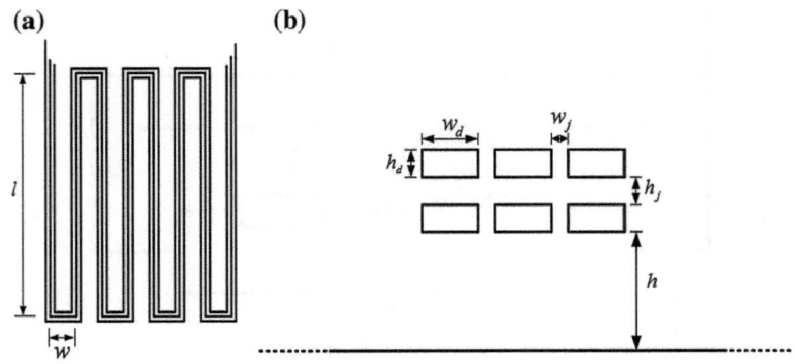

Fig. 1.51 The principle structure of the coil

The parameters of the coil are as follows: l = 60 mm; w = 6.5 mm; w_d = 1.96 mm; w_j = 0.2 mm; h_d = 0.1 mm; and h_j = 0.1 mm.

The coil was made by copper, and its material parameters are as follows: μ_r = 0.999991; σ = 2.67 × 10^7 S/m. The coil's direct current resistance and inductance are, respectively, R = 0.999991 Ω; L = 6.8 μH. The specimen is an aluminum plate, and its material parameters are μ_r = 1.000,021; σ = 3.8 × 10^7 S/m.

According to the given coil and aluminum specimen under test, the coil excitation is a tone burst with a frequency of 250 kHz. The coil's AC resistance, electrical inductance, and matching capacitance were calculated. When the matching capacitance was determined using the experimental method, the tone burst excitation signal was input into the parallel connection of the coil and capacitor. An oscilloscope was used to observe the voltage waveform at the two ends of the actual coil, and the value of the matching capacitance and the frequency of the excitation signal were adjusted, so that the value of the matching capacitance is the required value when the voltage amplitude value reaches its maximum under the excitation frequency of 250 kHz. The computation and experimental results are shown in Table 1.4.

In Table 1.4, C_{me} is the matching capacitance measured from the experiments. Other parameters are all calculated values. The error represents the difference between the experimental and calculated results. It is shown that no errors exceed 2 %, and therefore, the calculation method has a high accuracy.

From the calculated results, the skin effect and proximity effect between the coil conductors and the eddy current effect between coil and specimen all have a significant influence on the value of the coil's resistance and inductance. Compared to the voltaic resistance and inductance, the resistance will increase and the inductance will decrease. Meanwhile, the coil's liftoff value also has a significant effect on the coil's inductance. The influence of the coil's liftoff on the coil's parameters is illustrated in Fig. 1.52.

The calculated results of the parameters under different liftoff values are provided in Fig. 1.52. It shows that the coil's resistance value will decrease with increase in the liftoff value, and the coil's inductance value will increase with increase in the liftoff value. The corresponding changes all show an index characteristic. It is easy to give a physical explanation of the parameter's change with

Table 1.4 Calculation results of each parameter under different liftoff values

h/mm	0.5	1.0	1.5	2.0	2.5	3.0	3.5	4.0
R_{eq}/Ω	2.183	2.103	2.057	2.030	2.013	2.002	1.995	1.991
L_{eq}/μH	2.764	3.781	4.428	4.848	5.125	5.311	5.440	5.530
C_m/pF	117,050	95,250	84,170	78,060	74,430	72,150	70,660	69,630
C_{me}/pF	116,000	96,000	83,800	77,000	75,000	734,000	70,000	69,000
Error (%)	0.9	0.78	0.44	1.3	0.76	1.7	0.94	0.91

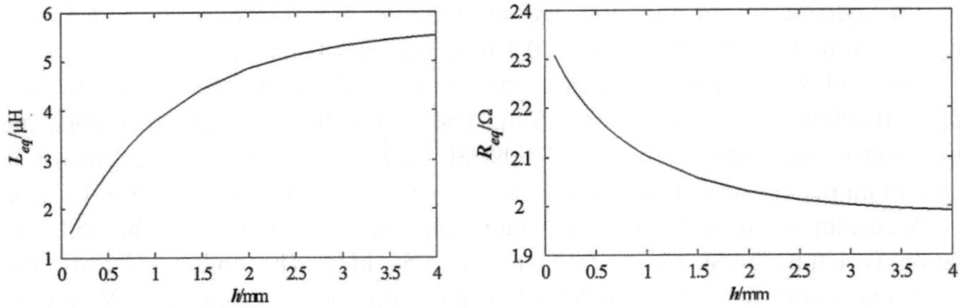

Fig. 1.52 The influence of the coil's liftoff value on the parameters

the variation in liftoff values: With an increase in the liftoff value, the skin effect and proximity effect between the coil conductor and specimen will die down; the distribution of the current density in the coil conductor will become average; and the conductor's area in which the same amount of current flows through will increase. Thus, the coil's resistance will decrease, while at the same time, with the increase in the liftoff value, the effect of the magnetic field produced by the eddy currents inside the specimen to the coil's magnetic field will decrease, and thus, the electrical inductance will increase.

References

1. Zhang, Y.: Development of oil and gas pipeline crack detector based on EMAT. Tsinghua University, Beijing (2008) (in Chinese)
2. Li, P.: Detection device based on electromagnetic ultrasonic guided wave in pipe. Tsinghua University, Beijing (2012) (in Chinese)
3. Zhang, Y., Huang, S., Zhao, W., et al.: Crack detection system of steel plate by electromagnetic ultrasonic testing. Nondestr. Test. **31**(4), 307–310 (2009) (in Chinese)
4. Huang, S., Zhao, W., Zhang, Y., et al.: Study on the lift-off effect of EMAT. Sens. Actuators **153**(2), 218–221 (2009)
5. Hao, K.S., Huang, S.L., Zhao, W.: Calculation of coil impedance and matching capacitance for EMAT. Chin. High Technol. Lett. **20**(8), 845–849 (2010). (in Chinese)

Chapter 2
Analytical Method of EMAT Based on Lorentz Force Mechanism

The energy conversion mechanism of EMAT that is used in non-ferromagnetic metal material is the *Lorentz force* mechanism. The corresponding EMAT is called the *Lorentz force*-based EMAT. For the *Lorentz force*-based EMAT, there is a relatively complete ultrasonic mathematic model that is used to describe the excitation and receiving of EMAT. But because of the coupling between the electromagnetic field and mechanical field in the complete energy conversion process, it can be very difficult to achieve its numerical simulation. In EMAT, the spiral coil and meander coil are widely used, and with the improvement of the coil's production process, those two kinds of coils are generally designed as the complex structures of multilayer or multisplit. Another alternative approach for theoretical analysis is making analytical models of EMAT with those two kinds of coil structure, and calculating its coil impedance and pulse response. Considering that most excitation power supplies of EMAT are voltage-sourced, while almost all the analysis of the existing mathematical models use current excitation to do the analysis of EMAT, it is necessary to propose the analytical method of EMAT under the excitation of impulse voltage. This chapter mainly concerns the simulation analysis of the entire process of EMAT based on the *Lorentz force* mechanism finite element method. The analytical calculation method of the EMAT has the spiral coil and meander coil structure under the excitation of pulsed voltage.

2.1 Multifield Coupling Equation of EMAT Based on Lorentz Force Mechanism

The EMAT based on the *Lorentz force* mechanism consists of the biasing magnet, coil, and non-ferromagnetic specimen. The biasing magnet is used to provide a bias magnetic field, the coil is used to produce a pulsed magnetic field, and the specimen is used for the excitation and reception of the ultrasonic waves. The static magnetic

S. Huang et al., *Electromagnetic Ultrasonic Guided Waves*, Springer Series in Measurement Science and Technology, DOI 10.1007/978-981-10-0564-0_2

field, pulsed eddy current field, and mechanical field are involved in the multifield coupling. The numerical simulation analysis of the whole energy conversion process of EMAT means solving the multifield coupling problem. In this work, the equations used to describe the physical fields that include the excitation and receiving processes of EMAT ultrasonic waves are given first.

2.1.1 Magnetic Field Equation of a Permanent Magnet

The magnetic field generated by a permanent magnet belongs to the static magnetic field. In the magnetic field of a permanent magnet, there is no free current, which means the current density is $J = 0$, as expressed in (2.1).

$$\nabla \times H = 0 \tag{2.1}$$

In the above equation, H is the magnetic field intensity and can be represented as the gradient of a scalar function, as shown in (2.2).

$$H = -\nabla \varphi_m \tag{2.2}$$

In the above equation, φ_m is the scalar magnetic potential.

Considering the existence of the permanent magnet, (2.3) can be obtained.

$$B = \mu H + \mu_0 M_0 = \mu H + B_r \tag{2.3}$$

In the above equation, μ is the magnetic permeability of the magnetic medium; μ_0 is the magnetic permeability of the vacuum, M_0 is the residual magnetization; B is the intensity of the magnetic induction, and B_r is the residual magnetic intensity in the permanent magnet.

Substituting (2.2) into (2.3), from $\nabla \cdot B = 0$, (2.4) can be obtained.

$$\mu \nabla^2 \varphi_m = -\mu_0 \nabla \cdot M_0 \tag{2.4}$$

According to the magnetic charge model of the computational electromagnetism, the spatial magnetic field is produced by magnetic charge. The magnetic field generated in the space by a permanent magnet with a limited size can be taken as the superposition of the magnetic field generated in the space by the magnetic charge that is distributed with a certain rule; that is, the permanent magnet with a limited size can be taken as the magnetic charge distribution with a certain rule. The relationship is shown in (2.5).

$$\rho_m = -\mu_0 \nabla \cdot M_0 \tag{2.5}$$

In the above equation, ρ_m is the magnetic charge density; for the permanent magnet magnetizing uniformly with a certain direction, M_0 is a constant vector. Therefore, (2.6) can be obtained.

$$\rho_m = 0 \tag{2.6}$$

For the permanent magnet with a uniform magnetization, there is no body magnetic charge density, only the surface magnetic charge density σ_m exists. Then, (2.4) can be expressed by (2.7).

$$\mu \nabla^2 \varphi_m = 0 \tag{2.7}$$

The interface condition between the magnetic pole surface of the permanent magnet and any other medium is listed in (2.8).

$$\sigma_m = \mu_0 M_0 \cdot n \tag{2.8}$$

2.1.2 Dynamic Magnetic Field Equation of the Pulsed Eddy Current [1]

There is no free charge in a system that consists of an EMAT coil and a specimen. If the effect of displacement current density is overlooked, the dynamic magnetic field of the pulsed eddy current can be expressed by (2.9).

$$\frac{1}{\mu} \nabla^2 A - \sigma \frac{\partial A}{\partial t} = -J_s \tag{2.9}$$

In the above equation, A is the vector magnetic potential; σ represents the material's conductivity, and J_s is the density of the source current.

The right part of (2.9) is the density of the source current. In the simulation of EMAT, the excitation current and voltage are generally provided. The current density should be determined by the given value of electric current or voltage. When the coil's skin effect and proximity effect can be neglected, the distribution of the average current density in the coil can be approximately solved by electric current, as shown in (2.10).

$$J_S = \frac{i}{S} \tag{2.10}$$

In the above equation, i is the total current and S is the cross-area of the coil conductor.

Because the excitation frequency applied in the EMAT's coil can be very high from KHz to MHz and the distance between the coil conductors, as well as that from the coil conductor to specimen, are all very small, the coil's skin effect, proximity effect, and eddy effect are all very significant. It is unnecessary to figure out the accuracy of the theoretical calculation that can be reduced if those effects discussed above are overlooked. Actually, the total current density is expressed in (2.11).

$$J_t = J_s + J_e = J_s - \sigma \frac{\partial A}{\partial t} \tag{2.11}$$

The total current can be expressed by (2.12).

$$\iint\limits_S J_t \mathrm{d}s = i \tag{2.12}$$

Then, (2.13) can be obtained.

$$J_s = \frac{i}{S} + \frac{1}{S} \iint\limits_S \sigma \frac{\partial A}{\partial t} \mathrm{d}s \tag{2.13}$$

And (2.14) can be obtained by substituting (2.13) into (2.9).

$$\frac{1}{\mu} \nabla^2 A - \sigma \frac{\partial A}{\partial t} + \frac{1}{S} \iint\limits_S \sigma \frac{\partial A}{\partial t} \mathrm{d}s = -\frac{i}{S} \tag{2.14}$$

The relationship between E (the electric field intensity of each location) and vector magnetic potential is expressed in (2.15).

$$E = -\frac{\partial A}{\partial t} \tag{2.15}$$

The eddy current density inside the coil and specimen is expressed in (2.16).

$$J_e = -\sigma \frac{\partial A}{\partial t} \tag{2.16}$$

From the definition of *Lorentz force*, the force density in the skin depth of the surface of a non-ferromagnetic specimen is related to the magnitude of the static magnetic field provided by the biasing magnet and the eddy current in the surface of the non-ferromagnetic conductor; it can be expressed by (2.17).

$$f_L = B_0 \times J_e \tag{2.17}$$

In the previous literature, the static magnetic field has uniform distribution. So the force on the surface of the non-ferromagnetic specimen extends in one single-direction, but this does not conform to the specific circumstances. In the practical application of EMAT, the value and direction of the static magnetic field provided by the permanent magnet are not uniform nor in a single-direction; thus, it will inevitably bring errors if the static magnetic field is assumed to have uniform distribution and one single direction.

The boundary condition must be satisfied on the outer boundary. That is the vector magnetic potential $A_z = 0$.

In the interface of two kinds of magnetic medium, the following interface conditions listed in (2.18) are satisfied.

$$A_i = A_{i+1} \tag{2.18}$$

$$\frac{1}{\mu_i}(\nabla \times A_i)_t - \frac{1}{\mu_{i+1}}(\nabla \times A_{i+1})_t = J_{si} \tag{2.19}$$

In the above equations, subscript i represents the adjacent magnetic medium; subscript t represents the tangential direction of the interface, and J_{si} is the surface current density in the adjacent interface.

2.1.3 Motion Equation of Particle in the Specimen

The tested specimen is isotropic material, so it can be assumed as having linear elasticity and continuity. The elastic deformation occurs inside the specimen under the *Lorentz force*. The motion equation in the specimen is expressed in (2.20).

$$\nabla \cdot \sigma + f_L = \rho \frac{\partial^2 u}{\partial t^2} \tag{2.20}$$

In the above equation, σ is the stress tensor; u is the displacement matrix, and ρ is the specimen's bulk density.

Considering the relationship between σ and u, (2.20) can be expressed using the displacement, as in (2.21).

$$G\nabla^2 u + (G + k)\nabla(\nabla \cdot u) + f_L = \rho \frac{\partial^2 u}{\partial t^2} \tag{2.21}$$

In the above equation, G and k are *Lame constants*.

The traction-free boundary condition should be satisfied such that stress is zero and the strain is not zero on the surface of the specimen.

2.1.4 Receiving Equation of Ultrasonic Signal

When the ultrasonic wave in the specimen propagates to the receiving coil of EMAT, the charged particle will generate a dynamic current under the function of the external bias magnetic field; its current density is expressed in (2.22).

$$J_L = \sigma v \times B_0 \tag{2.22}$$

The current density in the specimen will generate a dynamic magnetic field inside and around the specimen. The coil of EMAT in the dynamic magnetic field will generate an induced electromotive force, which is the coil's receiving signal.

In the signal receiving process of the coil, the magnetic field in the solving area is provided by the eddy current density in the specimen and the source current density. Generally, the receiving coil is in an open circuit state and the total current of the receiving coil is zero. The governing equation satisfied in the area of the receiving coil and specimen can be expressed in (2.23).

$$-\frac{1}{\mu}\nabla^2 A + \sigma\frac{\partial A}{\partial t} - \frac{\sigma}{S}\frac{\partial}{\partial t}\iint\limits_{\Omega_c} A\mathrm{d}s = J_L \tag{2.23}$$

After solving the vector magnetic potential of each region using the above equation, the induced electromotive force of the coil can be calculated. The induced electric field in the coil can be expressed by (2.24).

$$E = -\frac{\partial A}{\partial t} \tag{2.24}$$

The conductor's electromotive force at some point of the coil can be obtained by using the line integral of the electric field intensity, expressed in (2.25).

$$V_{\mathrm{pout}} = \int\limits_{l} -\frac{\partial A}{\partial t}\cdot \mathrm{d}l \tag{2.25}$$

Thus, the output voltage of the coil can be obtained by averaging the electromotive force of the point conductors in the coil, expressed in (2.26).

$$V_{\mathrm{out}} = \frac{\int_{\Omega} V_{\mathrm{pout}}\mathrm{d}\Omega}{\int_{\Omega}\mathrm{d}\Omega} \tag{2.26}$$

2.2 The Weak Form of the Coupling Field Equations

The analysis and simulation of the EMAT based on the Lorentz force mechanism aim at the specific physical model in combination with the corresponding boundary condition, solving the partial differential equations (2.7), (2.14), (2.21), and (2.23). The finite element method is an effective numerical method to solve this problem. When it comes to solving those equations using the finite element method, in fact, the solutions are the weak solutions, also referred to as the "weak form." Thus, it is necessary to derive the weak form of the above equations, that is, to derive the weak forms of (2.7), (2.14), and (2.21) based on the principle of virtual displacement. Since the weak forms of (2.23) and (2.14) are similar to each other, they will not be given any more. According to the differences in the EMAT's structure in the analytical process, the weak form of various field equations under the two-dimensional *Cartesian coordinate system* and axisymmetric coordinate system is needed, which are derived, respectively, as below.

2.2.1 The Weak Form of Coupled Equations Under Two-Dimensional Cartesian Coordinates

In the finite element analysis of EMAT, where the coil consists of linear conductors (such as the meander coil EMAT and rectangular spiral coil EMAT), all the factors in the conductor's length direction (assumed to be the direction of the Z axis), including the difference of the field characteristics among the permanent magnet, coil, and specimen, can be neglected. So the EMAT can be simplified into a two-dimensional model on the coil's cross section. Thus, among the field variables, the components of the x axis and y axis are the only factors that need to be considered.

2.2.1.1 The Weak Form of the Static Magnetic Field Equation of the Permanent Magnet

In the two-dimensional *Cartesian coordinate system*, the static magnetic field equation of the permanent magnet is expressed in (2.27).

$$\frac{\partial}{\partial x}\left(\mu\frac{\partial\varphi_m}{\partial x}\right)+\frac{\partial}{\partial y}\left(\mu\frac{\partial\varphi_m}{\partial y}\right)=0 \tag{2.27}$$

The virtual displacement function $\delta\varphi_m$ is multiplied at both sides of the equation at the same time, and the integral is calculated to obtain (2.28).

$$\int_{\Omega} \left[\frac{\partial}{\partial x}\left(\mu \frac{\partial \varphi_m}{\partial x}\right) + \frac{\partial}{\partial y}\left(\mu \frac{\partial \varphi_m}{\partial y}\right) \right] \delta\varphi_m \mathrm{d}A = 0 \tag{2.28}$$

In the above equation, Ω indicates the solution domain and dA indicates the surface element.

Carrying out partial integration for the left-hand side of the equation, we obtain (2.29).

$$\int_{\Omega} \mu \left(\frac{\partial \varphi_m}{\partial x}\frac{\partial \delta\varphi_m}{\partial x} + \frac{\partial \varphi_m}{\partial y}\frac{\partial \delta\varphi_m}{\partial y} \right) \mathrm{d}A = \int_{\Gamma} \mu \frac{\partial \varphi_m}{\partial n} \delta\varphi_m \mathrm{d}l \tag{2.29}$$

In the above equation, Γ indicates the boundary of the solution domain and dl is the boundary element.

2.2.1.2 The Weak Form of the Coil's Pulsed Eddy Current Field Equation

In the two-dimensional *Cartesian coordinate system*, the vector magnetic potential only has the component of the Z direction. The equation of the coil's pulsed eddy current field can be expressed in (2.30).

$$-\frac{\partial}{\partial x}\left(\frac{1}{\mu}\frac{\partial A_Z}{\partial x}\right) - \frac{\partial}{\partial y}\left(\frac{1}{\mu}\frac{\partial A_Z}{\partial y}\right) + \sigma \frac{\partial A_Z}{\partial t} + \frac{1}{S}\iint_S \sigma \frac{\partial A_Z}{\partial t} \mathrm{d}s = J_{sz} \tag{2.30}$$

Multiplying the virtual displacement function δA_Z of this equation and carrying out the integration is expressed by (2.31).

$$\begin{aligned} &\int_{\Omega} \left[-\frac{\partial}{\partial x}\left(\frac{1}{\mu}\frac{\partial A_z}{\partial x}\right) - \frac{\partial}{\partial y}\left(\frac{1}{\mu}\frac{\partial A_z}{\partial y}\right) + \sigma \frac{\partial A_z}{\partial t} + \frac{1}{S}\iint_S \sigma \frac{\partial A_z}{\partial t} \mathrm{d}s \right] \delta A_z \mathrm{d}A \\ &= \int_{\Omega} J_{sz} \delta A_z \mathrm{d}A \end{aligned} \tag{2.31}$$

Carrying out partial integration for the above equation using Green's function as in (2.32).

$$\int_{\Omega}\left(\frac{1}{\mu}\frac{\partial A_Z}{\partial x}\frac{\partial \delta A_Z}{\partial x}+\frac{1}{\mu}\frac{\partial A_z}{\partial y}\frac{\partial \delta A_z}{\partial y}\right)\mathrm{d}A$$
$$=-\int_{\Omega}\sigma\frac{\partial A_z}{\partial t}\delta A\mathrm{d}A+\int_{\Omega}\frac{1}{S}\iint_{S}\sigma\frac{\partial A_Z}{\partial t}\mathrm{d}s\delta A_z\mathrm{d}A$$
$$+\int_{\Omega}J_{sz}\delta A_z\mathrm{d}A+\int_{\Gamma}\frac{1}{\mu}\frac{\partial A_z}{\partial n}\delta A_z\mathrm{d}l \tag{2.32}$$

2.2.1.3 The Weak Form of the Specimen's Wave Equation

The problem of the specimen's particle displacement caused by the *Lorentz force* can be simplified as a plane-strain problem in the elasticity mechanics. For the plane-strain problem, the desired characteristics of the geometry and external forces are listed as follows: The sample is very long along one direction and all of the cross sections that are perpendicular to this direction are the same; the conditions of displacement and support are also same; the surface traction and body force are both perpendicular to this direction and will not change with the length.

For the problem of a plane-strain condition, the direction of the length is supposed to be along the *Z* coordinate axis. Because the sample is very long, the sectional dimension and external force will not change with the length. The stress and strain also have no change along the *Z* coordinate axis. Thus, the factors to be concerned with are the stress, strain, and displacement components including ε_x, ε_y, ε_{xy}, σ_x, σ_y, σ_{xy}, u, and v.

In the two-dimensional *Cartesian coordinate system*, the equilibrium equations of each point along the direction of x and y in the specimen are expressed in (2.33) and (2.34).

$$\frac{\partial \sigma_x}{\partial x}+\frac{\partial \sigma_{xy}}{\partial y}+f_x=\rho\frac{\partial^2 u}{\partial t^2} \tag{2.33}$$

$$\frac{\partial \sigma_{xy}}{\partial x}+\frac{\partial \sigma_y}{\partial y}+f_y=\rho\frac{\partial^2 v}{\partial t^2} \tag{2.34}$$

The geometric equation is expressed in (2.35).

$$\varepsilon_x=\frac{\partial u}{\partial x};\ \varepsilon_y=\frac{\partial v}{\partial y};\ \varepsilon_{xy}=\frac{\partial u}{\partial y}+\frac{\partial v}{\partial x} \tag{2.35}$$

The physical equation is expressed in (2.36).

$$\left.\begin{aligned} \varepsilon_x &= \frac{1-\mu_p^2}{E}\left(\sigma_x - \frac{\mu_p}{1-\mu_p}\sigma_y\right) \\ \varepsilon_x &= \frac{1-\mu_p^2}{E}\left(\sigma_y - \frac{\mu_p}{1-\mu_p}\sigma_x\right) \\ \varepsilon_{xy} &= \frac{2(1+\mu_p)}{E}\sigma_{xy} \end{aligned}\right\} \tag{2.36}$$

In the above equation, E represents the material's elastic modulus and μ_p is the *Poisson* ratio.

For the equilibrium equation, the virtual displacement functions δu and δv were used to obtain (2.37).

$$\begin{aligned} &\int_{\Omega} \left[\left(\frac{\partial \sigma_x}{\partial x} + \frac{\partial \sigma_{xy}}{\partial y} + f_x\right)\delta u + \left(\frac{\partial \sigma_{xy}}{\partial x} + \frac{\partial \sigma_y}{\partial y} + f_y\right)\delta v\right] \mathrm{d}A \\ &= \int_{\Omega} \left(\rho\frac{\partial^2 u}{\partial t^2}\delta u + \rho\frac{\partial^2 v}{\partial t^2}\delta v\right)\mathrm{d}A \end{aligned} \tag{2.37}$$

Carrying out partial integration for the above equation and submitting it into the geometric equation as in (2.38).

$$\begin{aligned} &\int_{\Omega} (\sigma_x\delta\varepsilon_x + \sigma_y\delta\varepsilon_y + \sigma_{xy}\delta\varepsilon_{xy})\mathrm{d}V + \int_{V} \left(\rho\frac{\partial^2 u}{\partial t^2}\delta u + \rho\frac{\partial^2 v}{\partial t^2}\delta v\right)\mathrm{d}A \\ &= \int_{\Omega} (f_x\delta u + f_y\delta v)\mathrm{d}A + \int_{\Gamma} (T_x\delta u + T_y\delta v)\mathrm{d}l \end{aligned} \tag{2.38}$$

In (2.38), T_x and T_y indicate the elastic sample's internal forces on the boundary. Because the specimen satisfies the boundary condition of traction free, the integral value on the right hand of the equation is zero.

2.2.2 The Weak Form of Coupled Equations in the Axisymmetric Coordinate System

For the spiral coil or EMAT used for the testing of the pipeline in the axial direction, the coil's structure and specimen satisfy the axisymmetric condition. In this case, the problem of three-dimensional analysis can be transformed into a two-dimensional problem in the axisymmetric coordinates. In the axisymmetric coordinates system, the vector magnetic potential and eddy current only have the component of θ direction and the other variables of various fields only have the components of r and z directions.

2.2.2.1 The Weak Form of the Permanent Magnet's Static Magnetic Field Equation

In the axisymmetric coordinates, the permanent magnet's static magnetic field equation is expressed in (2.39).

$$\frac{1}{r}\frac{\partial}{\partial r}\left(\mu r\frac{\partial \varphi_m}{\partial r}\right)+\frac{\partial}{\partial z}\left(\mu\frac{\partial \varphi_m}{\partial z}\right)=0 \tag{2.39}$$

In this equation, the virtual displacement function $\delta\varphi_m$ is multiplied and the integral is calculated to obtain (2.40).

$$\int_{\Omega}\left[\frac{1}{r}\frac{\partial}{\partial r}\left(\mu r\frac{\partial \varphi_m}{\partial r}\right)+\frac{\partial}{\partial z}\left(\mu\frac{\partial \varphi_m}{\partial z}\right)\right]\delta\varphi_m \mathrm{d}A=0 \tag{2.40}$$

Using *Green's equation*, partial integration is carried out on the left part of the above equation as in (2.41).

$$\int_{\Omega}\mu\left(r\frac{\partial \varphi_m}{\partial r}\frac{\partial \delta\varphi_m}{\partial r}+r\frac{\partial \varphi_m}{\partial z}\frac{\partial \delta\varphi_m}{\partial z}\right)\mathrm{d}A=\int_{\Gamma}\mu r\frac{\partial \varphi_m}{\partial n}\delta\varphi_m \mathrm{d}l \tag{2.41}$$

2.2.2.2 The Weak Form of the Coil's Pulsed Eddy Current Field

Under the axisymmetric coordinates, the equation of the coil's pulsed eddy current field can be expressed by (2.42).

$$-\frac{\partial}{\partial r}\left(\frac{1}{r\mu}\frac{\partial A_\theta}{\partial r}\right)-\frac{\partial}{\partial z}\left(\frac{1}{\mu}\frac{\partial A_\theta}{\partial z}\right)+\sigma\frac{\partial A_\theta}{\partial t}+\frac{1}{S}\iint_{s}\sigma\frac{\partial A_\theta}{\partial t}\mathrm{d}s=J_{s\theta} \tag{2.42}$$

Multiplying the virtual displacement function $\delta\varphi_m$ into the above equations, and carrying out the integration, (2.42) can be stated using (2.43).

$$\begin{aligned}&\int_{\Omega}\left[-\frac{\partial}{\partial r}\left(\frac{1}{r\mu}\frac{\partial A_\theta}{\partial r}\right)-\frac{\partial}{\partial z}\left(\frac{1}{\mu}\frac{\partial A_\theta}{\partial z}\right)+\sigma\frac{\partial A_\theta}{\partial t}+\frac{1}{S}\iint_{S}\sigma\frac{\partial A_\theta}{\partial t}\mathrm{d}s\right]\delta A_\theta \mathrm{d}A\\&=\int_{\Omega}J_{S\theta}\delta A_\theta \mathrm{d}A\end{aligned} \tag{2.43}$$

Using *Green's equation*, partial integration is carried out on the left part of the above equation, to obtain (2.44).

$$\int_{\Omega} \frac{1}{\mu}\left(r\frac{\partial A_\theta}{\partial r}\frac{\partial \delta A_\theta}{\partial r} + r\frac{\partial A_\theta}{\partial z}\frac{\partial \delta A_\theta}{\partial z}\right)\mathrm{d}A = -\int_{\Omega} r\sigma\frac{\partial A_\theta}{\partial t}\delta A_\theta \mathrm{d}A + \int_{\Omega} rJ_{S\theta}\delta A_\theta \mathrm{d}A + \int_{\Omega} r\frac{1}{S}\iint_{S} \sigma\frac{\partial A_\theta}{\partial t}\mathrm{d}s\delta A_\theta \mathrm{d}A + \int_{\Gamma} r\frac{1}{\mu}\frac{\partial A_\theta}{\partial n}\delta A_\theta \mathrm{d}l \tag{2.44}$$

2.2.2.3 The Weak Form of the Specimen's Wave Equation

In the axisymmetric coordinate system, the deformation states are all the same in each radial plane of the object, that is, both of them have no relation with θ; thus, the displacement components in the radial, tangential, and axial direction of each point in the object are expressed in (2.45).

$$u = u(r, z); v = 0; w = w(r, z) \tag{2.45}$$

In this way, the geometric equation satisfied by the elastic body is expressed in (2.46).

$$\varepsilon_r = \frac{\partial u}{\partial r}; \varepsilon_z = \frac{\partial u}{\partial z}; \varepsilon_\theta = \frac{u}{r}; \varepsilon_{rz} = \frac{\partial u}{\partial z} + \frac{\partial w}{\partial r} \tag{2.46}$$

According to the generalized *Hooke*'s *law*, the stress–strain relationship can be expressed by (2.47), (2.48), (2.49), and (2.50).

$$\varepsilon_r = \frac{1}{E}[\sigma_r - \mu_p(\sigma_\theta + \sigma_z)] \tag{2.47}$$

$$\varepsilon_\theta = \frac{1}{E}[\sigma_\theta - \mu_p(\sigma_r + \sigma_z)] \tag{2.48}$$

$$\varepsilon_z = \frac{1}{E}[\sigma_z - \mu_p(\sigma_r + \sigma_\theta)] \tag{2.49}$$

$$\varepsilon_{rz} = \frac{2(1-\mu_p)}{E}\sigma_{rz} \tag{2.50}$$

In the above equations, σ_r, σ_θ, and σ_z represent the positive stress of each axis direction, respectively. σ_{rz} is the shear stress.

The equilibrium equations of particle motion are listed in (2.51) and (2.52).

$$\frac{\partial \sigma_r}{\partial r} + \frac{\partial \sigma_{rz}}{\partial z} + \frac{\sigma_r - \sigma_\theta}{r} + f_r = \rho\frac{\partial^2 u}{\partial t^2} \tag{2.51}$$

$$\frac{\partial \sigma_{rz}}{\partial r} + \frac{\partial \sigma_{z}}{\partial z} + \frac{\sigma_{rz}}{r} + f_z = \rho \frac{\partial^2 w}{\partial t^2} \tag{2.52}$$

In the above equations, f_r and f_z represent the body forces existing on the specimen and ρ represents the bulk density of the specimen.

The principle of virtual displacement is used on the equilibrium equation to get (2.53).

$$\begin{aligned} &\int_V \left[\left(\frac{\partial \sigma_{rr}}{\partial r} + \frac{\partial \sigma_{rz}}{\partial z} + \frac{\sigma_{rr} - \sigma_{\theta\theta}}{r} + f_r \right) \delta u + \left(\frac{\partial \sigma_{rz}}{\partial r} + \frac{\partial \sigma_{zz}}{\partial z} + \frac{\sigma_{rz}}{r} + f_z \right) \delta w \right] \mathrm{d}V \\ &= \int_V \left(\rho \frac{\partial^2 u}{\partial t^2} \delta u + \rho \frac{\partial^2 w}{\partial t^2} \delta w \right) \mathrm{d}W \end{aligned} \tag{2.53}$$

Partial integration is carried out on the above equation and the geometric equation is used to obtain (2.54).

$$\begin{aligned} &\int_V (\sigma_{rr} \delta \varepsilon_{rr} + \sigma_{zz} \delta \varepsilon_{zz} + \sigma_{rz} \delta \varepsilon_{rz}) \mathrm{d}V + \int_V \left(\rho \frac{\partial^2 u}{\partial t^2} \delta u + \rho \frac{\partial^2 v}{\partial t^2} \delta v \right) \\ &= \int_V \left[\left(f_r + \frac{\sigma_{rr} - \sigma_{\theta\theta}}{r} \right) \delta u + \left(\frac{\sigma_{rz}}{r} + f_z \right) \delta v \right] \mathrm{d}V + \int_\Gamma (T_r \delta u + T_z \delta v) \mathrm{d}\Gamma \end{aligned} \tag{2.54}$$

In the above equation, T_r and T_z indicate the internal force of the specimen's boundary. Since the specimen satisfies the boundary conditions, both T_r and T_z are zero.

2.3 Finite Element Simulation of EMAT by COMSOL Multiphysics [2]

2.3.1 *Simulation Procedure of EMAT by COMSOL Multiphysics*

The *COMSOL Multiphysics* was used to do the modeling analysis of the problem described by the PDEs. There are two optional ways and one of them is the direct modeling for the PDEs. According to the expression form of the PDEs, it can be divided into three kinds: coefficient form, general form, and weak form. At the condition for which the parameters of the varied forms are given and their boundary conditions are satisfied, inputting those parameters and values of the boundary conditions means the modeling work of the physical model can be completed.

Another way is in using the analytical modules of the physical fields that are generally used, which are built into the *COMSOL Multiphysics*. Eight modules were included in the 3.5a edition of *COMSOL Multiphysics*, including AC/DC module, radio frequency module, acoustics module, chemical engineering module, geoscience module, heat transfer module, microelectromechanical system (MEMS) module, and structural mechanics module. No matter which module is selected, the essence of the finite element method using *COMSOL Multiphysics* is to solve the partial differential equations of the actual physical problems by transforming the equation set into the weak form with interface and boundary conditions. Therefore, the weak form is more widely used in carrying out the modeling work for the multiphysics coupling problem, for the reason that it is not only the clear essence of the FEM calculation of the multiphysics field, but also that it can be used to resolve the insoluble problem with the built-in module.

By using the *COMSOL Multiphysics*, the numerical methods and steps for the whole energy conversion process of EMAT based on the Lorentz force mechanism are listed as follows:

(1) Selecting the coordinate system that needs to be solved. If the two-dimensional analysis is adopted, the 2D rectangular coordinate system or axisymmetric coordinate system can be selected. It is necessary to note that the 2D axisymmetric coordinate system does not support the modeling of the weak form by itself. In the process of weak form modeling, the coordinate system is transformed into a rectangular coordinate system.
(2) The solving form was selected as the weak form. In this way, 4 set of equations should be selected.
(3) Setting the geometric model of each component of EMAT. A solution domain should be given when the numerical simulation of the electromagnetic field is done. Thus, a solution domain should be defined beside the model.
(4) According to those weak forms and boundary conditions of the various fields, the solution domain and boundary condition are set, respectively. For the EMAT based on the *Lorentz force* mechanism, besides the calculation of the coil's magnetic field when the skin effect and proximity effect are considered, the built-in modules of *COMSOL Multiphysics* can also be used in the calculation of the static magnetic field and mechanical field. Because the multifield coupling is involved, the coupling variable needs to be set. The product of the static magnetic induction intensity obtained from the solution of the static bias magnetic field, the specimen's magnetic permeability, and the particle's velocity is set as the density of the pulsed eddy current source when calculating the coil's ultrasonic wave. Because the induced electromotive force of each coil conductor should be calculated when calculating the ultrasonic signal received by the coils, the variables of the integral coupling should be set to calculate the induced electromotive force of each step length of the solution.
(5) Meshing for the solution domain. In the process of meshing, two points are important in improving the accuracy of the calculation: One is that the

elements of the mesh should be more than 2 under the skin depth of the specimen's upper surface; and the other is guaranteeing that there are more than 7 elements of mesh within the wavelength of the excited ultrasonic wave inside the specimen.

(6) Solving the model. The solution of the static bias magnetic field is a steady-state solution, while other fields are transient solutions. The transient solution needs to be set to the step length and the relative and absolute error of the solutions. Because the adoptive pulsed excitation signal is a tone-burst signal with high frequency, the step length should be set smaller to ensure the smoothness and stability of the solutions. Since the value of the displacement is small, the absolute error of the solution should be small enough during the process of solving the displacement of the internal points to ensure the correctness of the solution.

(7) Postprocessing of the results, including the exhibition of the variables' surface plots of the various fields and transient waveforms.

2.3.2 *Example of the Numerical Simulation and Experimental Verification*

To verify the correctness of the derived mathematical model that is used to represent the EMAT's character and the effectiveness of the *COMSOL Multiphysics* based on the numerical simulation of EMAT's working process, an example of the modeling of EMAT was conducted and the calculation result was verified by experimental results.

2.3.2.1 The Parameters of the Simulation

The meander coil was used to excite *Lamb waves* within an aluminum plate. Because of the dispersive nature of *Lamb waves*, before the fabrication of the EMAT the phase velocity and group velocity dispersion curves of *Lamb waves* should be calculated based on the elastic modulus, *Poisson ratio*, and the thickness of the aluminum plate. The appropriate wave mode was selected according to the dispersion curves. The EMAT coil was designed based on the chosen point. The parameters of the aluminum plate are listed in Table 2.1.

Using the calculation software of the dispersion curves, the dispersion curves of phase velocity and group velocity for the *Lamb waves* in the aluminum plate under different frequency-thickness products are listed, respectively, as follows (Fig. 2.1):

The frequency-thickness product of the aluminum plate with a thickness of 3 mm was selected as 1100 m × Hz. That is, the exciting frequency is 366.7 kHz. It is found that there are two modes of *Lamb waves* that can be generated at this condition: A_0 mode and S_0 mode. The phase velocity and group velocity of the A_0

Table 2.1 The dimension and parameters of the tested aluminum plate

Name	Value
Length	500 mm
Width	350 mm
Thickness	3 mm
Conductivity	$3.5 \times e^{-7}$ S/m
Elasticity modulus	70 Gpa
Poisson ratio	0.33

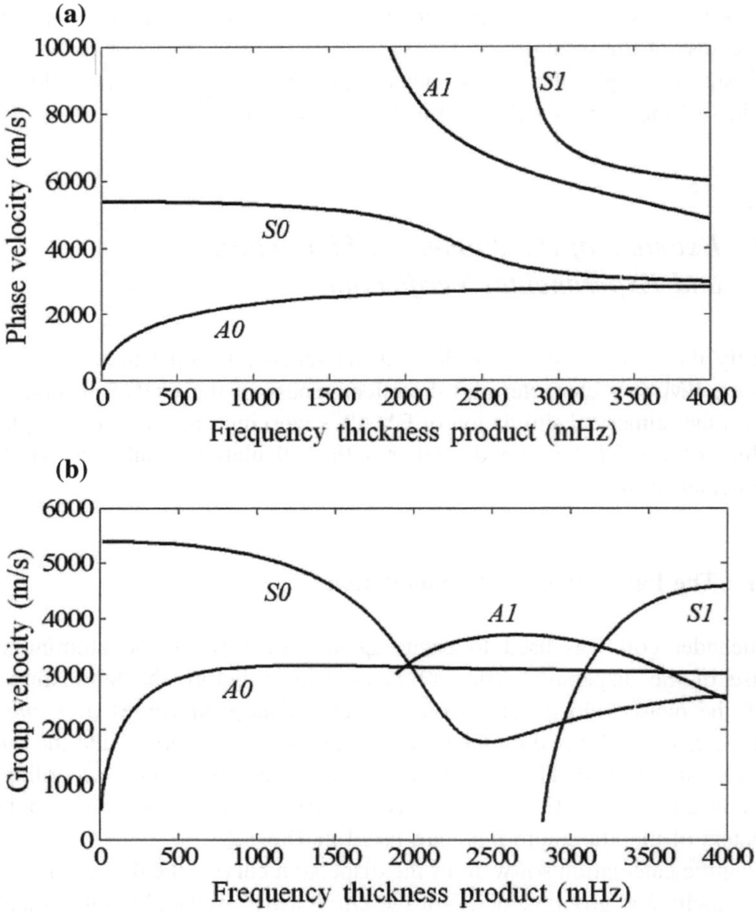

Fig. 2.1 The phase velocity (**a**) and group velocity (**b**) dispersion curves of Lamb waves in the aluminum plate

Fig. 2.2 The double-layer and double-split meander coil

Table 2.2 The coil's dimensions and material's parameters

Name	Value
Substrate thickness	0.500 mm
The width of the copper layer	0.720 mm
The thickness of the copper layer	0.035 mm
The line spacing	0.905 mm
The fold spacing	3.25 mm
The magnetic permeability of the copper layer	$4\pi \times 10^{-7}$ H/m
The electric conductivity of the copper layer	2.667×10^{7} S/m

mode are 5287 and 5044 m/s, respectively. The adopted double-layer and double-split meander coil is shown in Fig. 2.2, with the distance between the adjacent folds being 6.5 mm (Table 2.2).

In the experiment, as shown in Fig. 2.3, excitation and receiving probes of EMAT were used for the excitation and receiving of ultrasonic waves. The excitation probe was installed 90 mm away from the left end of the aluminum plate, the receiving probe was installed 120 mm away from the left end, and the center-to-center spacing of the two probes was 140 mm. The liftoff distance between the two probe's coil and the aluminum plate was 1 mm. A permanent magnet with a residual magnetic flux density of 1 T was located above the bending coil and the distance between them was 0.5 mm. The arrangement of the probe is shown in Fig. 2.3:

2.3.2.2 Experimental Setup

In order to verify the effectiveness of the simulation results, the pulse excitation and receiving device RPR-4000, produced by RITEC, was employed as the exciting source of the EMAT to produce the RF tone-burst signal. The amplitude and frequency of the signal can be adjusted arbitrarily to meet the requirements. Between the RPR-4000 and the excitation coil, an impedance matching device was used to complete the matching of the coil impedance and that of the exciting

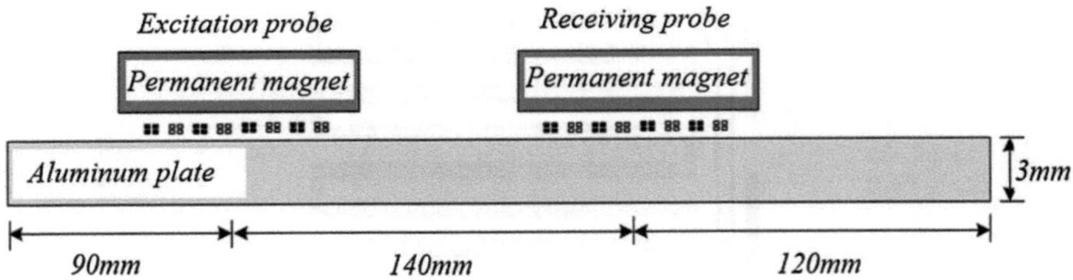

Fig. 2.3 A diagram of the arrangement of the probes for Lamb waves

source's output. The filtering and amplification of the receiving signal can be achieved using the RPR-4000. The bandwidth and amplification factor of the filter are also adjustable. The matching of the coil impedance and the input impedance of the RPR-4000 is achieved by the connection of the receiving coil and the impedance matching device, so that a greater power output can be achieved. The signal received by the RPR-4000 was displayed on the oscilloscope and was processed by the data acquisition software *Wave Star for Oscilloscopes* to collect the data. The final waveform was displayed on the PC. The setup of the experimental system is shown in Fig. 2.4.

2.3.2.3 The Process and Result of Calculations

The waveform of the tone-burst current excitation signal with a frequency of 366.7 kHz and 5 cycles is shown in Fig. 2.5.

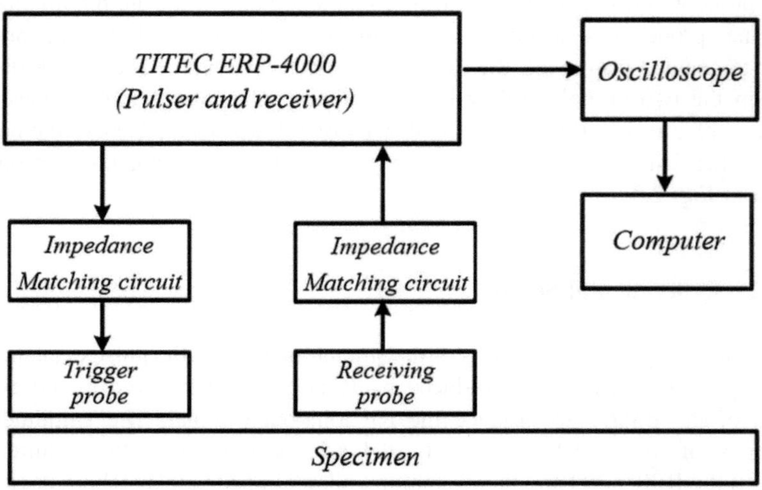

Fig. 2.4 The setup of the experimental system

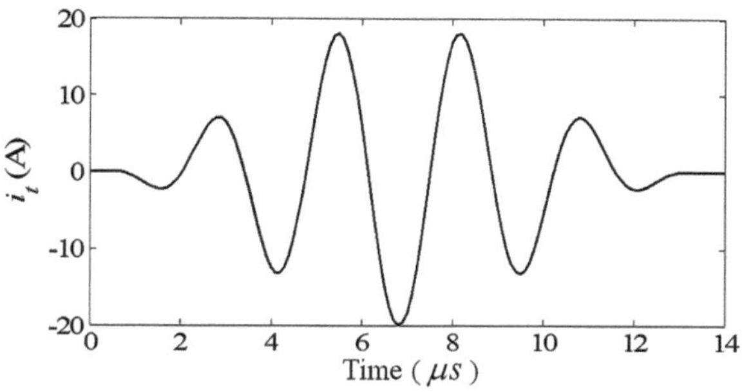

Fig. 2.5 The waveform of the excitation signal

The excitation current was input into *COMSOL Multiphysics* for simulation. The maximum element size in the meshing process inside the specimen was set as one-eighth of the wavelength of the *Lamb waves*, that is, 0.8125 mm. The time step is 0.00000002 s. In order to observe the obtained result, the observation points A and B inside the specimen were selected, with the location of point A just below the left-most conductor of the excitation coil, 0.01 mm below the top surface of the specimen under test. The location of point B is just below the center point of the receiving coil, 0.01 mm below the top surface of the specimen. A viewing zone *C* was selected, with its line segment located just below the exciting probe, 0.01 mm below the specimen's top face. Its width is 60 mm along the direction of the specimen's surface—the center of this line segment—in the horizontal coordinate, corresponding to the center of the meander coil and the permanent magnet.

Figures 2.6 and 2.7 show the *x* and *y* components of the static magnetic induction intensity generated by the permanent magnet in area *C*. In Fig. 2.6, the *x* components of the magnetic induction intensity are distributed symmetrically along the center of the area, with the values increasing from the center to both sides. The maximum value is located at the edge of the magnet. In Fig. 2.7, the *y* components of the magnetic induction intensity have a saddle-type distribution across the region. It is approximately a constant in the center of the area, with the values increasing from the center to both sides. The maximum value is also located at the edge of the magnet. It is found that, when the permanent magnet is reasonably well

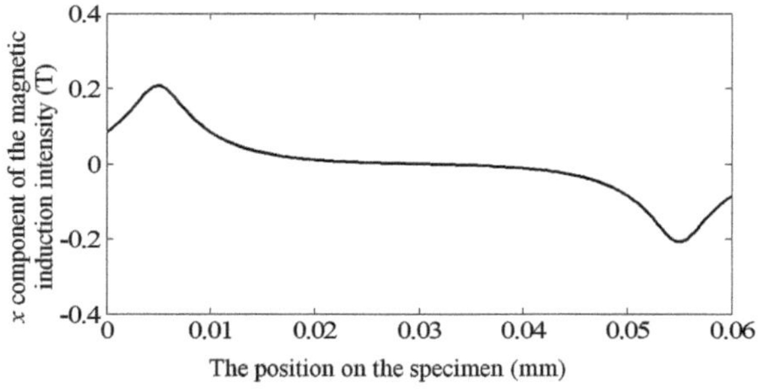

Fig. 2.6 The *x* component of the static magnetic induction intensity

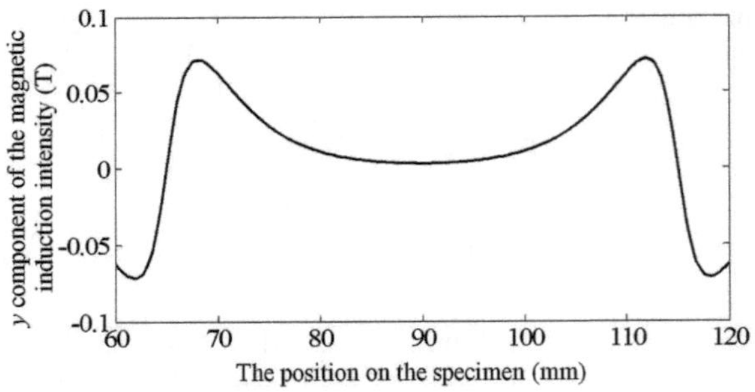

Fig. 2.7 The *y* component of the static magnetic induction intensity

designed, the *y* component of the bias magnetic field within the coil's area can be taken as a constant.

Figure 2.8 shows the equipotential line plot of the vector magnetic potential generated by the meander coil at a time of 10 μs when the tone burst is applied on the coil. With the pulsed magnetic field, the coil's induced eddy current and its distribution within the skin depth of the specimen's top surface are shown in Figs. 2.9 and 2.10. It is illustrated that, since the current direction of the two adjacent wires of the meander coil is opposite to each other, the direction of the induced eddy current inside the specimen is also opposite, below the meander coil, and the values of the eddy current density below each of the coil's line segments are approximately equal. The bias magnetic field, as shown in Figs. 2.6 and 2.7, which is affected by the pulsed eddy current, as shown in Fig. 2.10, can be used to generate *Lorentz force* inside the specimen. The force will be applied to the internal point of the specimen. At the moment of 10 μs, the distributions of *Lorentz force* in area *C* are shown in Figs. 2.11 and 2.12. Figure 2.11 shows the *x* component of the *Lorentz force* and Fig. 2.12 shows the *y* component of the *Lorentz force*. It is shown that, even if the densities of the eddy currents below each bending section of the

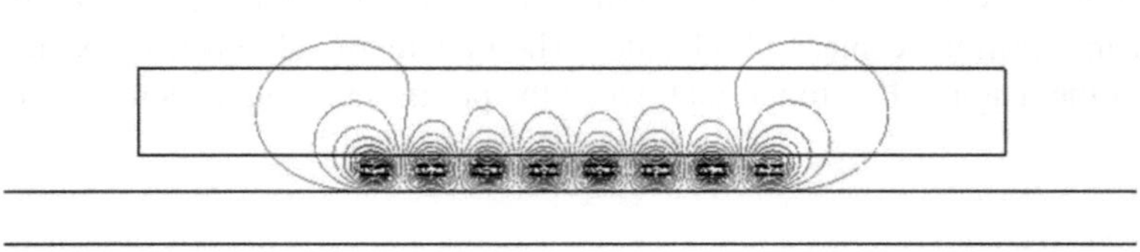

Fig. 2.8 The equipotential line of the vector magnetic potential

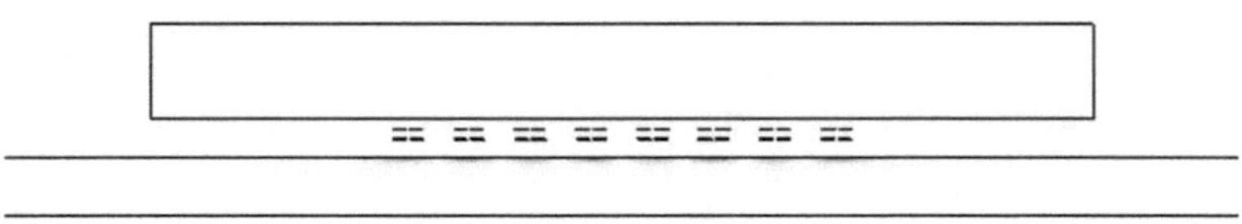

Fig. 2.9 The eddy current distribution inside the specimen at a time of 10 μs

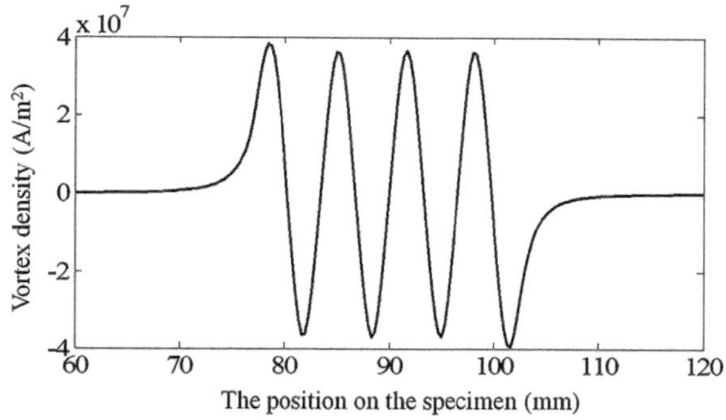

Fig. 2.10 The eddy current distribution on area *C* at a time of 10 μs

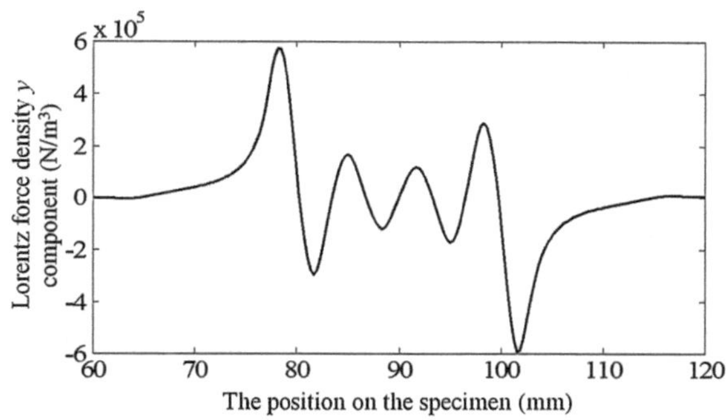

Fig. 2.11 The *x* component of *Lorentz force* on area *C* at a time of 10 μs

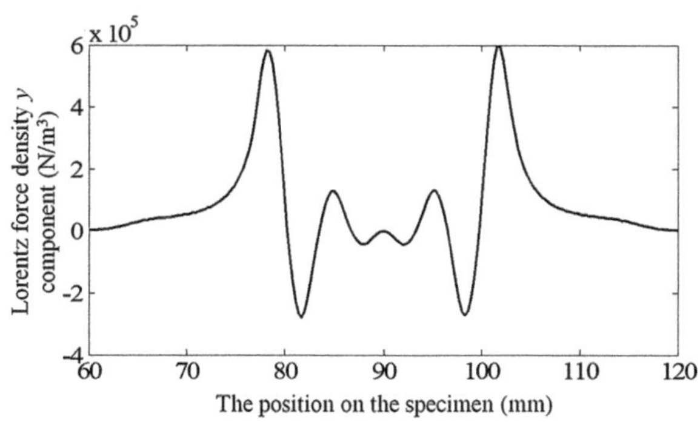

Fig. 2.12 The *y* component of *Lorentz force* on area *C* at the time of 10 μs

coil are equal to each other, there is a big difference in the *Lorentz force* inside the specimen below each fold line of the coil for the reason that the force is generated under the bias magnetic field with a non-uniform distribution. Therefore, if the magnet is poorly designed, the approximate average area of the magnetic field generated by the magnet will be smaller than the coil's width. It will bring a large calculation error with the assumption that the bias magnetic field is of uniform distribution.

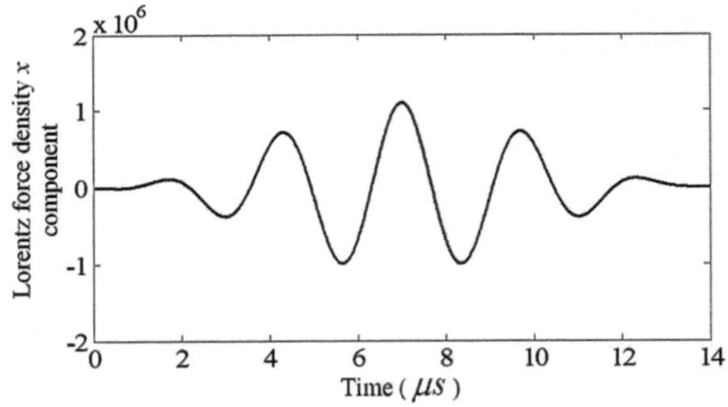

Fig. 2.13 The temporal variation of the *Lorentz force's x* component at point A

Figures 2.13 and 2.14 show the characteristics of a temporal variation of the *Lorentz force's* components *x* and *y* at the viewpoint A. Under the effect of the pulsed eddy current and bias magnetic field, the duration of the *Lorentz force* generated inside the specimen and that of the pulse are the same. The particles inside the specimen will generate vibration, exciting the ultrasonic wave, and the ultrasonic waves will propagate along the aluminum plate to be tested.

Figure 2.15 shows the *x* component of the internal particle's displacement of the specimen at the moment of 0, 10, 30, and 50 μs. It shows the process of the ultrasonic wave propagation in the specimen. At the moment of 0 μs, the excitation is still not applied on the coil, and there is no ultrasonic wave generated; at the time of 10 μs, the ultrasonic wave has been already generated, and it begins to propagate in two different directions; at the moment of 30 μs, the coil's excitation has already completed and the ultrasonic wave continues to propagate in two different directions. At this moment, two kinds of ultrasonic wave (A_0 and S_0) are excited inside the specimen. At the moment of 50 μs, the ultrasonic wave of the S_0 mode that propagates to the right side has already passed the receiving probe, while the ultrasonic wave of the A_0 mode has also arrived at the receiving probe. Meanwhile, the ultrasonic waves of the A_0 and S_0 mode, which propagate to the left side at first, are reflected from the left end of the specimen and change their direction of propagation to the right side of the specimen.

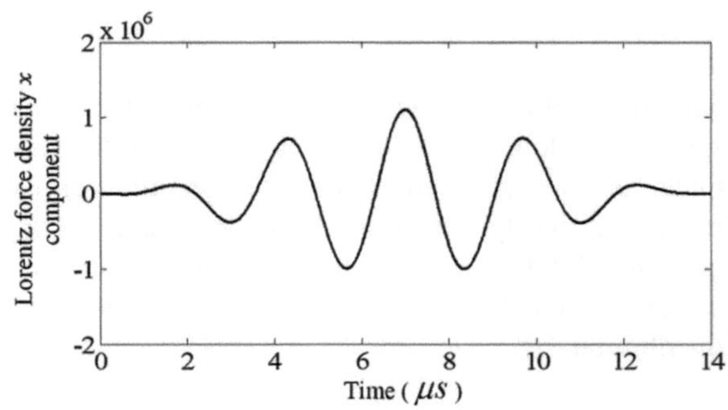

Fig. 2.14 The temporal variation curve of the *Lorentz force's y* component at point A

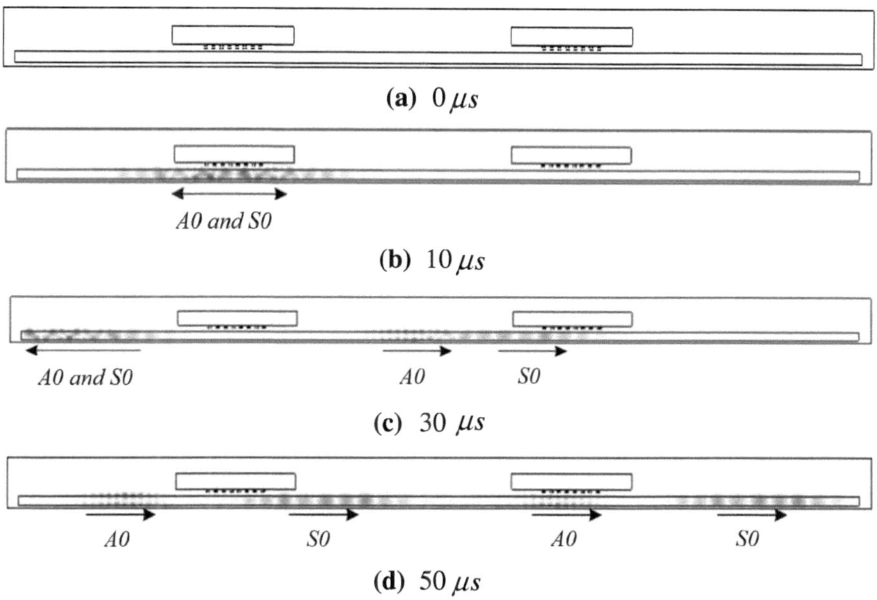

Fig. 2.15 The temporal variation of particle displacement component x

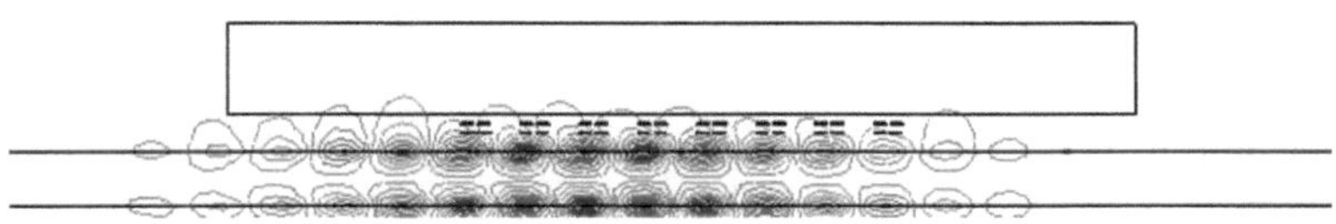

Fig. 2.16 The equipotential line of the magnetic vector potential generated by the particle motion inside the specimen

When the ultrasonic waves propagate to the receiving probe, the particle motion inside the specimen will generate a dynamic magnetic field under the effect of the bias magnetic field. An induced electromotive force will be generated inside the receiving coil, thus realizing the signal reception. At the moment of 50 μs, the equipotential lines of the magnetic vector potential are generated by the particle motion inside the specimen, as shown in Fig. 2.16.

Figure 2.17 shows the normalized waveform of the receiving signal from the receiving coil within 120 μs after the excitation from the probe. As illustrated in Fig. 2.17, the receiving probe receives 5 wave packets within the first 120 μs. For convenience, in the analysis each wave packet is named separately as S_{0R}, A_{0R}, S_{0LR}, S_{0L}, and A_{0LR}, according to the ultrasonic mode and the direction of their propagation. S_{0R} is the *Lamb wave* of the S_0 mode that propagates to the right side in the earliest stage; A_{0R} is the *Lamb wave* of the A_0 mode that propagates to the right side in the earliest stage; S_{0LR} represents the *Lamb* wave of the S_0 mode that propagates to the right side in the earliest stage and then propagates to the left side after the reflection at

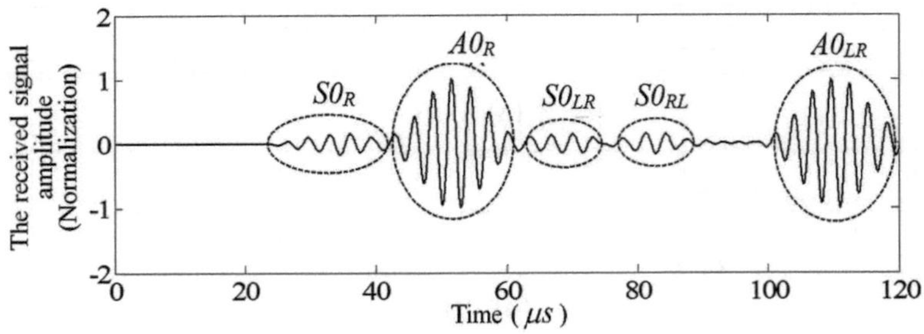

Fig. 2.17 The induced electromotive force of the receiving coil

the right end of the specimen; S_{0RL} represents the *Lamb wave* of the S_0 mode that propagates to the left side in the earliest stage and then propagates to the right side after the reflection at the left end of the specimen; and A_{0LR} is the *Lamb wave* of the A_0 mode that propagates to the left side in the earliest stage and then propagates to the right side after the reflection at the left end of the specimen.

In the experiment, the parameters used for EMAT are the same as those of the computational model, and the impulse excitation of the same frequency and number of periods is applied in the experiment. The normalized waveform of the received signal from the receiving probe in the experiment is shown in Fig. 2.18. Comparing Fig. 2.17 with 2.18, it can be found that the corresponding time of either the ultrasonic waves or of the wave packet received in the experiment is relatively comparable with that of the simulation results. This verifies the correctness and effectiveness of the numerical simulation method of the whole energy conversion process of the *Lorentz force*-based EMAT discussed in this book.

Based on the ultrasonic wave propagation distance for a certain time period, the group velocities of the *Lamb wave* of the S_0 mode and A_0 mode obtained from the numerical simulation and experiment can be calculated. A comparison of these values with the theoretical values is given in Table 2.3.

Table 2.3 shows that the differences among the theoretical, simulation, and experimental values of wave velocity are negligible and so proves the correctness and effectiveness of the simulation results and the experimental results.

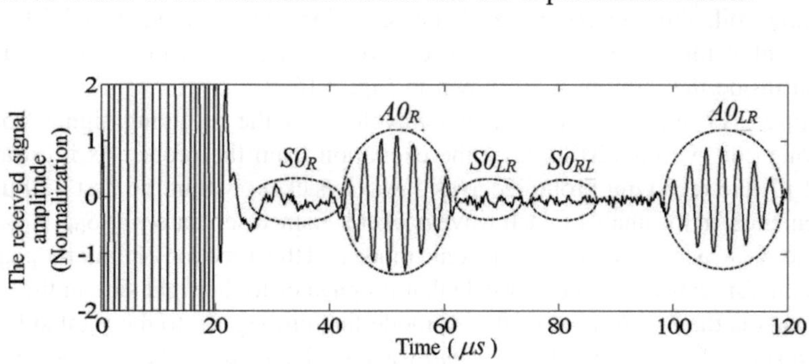

Fig. 2.18 The measured results from the receiving probe

Table 2.3 The comparisons of wave velocities

Wave mode	Theoretical wave velocity (m/s)	Simulation wave velocity (m/s)	Experimental wave velocity (m/s)
S_0	3148	3102	3147
A_0	5044	5291	5263

2.4 Analytical Modeling and Calculation of EMAT with Spiral Coil [3]

2.4.1 Configurations of the EMAT with Spiral Coils

Spiral coils can be used to excite and receive ultrasonic bulk waves inside the specimen. The typical configuration of the bulk wave EMAT is shown in Fig. 2.19. The permanent magnet or electromagnet is used to provide a static bias magnetic field which is perpendicular to the spiral coil and specimen. The spiral coil is placed above the specimen. When the spiral coil is supplied by the pulsed excitation with the desired center frequency, the ultrasonic wave could be generated by *Lorentz force* or magnetostrictive effect and propagates perpendicularly to the surface of the specimens. The echo wave could be detected in terms of the inverse process of generation.

In practical applications, the spiral coils are made by a printed circuit board (PCB) or flexible plate (FPC) technique for accurate design of the dimensions, compact configuration, and convenient application. The coil could be designed with single-layer or double-layer configurations, which consist of rectangular spiral copper wires on a fiberglass substrate. The sectional view schematic diagrams of the coils are shown in Figs. 2.20 and 2.21.

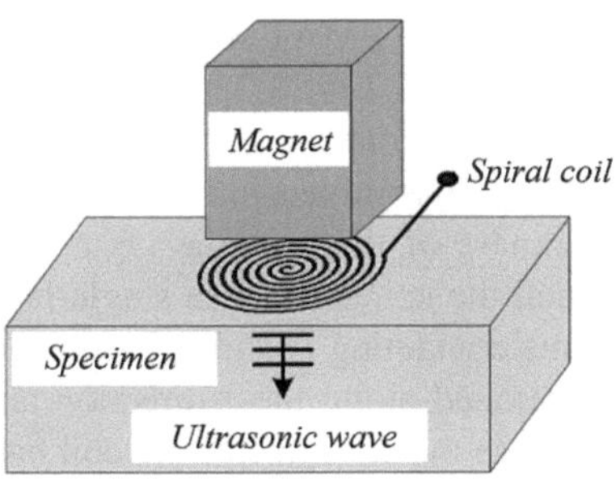

Fig. 2.19 The structure of EMAT: permanent magnet, spiral coil, and specimen

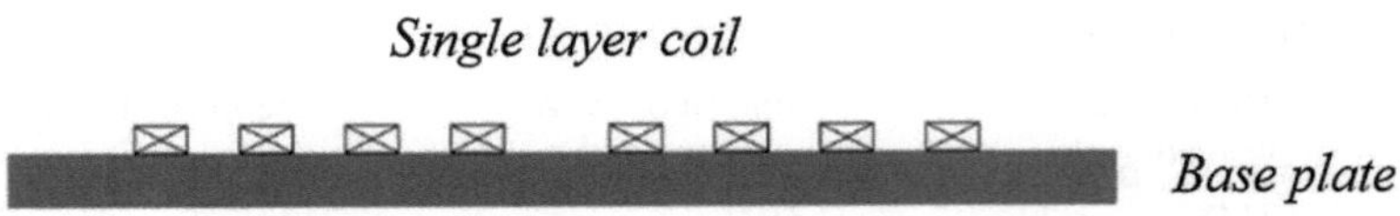

Fig. 2.20 The single-layer planar spiral coil

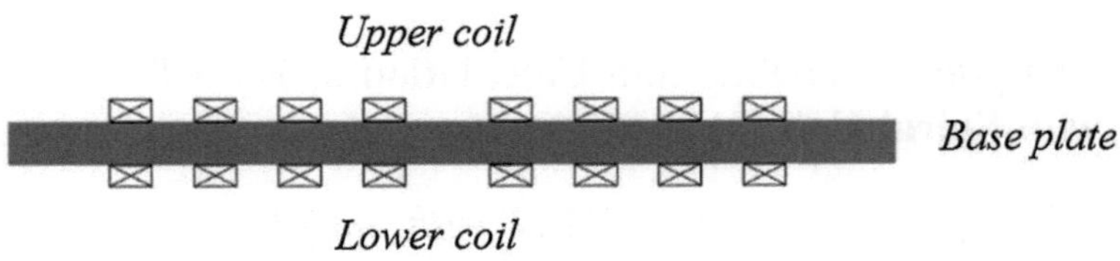

Fig. 2.21 The double-layer planar spiral coil

2.4.2 Frequency-Domain Solution

To analytically express the solutions, the following derivations and calculations are based on some assumptions:

(1) The mediums in all solution domains are linear, isotropic, and homogeneous.
(2) The proximity effect and skin effect for the copper wires are neglected, i.e., the distribution of current density is uniform in the copper wires.
(3) The displacement current is neglected because of its small value.
(4) The spiral coils are viewed as a superposition of concentric circles carrying the same current.

For the frequency-domain calculation of the magnetic field, the cylindrical coordinate system is used to model the EMAT. Thus, the magnetic vector potential (MVP) in each solution domain only has circumferential components. The physical model of the EMAT with a double-layer spiral coil in the cylindrical coordinates is illustrated in Fig. 2.22. The spiral coil is viewed as the arrays of N-concentric copper wire circles. The radiuses of the circles are as follows: r_{11}, r_{21}, r_{12}, r_{22}, …, r_{1N}, r_{2N}. The dimensions of the spiral coil in the z direction are l_1, l_2, l_3, and l_4. The thickness of the specimen is c. A conducting half-space is below the specimen. When the spiral coil is a single layer, the model can be simplified by deleting the terms containing l_3 and l_4 in the analytical expression. Seven solution domains are partitioned along the z direction and indicated as $a = 1, 2, \ldots, 7$, for calculation. Because the substrate of the coil has the same permeability and conductivity as air, we view it as air in the derivation process. The conductive specimens are non-magnetic.

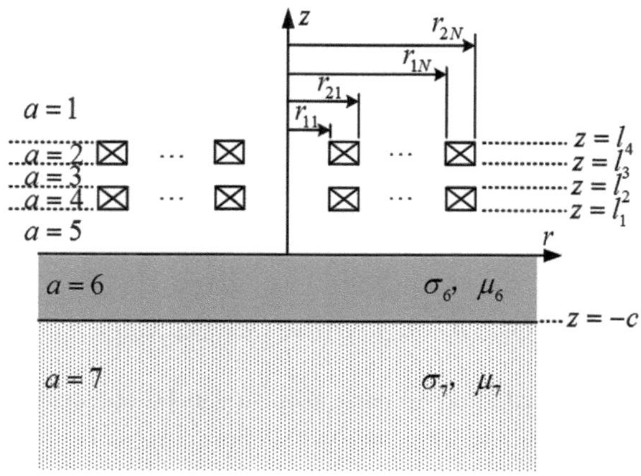

Fig. 2.22 Double-layer spiral coil at a distance above a conductive specimen and an infinite half-space

2.4.2.1 The Magnetic Vector Potential of a Delta Function Coil

To begin with, the MVP of a delta function coil $\delta(r - r_0)\delta(z - z_0)$ above a specimen is calculated. The schematic diagram is shown in Fig. 2.23, in which four solution domains are partitioned. The frequency-domain differential equation of the delta function coil above the specimen can be expressed as shown in (2.55); the solution area is 4.

$$\left(\frac{\partial^2}{\partial r^2} + \frac{1}{r}\frac{\partial}{\partial r} + \frac{\partial^2}{\partial z^2} - \frac{1}{r^2} - j\omega\mu_a\sigma_a\right)A_a(\omega, r, z) \\ = -\mu_a i(\omega)\delta(r - r_0)\delta(z - z_0) \tag{2.55}$$

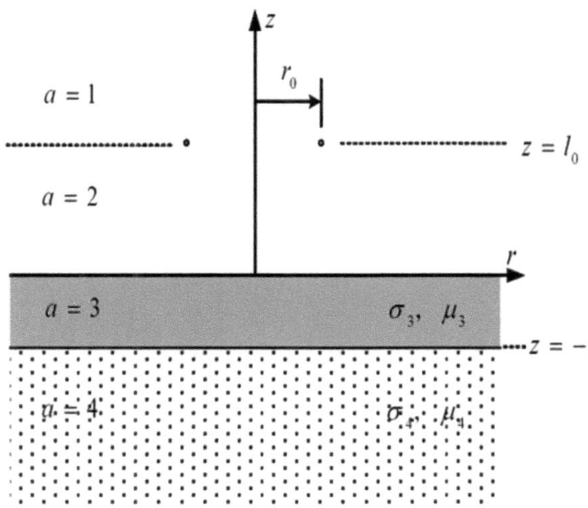

Fig. 2.23 Delta function coil at (r_0, z_0) above a conductive specimen and an infinite half-space

In the above equation, $A_a(\omega, r, z)$ is the MVP; $i(\omega)$ is the excitation current density; μ_a and σ_a are the permeability and conductivity, respectively, and $a = 1, 2, 3, 4$ represent the solution domains.

At the boundaries between adjacent domains, the boundary conditions as shown in (2.56) should be satisfied.

$$A_a(\omega, r, z_a) = A_{a+1}(\omega, r, z_a) \tag{2.56}$$

$$\frac{1}{\mu_a}\frac{\partial A_a}{\partial z}\bigg|_{z=z_a} = \frac{1}{\mu_{a+1}}\frac{\partial A_{a+1}}{\partial z}\bigg|_{z=z_a} - i(\omega)\delta(r - r_0)\delta(z - z_0) \tag{2.57}$$

The solutions in each domain can be obtained by the method of separation of variables, and they can be expressed by (2.58), (2.59), (2.60), and (2.61).

$$A_1(r, z) = \frac{1}{2}\mu_0 i(\omega) r_0 \int_0^\infty J_1(kr_0) J_1(kr) e^{-kl-kz}\left(e^{2kl} + P_1\right) \mathrm{d}k \tag{2.58}$$

$$A_2(r, z) = \frac{1}{2}\mu_0 i(\omega) r_0 \int_0^\infty J_1(kr_0) J_1(kr) e^{-kl}\left(e^{kz} + P_1 e^{-kz}\right) \mathrm{d}k \tag{2.59}$$

$$A_3(r, z) = \mu_0 i(\omega) r_0 \int_0^\infty J_1(kr_0) J_1(kr) e^{-kl}\left(P_2 e^{K_3 z} + P_3 e^{-K_3 z}\right) \mathrm{d}k \tag{2.60}$$

$$A_4(r, z) = \mu_0 i(\omega) r_0 \int_0^\infty J_1(kr_0) J_1(kr) e^{-kl} P_4 e^{-K_4 z} \mathrm{d}k \tag{2.61}$$

In the above equations, $J_1(x)$ is the first kind and first order of the *Bessel function*. Other related parameters can be expressed as follows:

$$K_a = \sqrt{k^2 + j\omega\mu_a\sigma_a}$$

$$P_1 = \frac{(k + K_3)(K_3 - K_4) + (k - K_3)(K_4 + K_3)e^{2K_3 e}}{(k - K_3)(K_3 - K_4) + (k + K_3)(K_4 + K_3)e^{2K_3 e}}$$

$$P_2 = \frac{k(K_3 + K_4)e^{2K_3 e}}{(k - K_3)(K_3 - K_4) + (k + K_3)(K_4 + K_3)e^{2K_3 e}}$$

$$P_3 = \frac{k(K_3 - K_4)}{(k - K_3)(K_3 - K_4) + (k + K_3)(K_4 + K_3)e^{2K_3 e}}$$

$$P_4 = \frac{2K_3 k e^{(K_3 + K_4)e}}{(k - K_3)(K_3 - K_4) + (k + K_3)(K_4 + K_3)e^{2K_3 e}}$$

2.4.2.2 The MVP of the Spiral Coil

For the rectangular section planar spiral coil, we choose one circle to consider first. Then, the overall MVP can be calculated by the superposition theorem. As far as the circles with the radiuses r_{11}, r_{12} and heights l_1, l_2 are concerned, the MVP in each solution domain is the superposition of those of the delta function coils and can be calculated by (2.62).

$$A_a(\omega, r, z) = \int_S A_a(\omega, r, z, r_0, l_0)\mathrm{d}s = \int_{r1}^{r2}\int_{l1}^{l2} A_a(\omega, r, z, r_0, l_0)\mathrm{d}r_0\mathrm{d}l \tag{2.62}$$

In the above equation, $A_a(\omega, r, z)$ is the MVP of the delta function coil positioned at (r_0, l_0).

The integration is performed and the MVP of the chosen circle is expressed in (2.63), (2.64), (2.65), and (2.66).

$$A^c_{1,2,3}(\omega, r, z) = \frac{1}{2}\mu_0 i(\omega)\int_0^\infty \frac{1}{k^3} I(kr_1, kr_2)J_1(kr)e^{-kz}\left[e^{kl_2} - e^{kl_1} - P_1\left(e^{-kl_2} - e^{-kl_1}\right)\right]\mathrm{d}k \tag{2.63}$$

$$A^c_5(\omega, r, z) = \frac{1}{2}\mu_0 i(\omega)\int_0^\infty \frac{1}{k^3} I(kr_1, kr_2)J_1(kr)\left[\left(e^{-kl_1} - e^{kl_2}\right)\left(e^{kz} + P_1 e^{-kz}\right)\right]\mathrm{d}k \tag{2.64}$$

$$A^c_6(\omega, r, z) = \mu_0 i(\omega)\int_0^\infty \frac{1}{k^3} I(kr_1, kr_2)J_1(kr)\left[\left(e^{-kl_1} - e^{kl_2}\right)\left(P_2 e^{K_1 z} + P_3 e^{-K_1 z}\right)\right]\mathrm{d}k \tag{2.65}$$

$$A^c_7(\omega, r, z) = \mu_0 i(\omega)\int_0^\infty \frac{1}{k^3} J(kr_1, kr_2)J_1(kr) \left(e^{-kl_1} - e^{-kl_2}\right)P_4 e^{K_2 z}\mathrm{d}k \tag{2.66}$$

In the above equations, the superscript c represents the MVP of the chosen circle and K_6 and K_7 take the place of K_3 and K_4 in P_1, P_2, P_3, and P_4.

$$\begin{aligned} I(x_1, x_2) &= \int_{x_1}^{x_2} xJ_1(x)\mathrm{d}x \\ &= \frac{\pi}{2}\{x_2[J_0(x_2)H_1(x_2) - J_1(x_2)H_0(x_2)] \\ &\quad - x_1[J_0(x_1)H_1(x_1) - J_1(x_1)H_0(x_1)]\} \end{aligned} \tag{2.67}$$

In (2.67), H_n represents the Struve function.

(2.65) and (2.65) are valid for the domains above the coil and below the coil, respectively. The MVP in domain 4 can be obtained by substituting $l_1 = z$ in (2.64), and adding (2.68).

$$\begin{aligned} & A_4^c(\omega, r, z) \\ & = \frac{1}{2}\mu_0 i(\omega) \int_0^\infty \frac{1}{k^3} I(kr_1, kr_2) J_1(kr) \left[2 - e^{k(z-l_2)} - e^{-k(z-l_1)} + P_1\left(e^{-kl_1} - e^{-kl_2}\right)e^{-kz}\right] \mathrm{d}k \end{aligned} \tag{2.68}$$

On the basis of the MVP of a single coil, the MVP of other $n - 1$ circles in the lower-layer coil could be obtained in the same way. We add all the MVPs of each circle to obtain the MVP of all the lower-layer coils, as shown in (2.69).

$$A_a^l(\omega, r, z) = \sum_i^N A_a^{lci}(\omega, r, z) \tag{2.69}$$

In the above equation, the superscript l represents the lower-layer coil. In the same way, replacing l_1 and l_2 with l_3 and l_4, we get the MVP of the upper-layer coil, as shown in (2.70).

$$A_a^u(\omega, r, z) = \sum_i^N A_a^{uci}(\omega, r, z) \tag{2.70}$$

In the above equation, the superscript u represents the upper-layer coil. Thus, the MVP of the double-layer spiral coil is expressed in (2.71).

$$A_a^d(\omega, r, z) = A_a^l(\omega, r, z) + A_a^u(\omega, r, z) \tag{2.71}$$

In the above equation, the superscript d represents the double-layer coil. Here, what we are concerned with are the MVPs in domains 2, 4, and 6 in Fig. 2.22. They are derived in (2.72).

$$A_6^d(\omega, r, z) = \mu_0 i(\omega) \int_0^\infty \frac{1}{k^3} \sum_i^N I(kr_{1i}, kr_{2i}) J_1(kr) \left(e^{-kl_1} - e^{-kl_2} + e^{-kl_3} - e^{-kl_4}\right)\left(P_2 e^{K_1 z} + P_3 e^{-K_1 z}\right) \mathrm{d}k \tag{2.72}$$

$$A_2^d(\omega, r, z) = \frac{1}{2}\mu_0 i(\omega) \int_0^\infty \frac{1}{k^3} \sum_i^N I(kr_{1i}, kr_{2i}) J_1(kr) \left[2 - e^{k(z-l_4)} - e^{-k(z-l_3)} + e^{-k(z-l_2)} - e^{-k(z-l_1)} + P_1\left(e^{-kl_1} - e^{-kl_2} + e^{-kl_3} - e^{-kl_4}\right) e^{-kz}\right] \mathrm{d}k \tag{2.73}$$

$$A_4^d(\omega, r, z) = \frac{1}{2}\mu_0 i(\omega) \int_0^\infty \frac{1}{k^3} \sum_i^N I(kr_{1i}, kr_{2i}) J_1(kr) \left[2 - e^{k(z-l_2)} - e^{-k(z-l_1)} + e^{-k(z-l_3)} - e^{-k(z-l_4)} + P_1\left(e^{-kl_1} - e^{-kl_2} + e^{-kl_3} - e^{-kl_4}\right) e^{-kz}\right] \mathrm{d}k \tag{2.74}$$

2.4.2.3 Analytical Expression of Magnetic Flux Intensity, Pulsed Eddy Current and Input Impedance

The dynamic magnetic flux density in the specimen can be calculated according to (2.75).

$$B = \nabla \times A \tag{2.75}$$

Because A only has the component φ, B will only have the r component and z component.

$$B_r(\omega, r, z) = -\mu_0 i(\omega) \int_0^\infty \frac{1}{k^3} \sum_i^N I(kr_{1i}, kr_{2i}) J_1(kr) \left(e^{-kl_1} - e^{-kl_2} + e^{-kl_3} - e^{-kl_4}\right)\left(P_2 K_1 e^{K_1 z} - P_3 K_1 e^{-K_1 z}\right) \mathrm{d}k \tag{2.76}$$

$$B_z(\omega, r, z) = \mu_0 i(\omega) \int_0^\infty \frac{1}{k^2} \sum_i^N I(kr_{1i}, kr_{2i}) J_0(kr) \left(e^{-kl_1} - e^{-kl_2} + e^{-kl_3} - e^{-kl_4}\right)\left(P_2 e^{K_1 z} + P_3 e^{-K_1 z}\right) \mathrm{d}k \tag{2.77}$$

The pulsed eddy current is induced in the skin depth of the specimen. In terms of the relationship between the eddy current J and the MVP in the frequency domain, i.e., $\boldsymbol{J} = -j\omega\sigma\boldsymbol{A}$, the pulsed eddy current can be obtained by substituting (2.72).

$$
\begin{aligned}
&J(\omega, r, z) = \\
&-j\omega\sigma_6\mu_0 i(\omega)\int_0^{\infty}\frac{1}{k^3}\sum_i^N I(kr_{1i}, kr_{2i})J_1(kr)\left(e^{-kl_1} - e^{-kl_2} + e^{-kl_3} - e^{-kl_4}\right)\left(P_2e^{K_1z} + P_3e^{-K_1z}\right)dk
\end{aligned}
\tag{2.78}
$$

Before calculating the input impedance, the induced voltage should be obtained first. The induced voltage in the rectangular cross-sectional circle coil is expressed in (2.79).

$$
V(\omega) = \frac{j\omega 2\pi}{\text{coilcross section}}\iint\limits_{\text{coilcrosssection}} rA(\omega, r, z) \tag{2.79}
$$

The induced voltage of the double-layer coil is obtained by adding all the circles and can be expressed by (2.80).

$$
\begin{aligned}
&V^d(\omega) \\
&= \sum_i^N\left[\frac{j\omega 2\pi}{(l_2 - l_1)(r_{2i} - r_{1i})}\int_{l_1}^{l_2}\int_{r_{1i}}^{r_{2i}} rA_2^d(\omega, r, z)\mathrm{d}r\mathrm{d}z + \frac{j\omega 2\pi}{(l_4 - l_3)(r_{2i} - r_{1i})}\int_{l_3}^{l_4}\int_{r_{1i}}^{r_{2i}} rA_4^d(\omega, r, z)\mathrm{d}r\mathrm{d}z\right]
\end{aligned}
\tag{2.80}
$$

Substituting (2.73) and (2.74) into (2.80) and adding the direct current resistance of the coil, the expression of the coil's input impedance in terms of *Ohm's law* can be obtained.

$$
\begin{aligned}
&Z(\omega) = \\
&Z_0 + \sum_{i=1}^N\left\{\frac{j\omega\mu_0 2\pi}{(l_2 - l_1)^2(r_{2i} - r_{1i})^2}\int_0^{\infty}\frac{1}{k^6}I(kr_{1i}, kr_{2i})\sum_{i=1}^N I(r_{1i}, r_{2i})\sum_{i=1}^N I(r_{1i}, r_{2i})\right. \\
&[2k(l_2 - l_1) + 2(e^{-k(l_2 - l_1)} - 1) + (e^{-kl_1} - e^{-kl_2})(e^{kl_4} - e^{kl_3}) \\
&+ P_1(e^{-kl_1} - e^{-kl_2} + e^{-kl_3} - e^{-kl_4})(e^{-kl_1} - e^{-kl_2})]\mathrm{d}k \\
&+ \frac{j\omega\mu_0 2\pi}{(l_4 - l_3)^2(r_{2i} - r_{1i})^2}\int_0^{\infty}\frac{1}{k^6}I(r_{1i}, r_{2i})\sum_{i=1}^N I(r_{1i}, r_{2i}) \\
&[2k(l_4 - l_3) + 2(e^{-k(l_4 - l_3)} - 1) + (e^{-kl_1} - e^{-kl_2})(e^{kl_4} - e^{kl_3}) \\
&\left. + P_1(e^{-kl_3} - e^{-kl_4} + e^{-kl_1} - e^{-kl_2})(e^{-kl_3} - e^{-kl_4})]\mathrm{d}k\right\}
\end{aligned}
\tag{2.81}
$$

In the above equation, Z_0 is the direct current resistance of the coil.

$$Z_0 = \frac{1}{\sigma_6}\sum_{i=1}^{N}\left[\frac{\pi(r_{2i}+r_{1i})}{(r_{2i}-r_{1i})(l_2-l_1)}+\frac{\pi(r_{2i}+r_{1i})}{(r_{2i}-r_{1i})(l_4-l_3)}\right] \tag{2.82}$$

2.4.2.4 Series Expressions of Flux Density, Eddy Current, and Input Impedance

In the expression of magnetic flux density, pulsed eddy current, and input impedance, the infinite integration is difficult to implement because the integrand is too complicated and contains double integrals of the one-order Bessel function. To lower the calculation difficulty, the approximate series expression of the magnetic flux density, pulsed eddy current, and input impedance is proposed by the truncated-domain eigenfunction expansion (TREE) method. Although the number of series terms and truncated distances bring about calculation errors, the errors can be readily controlled.

By setting the solution area as $0 \le r \le \infty$ instead of $0 \le r \le R$, the infinite integration can be transformed into the sum of an infinite series. When the chosen R is large enough, the calculation results will be consistent with the real results. According to the physical significance of MVP, it is bounded in the solution area and a magnetic insulation boundary condition is exerted.

$$|A|_{r<R}| < \infty \tag{2.83}$$

$$A|_{r<R} = 0 \tag{2.84}$$

Then, the general solution of MVP can be expressed by (2.85).

$$A_n(\omega, r, z) = \sum_{m=1}^{\infty}(A_m(k_m)e^{K_{n}z} + B_m(k_m)e^{-K_{n}z})C_m(k_m)J_1(k_m r) \tag{2.85}$$

According to (2.84), we can get (2.86).

$$J_1(k_m R) = 0 \tag{2.86}$$

It is found that $K_m R$ is the mth zero root x_m of $J_1(x)$. Thus, (2.87) can be obtained.

$$k_m = \frac{x_m}{R} \tag{2.87}$$

After a series of derivations, in a similar manner to the above sections, the ultimate series expressions of magnetic flux density, pulsed eddy current, and input impedance are expressed by (2.88), (2.89), (2.90), and (2.91).

$$B_r = -2\mu_0 i(\omega) \sum_{i=1}^{N} \sum_{m=1}^{\infty} \frac{I(kr_{1i}, kr_{2i}) J_1(k_m r)}{k_m^4 R^2 J_0^2(k_m R)} \left(e^{-k_m l_1} - e^{-k_m l_2} + e^{-k_m l_3} - e^{-k_m l_4}\right)\left(P_{2m} K_{1m} e^{K_{1m} z} - P_{3m} K_{1m} e^{-K_{1m} z}\right) \tag{2.88}$$

$$B_z(\omega, r, z) = 2\mu_0 i(\omega) \sum_{i=1}^{N} \sum_{m=1}^{\infty} \frac{I(kr_{1i}, kr_{2i}) J_0(k_m r)}{k_m^3 R^2 J_0^2(k_m R)} (e^{-kl_1} - e^{-kl_2} + e^{-kl_3} - e^{-kl_4})(P_{2m} e^{K_1 z} + P_{3m} e^{-K_1 z}) \tag{2.89}$$

$$J_e(\omega, r, z) = -2j\omega\sigma_6 \mu_0 i(\omega) \sum_{i=1}^{N} \sum_{m=1}^{\infty} \frac{I(kr_{1i}, kr_{2i}) J_0(k_m r)}{k_m^4 R^2 J_0^2(k_m R)} \left(e^{-k_m l_1} - e^{-k_m l_2} + e^{-k_m l_3} - e^{-k_m l_4}\right)\left(P_{2m} e^{K_{1m} z} + P_{3m} e^{-K_{1m} z}\right) \tag{2.90}$$

$$\begin{aligned} Z(\omega) = Z_0 + \sum_{i=1}^{N} \sum_{m=1}^{\infty} \Bigg\{ & \frac{j2\pi\omega\mu_0 I(r_{1i}, r_{2i}) \sum_{i=1}^{N} I(kr_{1i}, kr_{2i})}{(l_2 - l_1)^2 (r_{2i} - r_{1i})^2 k_m^7 R^2 J_0^2(k_m R)} \cdot \\ & [2(l_2 - l_1) + 2(e^{-k(l_2 - l_1)} - 1) + (e^{-kl_1} - e^{-kl_2})(e^{kl_4} - e^{kl_3}) \\ & + P_{1m}(e^{-kl_1} - e^{-kl_2} + e^{-kl_3} - e^{-kl_4})(e^{-kl_1} - e^{-kl_2})] \\ & + \frac{j2\pi\omega\mu_0 I(r_{1i}, r_{2i}) \sum_{i=1}^{N} I(kr_{1i}, kr_{2i})}{(l_4 - l_3)^2 (r_{2i} - r_{1i})^2 k_m^7 R^2 J_0^2(k_m R)} [2(l_4 - l_3)] \\ & + \frac{j2\pi\omega\mu_0 I(r_{1i}, r_{2i}) \sum_{i=1}^{N} I(kr_{1i}, kr_{2i})}{(l_4 - l_3)^2 (r_{2i} - r_{1i})^2 k_m^7 R^2 J_0^2(k_m R)} [2(l_4 - l_3) \\ & + 2(e^{-k(l_4 - l_3)} - 1) + (e^{-kl_1} - e^{-kl_2})(e^{kl_4} - e^{kl_3}) \\ & + P_{1m}(e^{-kl_3} - e^{-kl_4} + e^{-kl_1} - e^{-kl_2})(e^{-kl_3} - e^{-kl_4})] \Bigg\} \end{aligned} \tag{2.91}$$

2.4.3 *The Time-Domain Solutions*

Pulsed excitations and responses can be expressed as the superposition of a series of sinusoidal signals with different frequencies. When the frequency-domain expressions of magnetic flux density and pulsed eddy current are known, their time-domain expressions can be obtained by the inverse Fourier transformation.

For (2.88), (2.89), and (2.90), the corresponding time-domain expressions can be obtained.

$$B_r(t,k,z) = \frac{1}{2\pi}\int_{-\infty}^{\infty} B_r(\omega,k,z)e^{j\omega t}\mathrm{d}\omega \tag{2.92}$$

$$B_z(t,r,z) = \frac{1}{2\pi}\int_{-\infty}^{\infty} B_z(\omega,r,z)e^{j\omega t}\mathrm{d}\omega \tag{2.93}$$

$$J_e(t,r,z) = \frac{1}{2\pi}\int_{-\infty}^{\infty} J_e(\omega,r,z)e^{j\omega t}\mathrm{d}\omega \tag{2.94}$$

(2.92), (2.93), and (2.94) are difficult to analytically implement for the infinite integration of the integrands $B_r(\omega, r, z)$, $B_z(\omega, r, z)$, and $J_e(\omega, r, z)$ with such complex expressions. The FFT and IFFT methods are good choices for their calculation. The procedure is that the time-domain current excitation signal is transformed to the frequency domain by FFT, and then, the magnetic flux densities and eddy currents at chosen frequencies are calculated according to (2.88), (2.89), and (2.90) and summed. Lastly, the time-domain responses of the magnetic flux density and pulsed eddy current are obtained by IFFT of the summed frequency-domain signals.

When the pulsed eddy current in the skin depth of the specimen is known, the *Lorentz force* can be calculated using (2.95).

$$f_L(t,r,z) = B_0 \times J_e(t,r,z) \tag{2.95}$$

In the above equation, $f_L(t, r, z)$ represents the *Lorentz force* and B_0 represents the bias static magnetic flux density.

2.4.4 Results Comparison and Discussion

In order to validate the proposed analytical model and calculation method, an example is demonstrated. Moreover, the results are compared to those of the TSFEM method and experiment.

A double-layer PCB spiral coil, shown in Fig. 2.24, is used for the calculation. The specimen is an aluminum plate. The liftoff distance between the coil and specimen is 0.5 mm. The infinite half-space below the aluminum plate is air. The dimensions and properties of the coil and specimen employed in the simulation and experiment are shown in Tables 2.4 and 2.5.

Generally, the EMATs are supplied by narrow pulses or modulated tone-burst signals, which are generated by the function generator and amplified by the power amplifier.

The modulated tone-burst current signal can be expressed using (2.96).

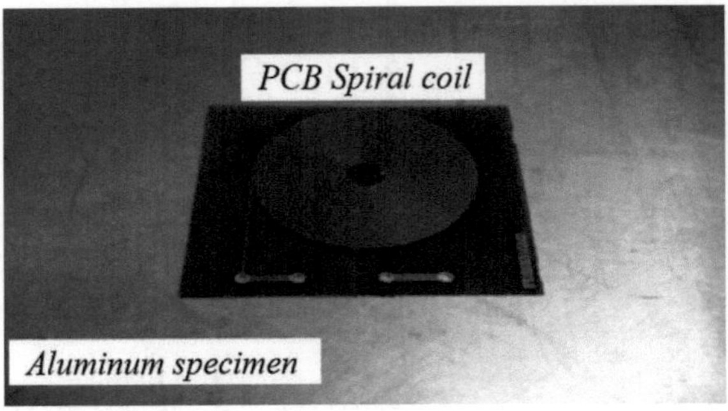

Fig. 2.24 Double-layer PCB spiral coil above the specimen

Table 2.4 The width of the double-layer spiral coil and the material parameters

Parameter name	Value
Substrate thickness	0.500 mm
Copper wire width	0.254 mm
Copper wire thickness	0.050 mm
r_{11}	2.921 mm
r_{21}	3.175 mm
The conductor's ring space	0.254 mm
Number of turns	29
Copper wire's magnetic permeability	$4\pi \times 10^{-7}$ H/m
Copper wire's electric conductivity	2.66×10^{7} S/m

Table 2.5 The specimen's dimension and the material's parameters

Parameter name	Value
Specimen's thickness	5.00 mm
Magnetic permeability	$4\pi \times 10^{-7}$ H/m
Electric conductivity	3.571×10^{-7} S/m

$$i(t) = \begin{cases} I_0\left(1 - \cos\frac{\omega t}{n}\right)\cos\omega t & 0 \le t \le \frac{2n\pi}{\omega} \\ 0 & t \ge \frac{2n\pi}{\omega} \end{cases} \tag{2.96}$$

In the above equation, $\omega = 2\pi f$. Here, $I_0 = 20A$, $n = 2$, and $f = 1$ MHz were selected; the waveform of the pulsed current signal is shown as follows (Fig. 2.25):

In the calculation process, the truncation radius is $R = 696$ mm and the summation number is 50. The sampling frequency and data number used in the FFT-IFFT method are chosen as 32 MHz and 64, respectively. Then, an impedance analyzer is used to measure the resistance and inductance of the coil above the specimen, so that the input impedance of the EMAT can be obtained. The time-domain response of the magnetic flux density and pulsed eddy current calculated by the TSFEM method is implemented through the FEM analysis software. To improve the accuracy, a denser mesh is used in the coil cross section and surface of the specimen. The number of the overall triangular elements is 13,193.

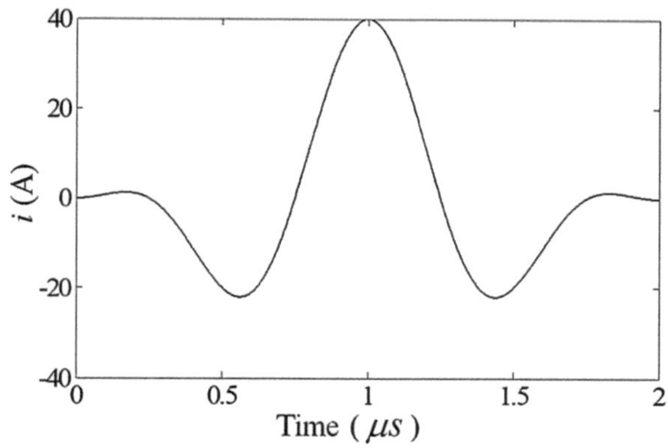

Fig. 2.25 The waveform of the tone-burst signal supplied to the spiral coil. The center frequency of the signal is 1 MHz and the number of periods is 2

To design the matching circuits for the EMATs, the impedance of the coil at the center frequency needs to be determined. Here, for validating the input impedance expressions as (2.81) and (2.91), the input impedance at frequencies 0.5, 1, and 1.5 MHz are calculated and then measured by the impedance analyzer. The results are shown in Table 2.6.

As shown in Table 2.6, the calculated results and measured results are in agreement, which validates the validity and accuracy of the input impedance expressions and the TREE method.

The transient time-domain signals of B_r, B_z, and J_e from both the proposed method and the TSFEM method are presented in Figs. 2.20, 2.21, and 2.22.

As shown in Figs. 2.26, 2.27, and 2.28, the results from the proposed method are consistent with those from the TSFEM method. The small discrepancies are that as follows: (1) B_r and B_z from the proposed method are a little larger than those from the TSFEM method, while J_e has the opposite relationship, and (2) the peak value times and the waveforms in the initial stage have certain differences. An evaluation of the errors of peak values and peak value times is shown in Table 2.7.

Assuming that the bias magnetic field is uniform and equals 1 T, we calculate the *Lorentz force* on the surface of the specimen according to Eq. (2.95). The *Lorentz force* inside the specimens under different liftoff values is shown in the following figure.

As shown in Fig. 2.29, it is found that *Lorentz force* will decrease with an increase in the liftoff distance. If the *Lorentz force* is given, the particle's

Table 2.6 Input impedance by calculation and measurement at different frequency

Frequency (MHz)	Calculation value (Ω)	Measuring value (Ω)
0.5	12.51 + j30.64	13.03 + j31.34
1	13.19 + j60.34	13.89 + j63.25
1.5	13.71 + j89.87	14.75 + j93.80

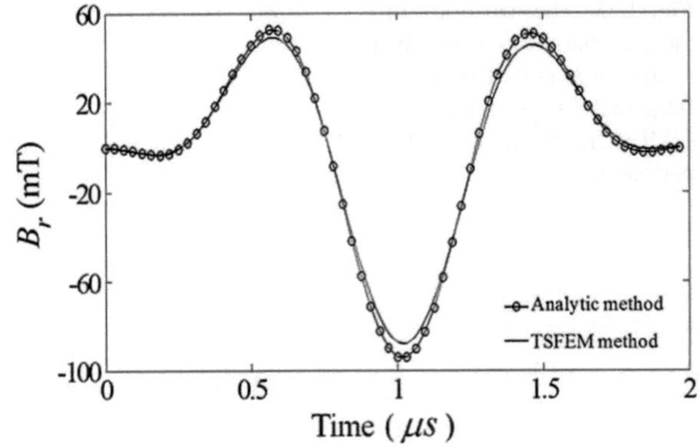

Fig. 2.26 Magnetic flux densities in the r direction at position (3.048 mm, −0.001 mm) on the surface of the specimen

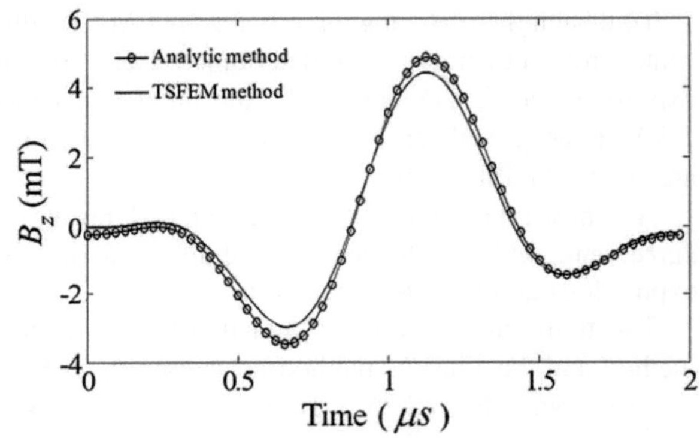

Fig. 2.27 Magnetic flux densities in the z direction at position (3.048 mm, −0.001 mm) on the surface of the specimen

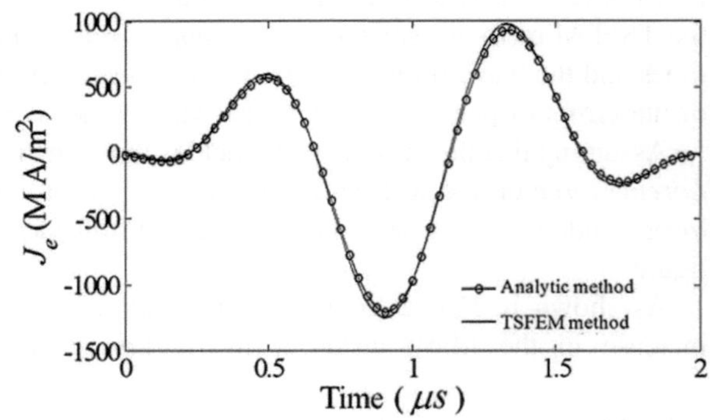

Fig. 2.28 Magnetic flux densities at position (3.048 mm, −0.001 mm) on the surface of the specimen

Table 2.7 Errors of peak value and peak value time comparing two methods

Error	B_r	B_z	J_e
Waveform peak	7.4	9.7	3.1
Waveform peak's time	0.78	1.3	0.41

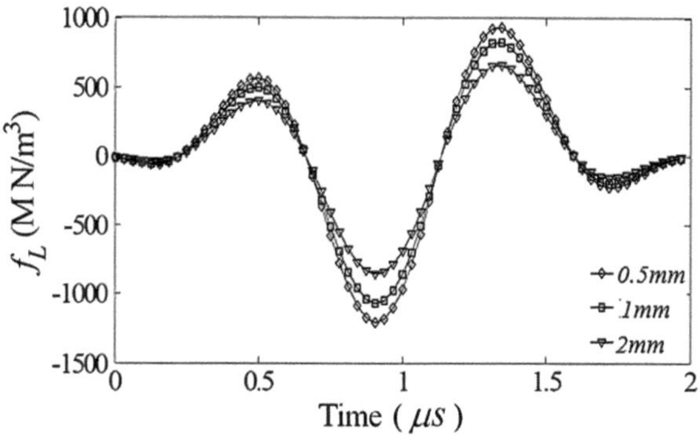

Fig. 2.29 *Lorentz force* at position (3.048 mm, −0.01 mm) on the surface of the specimen calculated by the proposed method when the liftoff distances are 0.5, 1, and 2 mm

displacement can be calculated according to the theory of elasticity for *Lorentz force* mechanism EMATs.

Although the analytical expressions of magnetic flux density, pulsed eddy current, and input impedance are complicated, they are valid and accurate when calculated by TREE and FFT-IFFT method, which is entirely much simpler than the TSFEM method. Moreover, the distinct mathematical expressions and faster calculation speed are also advantages over the TSFEM method.

The following causes could result in errors between calculation results and measurement results. The coil's skin effect and proximity effect were overlooked in the analytic model, which might cause certain errors in the circumstance of high frequency; in addition, some errors could be caused by the TREE method itself.

2.5 Analytical Modeling and Calculation of the Meander Coil EMAT [4]

In the field of nondestructive testing with EMAT under different static bias magnetic fields, the meander coil EMAT has an important position and is widely used for generating and detecting various types of ultrasonic waves, such as the *Lamb wave*, *Rayleigh wave*, and *shear horizontal* (SH) wave. The analytical model and analysis of EMAT with a meander coil structure will be illustrated in this section.

2.5.1 Meander Coil EMAT Configuration and Calculation Model

Nowadays, with the development of printed circuit boards (PCB) or flexible print circuit (FPC) techniques, the meander coil can be made more complex, accurate, and portable to improve the performance of EMAT and facilitate its actual application. The typical configurations of a meander coil are shown in Fig. 2.30.

The purpose of the multisplit configuration of a meander coil is to increase the turn number of each fold. For the same reason, sometimes, the coil is designed into a multilayer configuration, using the configuration of equidistance or non-equidistance to control the excited ultrasonic wave modes through controlling the coil's spacing.

When used for ultrasonic wave generation and detection, the meander coils are placed parallel above the surface of the specimen. If we neglect the ends connection of the meander coil, the section diagram of the EMAT is as shown in Fig. 2.31. We assume that the meander coil is an *M*-layer, *N*-conductor coil, and the number of bends is 2*Q*. This model is the general model of the meander coil. The coil with an equidistance structure is a special case.

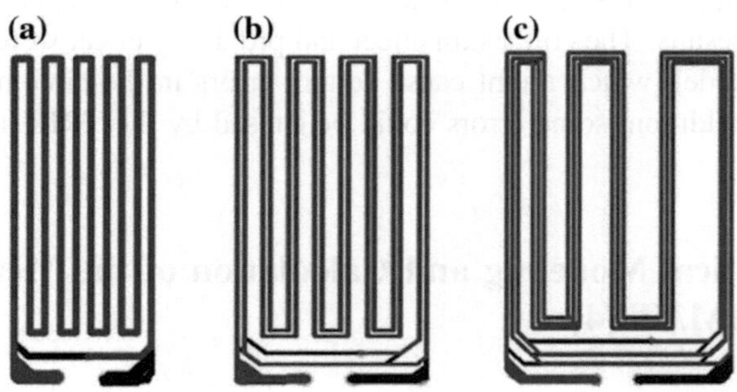

Fig. 2.30 Configuration of meander coils: **a** Isometric single-split coil; **b** Isometric multisplit coil; **c** Non-equidistance multisplit coil

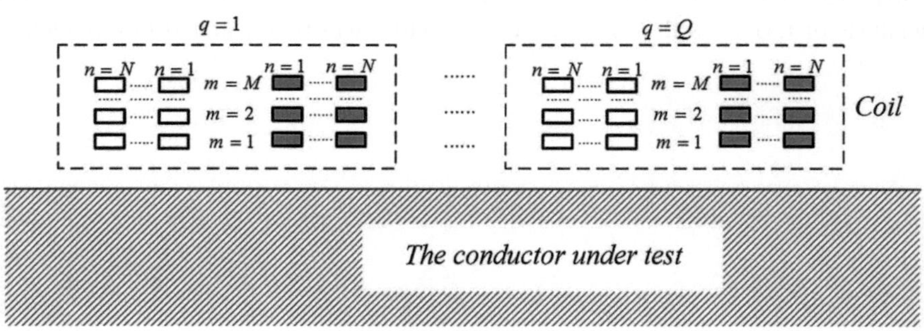

Fig. 2.31 Model of a meander coil EMAT

As shown in Fig. 2.31, according to the characteristics of the meander coil, the current directions in the two adjacent bends of the coil are opposite to each other. Therefore, the adjacent bends can be viewed as a closed coil. Thus, the *M*-layer, *N*-conductor coil can be viewed as the group of $M \times N \times Q$ single rectangular coils.

2.5.2 *The Frequency-Domain Calculation of the Coil's Impedance and Magnetic Field*

2.5.2.1 The Scalar Potential of the Rectangle δ Coil

The schematic diagram of a δ coil with current $\boldsymbol{I}$ above an infinite half-space is shown in Fig. 2.32. The solution domain is divided into three subdomains: the domain above the δ coil R_a, the domain between the δ coil and the conductor R_b, and the conductor domain R_c. In Fig. 2.32, x_0 and y_0 indicate the half width and length of the δ coil.

Because the frequencies in the EMAT nondestructive testing are generally lower than 10 MHz, the displacement current and velocity effect can be neglected; thus, the time-harmonic electromagnetic field governing equation is expressed in (2.97).

$$\nabla^2 A + \gamma^2 A = -\mu J_s \tag{2.97}$$

In the above equation, $\boldsymbol{A}$ is the magnetic vector potential; $\boldsymbol{J_s}$ is the source current density; $\Upsilon^2 = -j\omega\mu\sigma$; ω is the angular frequency; μ is the medium permeability; σ is the medium conductivity, and ∇^2 is the *Laplace operator.*

In terms of *Coulomb*'s criteria $\nabla \cdot A = 0$, there is a second vector potential $\boldsymbol{W}$, which satisfies $A = \nabla \times W$, and W can be decomposed as two scalar potentials. W can be expressed using (2.98).

$$W = uW_1 + u \times \nabla W_2 \tag{2.98}$$

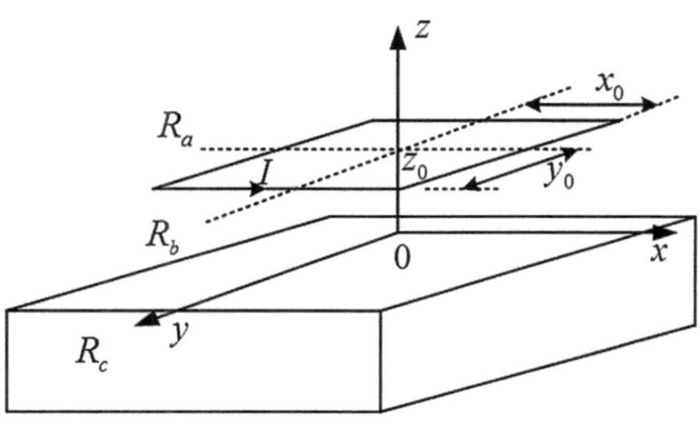

Fig. 2.32 Rectangle δ coil above the conductor

In the above equation, W_1 and W_2 are scalar potentials and u is the vector that meets the requirement.

We choose $u = \hat{z}$, A can be expressed by the second vector potential $\boldsymbol{W}$.

$$A = \nabla \times (W_1\hat{z} + \hat{z} \times \nabla W_2) \tag{2.99}$$

Substituting (2.99) into (2.97), we can obtain (2.100).

$$\nabla \times \left[\hat{z}\nabla^2 W_1 + \hat{z} \times \nabla(\nabla^2 W_2) + \gamma^2(W_1\hat{z} + \hat{z} \times \nabla W_2)\right] = -\mu J_s \tag{2.100}$$

In the air domains R_a and R_b, the source current density and conductivity are 0; thus, we have (2.101), (2.102), and (2.103).

$$\nabla^2 W_{1(a,b)} = 0 \tag{2.101}$$

$$\nabla^2 W_{2(a,b)} = 0 \tag{2.102}$$

$$A_{(a,b)} = \nabla \times (W_{1(a,b)}\hat{z}) - (\hat{z} \cdot \nabla)\nabla W_{2(a,b)} \tag{2.103}$$

In the above equations, the subscript (a,b) represents the variables in domains a and b; thus, we have (2.104).

$$B_{(a,b)} = \nabla\left(\frac{\partial W_{1(a,b)}}{\partial z}\right) \tag{2.104}$$

Therefore, we just need to calculate the scalar potential $W_{1(a,b)}$, which is the superposition of the coil's scalar potential $W_{1(a,b)s}$ and the eddy current's scalar potential $W_{1(a,b)e}$, as expressed in (2.105).

$$W_{1(a,b)} = W_{1(a,b)s} + W_{1(a,b)e} \tag{2.105}$$

Thus, we have (2.106).

$$\nabla^2\left(W_{1(a,b)s} + W_{1(a,b)e}\right) = 0 \tag{2.106}$$

In domain R_c, according to the interface conditions, in interface $z = 0$, the $W_{2(c)}$ should satisfy (2.107).

$$\begin{cases} \frac{\partial W_{2(e)}}{\partial z} = 0 \\ W_{2(e)} = 0 \end{cases} \tag{2.107}$$

In the above equation, the subscript (c) represents the variables in domain c. Since there is no source in domain R_c, $W_{2(e)} = 0$; thus, we have (2.108).

$$A_{(c)} = \nabla \times \left(W_{1(c)}\hat{z}\right) \tag{2.108}$$

According to (2.100), we obtain (2.109).

$$\nabla^2 W_{1(c)} + \Upsilon^2 W_{1(c)} = 0 \tag{2.109}$$

Therefore, when we calculate the magnetic field of the δ coil, we just need to calculate W_1 in terms of (2.106) and (2.109).

The general solutions of (2.106) and (2.109) can be expressed as the forms of the double *Fourier* integral, which are shown in (2.110), (2.111) and (2.112).

$$W_{1(b)s} = \int_{-\infty}^{+\infty}\int_{-\infty}^{+\infty} C(\alpha,\beta)e^{kz}e^{j\alpha x}e^{j\beta y}\mathrm{d}\alpha\mathrm{d}\beta \tag{2.110}$$

$$W_{1(b)e} = \int_{-\infty}^{+\infty}\int_{-\infty}^{+\infty} C(\alpha,\beta)\frac{k\mu_r - \lambda}{k\mu_r + \lambda}e^{-kz}e^{j\alpha x}e^{j\beta y}\mathrm{d}\alpha\mathrm{d}\beta \tag{2.111}$$

$$W_{1(e)} = \int_{-\infty}^{+\infty}\int_{-\infty}^{+\infty} C(\alpha,\beta)\frac{2k\mu_r}{k\mu_r + \lambda}e^{\lambda z}e^{j\alpha x}e^{j\beta y}\mathrm{d}\alpha\mathrm{d}\beta \tag{2.112}$$

In the above equations, α and β are integral variables; $C(\alpha, \beta)$ is the undetermined coefficient; $k = \sqrt{\alpha^2 + \beta^2}$; and $\lambda = \sqrt{k^2 + \gamma^2}$. From the above three equations, when the coefficients are determined, the scalar potential W_1 of the whole domain can be calculated.

According to the *Biot–Savart* theorem, we get (2.113).

$$B_{(a,b)s} = \nabla \times \frac{\mu_0 I}{4\pi}\oint_l \frac{\mathrm{d}\boldsymbol{l}}{r} \tag{2.113}$$

In the above equation, $r = \sqrt{(x - x_0)^2 + (y - y_0)^2 + (z - z_0)^2}$ is the distance between *point* (x, y, z) and *point* (x_0, y_0, z_0); μ_0 is the vacuum permeability, and l is the integral path.

The Stokes theorem is used for the above equation, so that we get (2.114).

$$B_{(a,b)s} = \nabla\left[-\frac{\mu_0 I}{4\pi}\iint_{\Omega} \mathrm{d}s \cdot \nabla\left(\frac{1}{r}\right)\right] \tag{2.114}$$

In the above equation, Ω represents the arbitrary curved surfaces encircled by l. Comparing (2.103) and (2.113), we get (2.115).

$$W_{1(a,b,s)} = -\frac{\mu_0 I}{4\pi} \iint\limits_S \mathrm{d}s \cdot \nabla\left(\int \frac{dz}{r}\right) \tag{2.115}$$

In the above equation, we have the following relationship, as shown in (2.116).

$$\frac{1}{r} = \frac{1}{2\pi} \int\limits_{-\infty}^{+\infty} \int\limits_{-\infty}^{+\infty} \frac{e^{\mp k(z-z_0)}}{k} e^{j[\alpha(x-x_0) + \beta(y-y_0)]} \mathrm{d}\alpha \mathrm{d}\beta \tag{2.116}$$

Substituting (2.116) into (2.115), we obtain (2.117) and (2.118).

$$W_{1(a)s} = \frac{\mu I}{2\pi^2} \int\limits_{-\infty}^{+\infty} \int\limits_{-\infty}^{+\infty} \frac{e^{-k(z-z_0)}}{k} \frac{\sin(\alpha y_0)}{\alpha} \times \frac{\sin(\beta y_0)}{\beta} e^{j\alpha x} e^{j\beta y} \mathrm{d}\alpha \mathrm{d}\beta \tag{2.117}$$

$$W_{1(b)s} = \frac{\mu I}{2\pi^2} \int\limits_{-\infty}^{+\infty} \int\limits_{-\infty}^{+\infty} \frac{e^{k(z-z_0)}}{k} \frac{\sin(\alpha y_0)}{\alpha} \times \frac{\sin(\beta y_0)}{\beta} e^{j\alpha x} e^{j\beta y} \mathrm{d}\alpha \mathrm{d}\beta \tag{2.118}$$

2.5.2.2 Scalar Potential of Rectangular Single-Turn Coil

Firstly, considering the single-turn coil, the configuration is as shown in Fig. 2.33. In Fig. 2.33, x_0 and y_0 represent the distance from the inner edge to the z axis in length and width directions, respectively; w represents the width of the coil conductor, and h_1 and h_2 represent the distance from the xoy plane to the upper and lower edges, respectively.

The scalar potential of the single-turn coil can be viewed as the superposition of those δ coils. If we assume that the current density in the section of the conductor is continuous, we can get (2.119).

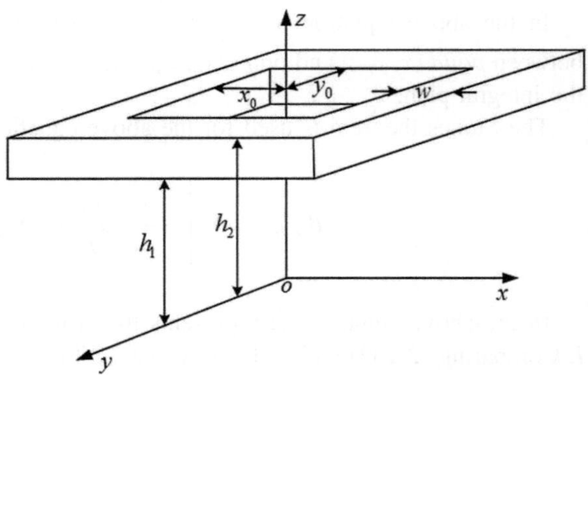

Fig. 2.33 Rectangular single-turn coil

$$W^S_{1(a,b)s} = \int_0^w \int_{h_1}^{h_2} W_{1(a,b)s} \mathrm{d}z_0 \mathrm{d}w \tag{2.119}$$

In the above equation, the superscript S represents the single-turn coil.

Substituting (2.117) and (2.118) into (2.119), and calculating the integral, we can get (2.120) and (2.121).

$$W^S_{1(a)s} = \frac{\mu_0 I}{2\pi^2(h_2 - h_1)w} \int_{-\infty}^{+\infty} \int_{-\infty}^{+\infty} \frac{e^{-kz}}{k^2}(e^{kh_2} - e^{kh_1}) \frac{P}{\alpha\beta} e^{j\alpha x} e^{j\beta y} \mathrm{d}\alpha \mathrm{d}\beta \tag{2.120}$$

$$W^S_{1(b)s} = \frac{\mu_0 I}{2\pi^2(h_2 - h_1)w} \int_{-\infty}^{+\infty} \int_{-\infty}^{+\infty} \frac{e^{kz}}{-k^2}(e^{-kh_2} - e^{-kh_1}) \frac{P}{\alpha\beta} e^{j\alpha x} e^{j\beta y} \mathrm{d}\alpha \mathrm{d}\beta \tag{2.121}$$

In the above two equations, there is some relationship, as follows.

$$P = \begin{cases} \dfrac{\sin[(\alpha-\beta)w + \alpha x_0 - \beta y_0] - \sin(\alpha x_0 - \beta y_0)}{2(\alpha - \beta)} & \\ -\dfrac{\sin[(\alpha+\beta)w + \alpha x_0 + \beta y_0] - \sin(\alpha x_0 + \beta y_0)}{2(\alpha+\beta)} & \alpha \neq \beta \\ \dfrac{w}{2}\cos[\alpha(x_0 - y_0)] & \\ -\dfrac{\sin[\alpha(2w + x_0 + y_0)] - \sin[\alpha(x_0 + y_0)]}{4\alpha} & \alpha = \beta \end{cases}$$

(2.120) and (2.121) give the scalar potentials above and under the coil. For the domain in the section of the coil, the scalar potential in an arbitrary *point* (x, y, z) can be obtained by superpositioning the scalar potential above and under the coil, and then substituting $h_2 = z$ and $h_1 = z$ into (2.120) and (2.121) separately, and summing them, to get (2.122).

$$\begin{aligned} W_{1\mathrm{coilsection}} = {} & \frac{\mu_0 I}{2\pi^2(h_2 - h_1)w} \int_{-\infty}^{+\infty} \int_{-\infty}^{+\infty} \frac{1}{k^2}[2 - e^{-k(z-h_1)}] \\ & - e^{k(z-h_2)} \frac{P}{\alpha\beta} e^{j\alpha x} e^{j\beta y} \mathrm{d}\alpha \mathrm{d}\beta \end{aligned} \tag{2.122}$$

When the center coordinates of the single-turn coil are (x', y', z_0), we can supplement the term $e^{-j\alpha x'} e^{-j\beta y'}$ in (2.120), (2.121), and (2.122) according to coordinate transformation.

2.5.2.3 Scalar Potential of Meander Coil

When the scalar potential of the single-turn coil is obtained, the scalar potential of the meander coil can be obtained by coordinate transformation and superposition.

For the configuration of the meander coil shown in Fig. 2.25, we assume the center coordinate of group ($q = 1$) is $(0, 0, z_m)$; thus, the center coordinates of group ($q = 2 \sim Q$) are $(x_q', 0, z_m)$. In each layer, there are $N \times Q$ single-turn coils; thus, the scalar potential in the m layer can be expressed by (2.123), (2.124), and (2.125).

$$W^m_{1(a)s} = \frac{\mu_0 I}{2\pi^2 (h_2 - h_1) w} \int\limits_{-\infty}^{+\infty} \int\limits_{-\infty}^{+\infty} \frac{e^{-kz}}{k^2} \left(e^{kh_m} - e^{kh_{1m}}\right) \frac{\sum_{n=1}^{N} P_n}{\alpha\beta} \sum_{q=1}^{Q} e^{-j\alpha x'} e^{j\alpha x} e^{j\beta y} \mathrm{d}\alpha \mathrm{d}\beta \tag{2.123}$$

$$W^m_{1(b)s} = \frac{\mu_0 I}{2\pi^2 (h_2 - h_1) w} \int\limits_{-\infty}^{+\infty} \int\limits_{-\infty}^{+\infty} \frac{e^{kz}}{k^2} \left(e^{-kh_{2m}} - e^{-kh_{1m}}\right) \frac{\sum_{n=1}^{N} P_n}{\alpha\beta} \sum_{q=1}^{Q} e^{-j\alpha x'_q} e^{j\alpha x} e^{j\beta y} \mathrm{d}\alpha \mathrm{d}\beta \tag{2.124}$$

$$\begin{aligned} W^m_{1\text{coilsection}} = & \frac{\mu_0 I}{2\pi^2 (h_2 - h_1) w} \int\limits_{-\infty}^{+\infty} \int\limits_{-\infty}^{+\infty} \frac{1}{k^2} \left(2 - e^{-k(z-h_{2m})} - e^{-k(z-h_{1m})}\right) \\ & \frac{\sum_{n=1}^{N} P_n}{\alpha\beta} \sum_{q=1}^{Q} e^{-j\alpha x'_q} e^{j\alpha x} e^{j\beta y} \mathrm{d}\alpha \mathrm{d}\beta \end{aligned} \tag{2.125}$$

In the above equations, there is some relationship, as follows, in which the superscript m represents the layer.

$$P_n = \begin{cases} \dfrac{\sin[(\alpha-\beta)w + \alpha x_{0n} - \beta y_{0n}] - \sin(\alpha x_{0n} - \beta y_{0n})}{2(\alpha-\beta)} & \\ \qquad & \alpha \neq \beta \\ -\dfrac{\sin[(\alpha+\beta)w + \alpha x_{0n} + \beta y_{0n}] - \sin(\alpha x_{0n} + \beta y_{0n})}{2(\alpha+\beta)} & \\ \dfrac{w}{2}\cos[\alpha(x_{0n} - y_{0n})] & \\ -\dfrac{\sin[\alpha(2w + x_{0n} + y_{0n})] - \sin[\alpha(x_{0n} + y_{0n})]}{4\alpha} & \alpha = \beta \end{cases}$$

The scalar potential in domain R_b can be obtained by the superposition of the scalar potential of each area, as shown in (2.126).

$$W_{1(b)s} = \frac{\mu_0 I}{2\pi^2(h_2 - h_1)w} \int_{-\infty}^{+\infty} \int_{-\infty}^{+\infty} \frac{e^{kz}}{-k^2} \sum_{m=1}^{M} \left(e^{-kh_{2m}} - e^{-kh_{1m}}\right) \times \frac{1}{\alpha\beta} \sum_{n=1}^{N} P_n \sum_{q=1}^{Q} e^{-j\alpha x'_q} e^{j\alpha x} e^{j\beta y} \mathrm{d}\alpha \mathrm{d}\beta \tag{2.126}$$

Comparing the above equation and (2.110), the coefficient $C(\alpha,\beta)$ can be obtained, as shown in (2.127).

$$C(\alpha, \beta) = -\frac{\mu_0 I}{2\pi^2(h_2 - h_1)wk^2\alpha\beta} \sum_{m=1}^{M} \left(e^{-kh_{2m}} - e^{-kh_{1}m}\right) \sum_{n=1}^{N} P_n \sum_{q=1}^{Q} e^{-j\alpha x'_q} \tag{2.127}$$

Then, $W_{1(b)s}$ and $W_{1(c)}$ can be obtained by substituting Eq. (2.127) into (2.111) and (2.112), separately.

For the meander coil, it is worth paying attention to the magnetic field of each layer of conductor coil. For each layer of coil, the scalar potential of the conductor's region is equal to the sum of the scalar potential generated by the layer itself and coils of other layers; thus, the equation for the area of the coil of the mth layer is listed in (2.128), in which the superscript m represents the layer.

$$W^{m'}_{l\text{coilsection}} = \frac{\mu_0 I}{2\pi^2(h_2 - h_1)w} \int_{-\infty}^{+\infty} \int_{-\infty}^{+\infty} \left\{ \frac{1}{k^2} \left[2 - e^{-k(z-h_{1m'})} - e^{k(z-h_{2m'})}\right] + \sum_{m=1}^{m'-1} \frac{e^{-kz}}{k^2} \left(e^{kh_{2m}} - e^{kh_{1m}}\right) - \sum_{m=m'+1}^{M} \frac{e^{kz}}{k^2} \left(e^{-kh_{2m}} - e^{-kh_{1m}}\right) \right\} \times \frac{1}{\alpha\beta} \sum_{n=1}^{N} P_n \sum_{q=1}^{Q} e^{-j\alpha x'_q} e^{j\alpha x} e^{j\beta y} \mathrm{d}\alpha \mathrm{d}\beta \tag{2.128}$$

2.5.2.4 Eddy Current, Magnetic Flux Density, and Coil's Impedance

The eddy current is induced into the tested conductor, which is necessary for the electromagnetic ultrasonic testing. According to electromagnetic field theory, the eddy current generated by the meander coil on the surface of the conductor can be calculated by (2.129), in which J_e represents the eddy current in the conductor.

$$J_e = -j\omega\sigma A_{(c)} \tag{2.129}$$

Substituting (2.102) into (2.129), we can obtain (2.130), in which $\hat{x}$ and $\hat{y}$ are direction vectors.

$$J_e = \frac{j\omega\sigma\mu_0 I}{\pi^2(h_2 - h_1)w} \int_{-\infty}^{+\infty} \int_{-\infty}^{+\infty} \frac{e^{\lambda z}}{k} \times \frac{\mu_r}{k\mu_r + \lambda} \times (j\beta\hat{x} - j\alpha\hat{y}) \sum_{m=1}^{M} \left(e^{-kh_{2m}} - e^{-kh_{1m}}\right) \times \frac{1}{\alpha\beta} \sum_{n=1}^{N} P_n \sum_{q=1}^{Q} e^{-j\alpha x'_q} e^{j\alpha x} e^{j\beta y} \mathrm{d}\alpha \mathrm{d}\beta \tag{2.130}$$

According to (2.104), (2.112), (2.125), and (2.126), the magnetic flux density can be obtained within domain R_b and R_c, as shown in (2.131) and (2.132).

$$B_{(b)} = \frac{\mu_0 I}{2\pi^2(h_2 - h_1)w} \int_{-\infty}^{+\infty} \int_{-\infty}^{+\infty} \left[\frac{1}{-k}\left(e^{kz} - \frac{k\mu_r - \lambda}{k\mu_r + \lambda} e^{-kz}\right) \times (j\alpha\hat{x} + j\beta\hat{y}) - \left(e^{kz} + \frac{k\mu_r - \lambda}{k\mu_r + \lambda} e^{-kz}\right)\hat{z}\right] \sum_{m=1}^{M} \left(e^{-kh_{2m}} - e^{-kh_{1m}}\right) \times \frac{1}{\alpha\beta} \sum_{n=1}^{N} P_n \sum_{q=1}^{Q} e^{-j\alpha x'_q} e^{j\alpha x} e^{j\beta y} \mathrm{d}\alpha \mathrm{d}\beta \tag{2.131}$$

$$B_{(c)} = \frac{\mu_0 I}{2\pi^2(h_2 - h_1)w} \int_{-\infty}^{+\infty} \int_{-\infty}^{+\infty} \left[\frac{e^{\lambda z}}{-k} \times \frac{2k\mu_r\lambda}{k\mu_r + \lambda} \times (j\alpha\hat{x} + j\beta\hat{y} + \lambda\hat{z})\right] \sum_{m=1}^{M} (e^{-kh_{2m}} - e^{-kh_{1m}}) \times \frac{1}{\alpha\beta} \sum_{n=1}^{N} P_n \sum_{q=1}^{Q} e^{-j\alpha x'_q} e^{j\alpha x} e^{j\beta y} \mathrm{d}\alpha \mathrm{d}\beta \tag{2.132}$$

The coil's impedance can be expressed as the sum of the coil's DC impedance, the coil's impedance in free space, and the impedance generated by the eddy current induced by the conductor, as shown in (2.133).

$$Z = Z_d + Z_0 + \Delta Z \tag{2.133}$$

In the above equation, Z_d is the DC impedance; Z_0 is the impedance in free space, and ΔZ is the impedance generated by the eddy current.

Impedance Z_0 can be calculated by (2.134).

$$Z_0 = \frac{V}{I} \tag{2.134}$$

Therefore, if we want to calculate Z_0, it is necessary to calculate the electromotive force first. According to the idea of scalar potential calculation, the electromotive force of the single-turn coil should be calculated first, as shown in (2.135).

$$V^S = \frac{j\omega}{(h_2 - h_1)w} \int_{\text{coilcrosssection}} \left(\int_S B \cdot \mathrm{d}s \right) \mathrm{dArea} \tag{2.135}$$

The electromotive force of the meander coil is the superposition of the electromotive force of all equivalent rectangular coils; thus, we have (2.136).

$$\begin{aligned} V = {} & \frac{2j\omega\mu_0 I}{\pi^2 (h_2 - h_1)^2 w^2} \sum_{m'=1}^{M} \int_{-\infty}^{+\infty} \int_{-\infty}^{+\infty} \left\{ 2(h_{2m'} - h_{1m'}) \right. \\ & + \frac{2}{k}\left[e^{-k(h_{2m'} - h_{1m'})} - 1 \right] + \sum_{m=1}^{m'-1} \frac{1}{k} (e^{kh_{2m}} - e^{kh_{1m}})(e^{-kh_{1m'}} - e^{-kh_{2m'}}) \\ & \left. + \sum_{m=m'+1}^{M} \frac{1}{k} (e^{-kh_{2m}} - e^{-kh_{1m}})(e^{kh_{1m'}} - e^{kh_{2m'}}) \right\} \\ & \times \frac{1}{(\alpha\beta)^2} \left(\sum_{n=1}^{N} P_n \right)^2 \left(\sum_{q=1}^{Q} e^{-j\alpha x'_q} + \sum_{q=1}^{Q} e^{j\alpha x'_q} \right) \mathrm{d}\alpha \mathrm{d}\beta \end{aligned} \tag{2.136}$$

Thus, according to (2.134), we can get (2.137).

$$\begin{aligned} Z_0 = {} & \frac{2j\omega\mu_0}{\pi^2 (h_2 - h_1)^2 w^2} \sum_{m'=1}^{M} \int_{-\infty}^{+\infty} \int_{-\infty}^{+\infty} \left\{ 2(h_{2m'} - h_{1m'}) \right. \\ & + \frac{2}{k}[e^{-k(h_{2m'} - h_{1m'})} - 1] + \sum_{m=1}^{m'-1} \frac{1}{k} (e^{kh_{2m}} - e^{kh_{1m}})(e^{-kh_{1m'}} - e^{-kh_{2m'}}) \\ & \left. + \sum_{m=m'+1}^{M} \frac{1}{k} (e^{-kh_{2m}} - e^{-kh_{1m}})(e^{kh_{1m'}} - e^{kh_{2m'}}) \right\} \\ & \times \frac{1}{(\alpha\beta)^2} \left(\sum_{n=1}^{N} P_n \right)^2 \left(\sum_{q=1}^{Q} e^{-j\alpha x'_q} + \sum_{q=1}^{Q} e^{j\alpha x'_q} \right) \mathrm{d}\alpha \mathrm{d}\beta \end{aligned} \tag{2.137}$$

(2.138) can be used to calculate the impedance ΔZ.

$$\Delta Z = \Delta R + j\Delta X = \frac{j8\pi^2\omega}{\mu_0 i^2} \int_{-\infty}^{+\infty} \int_{-\infty}^{+\infty} k^3 C(\alpha, \beta) \times C(-\alpha, -\beta) \frac{k\mu_r - \lambda}{k\mu_r + \lambda} \mathrm{d}\alpha \mathrm{d}\beta \tag{2.138}$$

Substituting $C(\alpha, \beta)$ into the above equation, we obtain (2.139).

$$\Delta Z = \frac{j2\omega\mu_0}{\pi^2 (h_2 - h_1)^2 w^2} \int_{-\infty}^{+\infty} \int_{-\infty}^{+\infty} \frac{1}{k(\alpha\beta)^2} \left[\sum_{m=1}^{M} \left(e^{-kh_{2m}} - e^{-kh_{1m}} \right) \right]^2 \left(\sum_{n=1}^{N} P_n \right)^2 \left(\sum_{q=1}^{Q} e^{-j\alpha x'_q} \times \sum_{q=1}^{Q} e^{j\alpha x'_q} \right)] \times \frac{k\mu_r - \lambda}{k\mu_r + \lambda} \mathrm{d}\alpha \mathrm{d}\beta \tag{2.139}$$

The meander coil's DC impedance Z_d is DC resistance, which can be calculated by the coil's dimension and electrical conductivity, as shown in (2.140).

$$Z_d = \frac{L}{\sigma_c (h_2 - h_1) w} \tag{2.140}$$

In the above equation, L is the total length of the meander coil conductor and σ_c is the electrical conductivity of the coil conductor.

2.5.3 The Calculation of the Time-Domain Pulsed Magnetic Field

In the electromagnetic ultrasonic testing, the applied excitation is a pulse signal. The pulsed magnetic field generated by the coil and the eddy current in the specimen should be calculated, that is, the pulsed response should be calculated.

The pulsed excitation and the corresponding pulsed response can be expressed as the superposition of a series of sinusoidal signals of different frequencies. For the conditions of the expressions of magnetic induction intensity and eddy current, the time-domain response can be obtained by the inverse Fourier transformation (IFT), as shown in (2.141), (2.142), and (2.143).

$$J_e(t) = \frac{1}{2\pi} \int_{-\infty}^{\infty} J_e e^{j\omega t} \mathrm{d}\omega \tag{2.141}$$

$$B_{(b)}(t) = \frac{1}{2\pi} \int_{-\infty}^{\infty} B_{(b)} e^{j\omega t} \mathrm{d}\omega \tag{2.142}$$

$$B_{(c)}(t) = \frac{1}{2\pi} \int_{-\infty}^{\infty} B_{(c)} e^{j\omega t} \mathrm{d}\omega \tag{2.143}$$

In the above equations, $J_e(t)$, $B_{(b)}(t)$, and $B_{(c)}(t)$ represent the time-domain signals of the eddy current and magnetic flux density, respectively.

However, it is hard to obtain the expressions of magnetic flux density and eddy current in the time domain by calculating the IFT of (2.141) to (2.143) directly.

2.5.4 Example and Comparison of Results

In order to verify the effectiveness of the analytical expression and calculation, a complete EMAT model, comprising a square permanent magnet with a perpendicular magnetic field, was set up. The two-layer structure coil, with each layer having eight meander belts, is taken as an example to perform the calculation. The coil is placed above the specimen and the liftoff distance between the coil and specimen is achieved using the non-magnetic and non-conductive thin liftoff sheet. The value of the liftoff can be adjusted, as shown in Fig. 2.34.

In Fig. 2.34, the coil's dimension and material's parameters are shown in Tables 2.1 and 2.2. In the experiment, a radio frequency (RF) power amplifier (AG1024), which can provide a continuous sine signal and tone-burst signal, is supplied to the meander coil. This radio frequency power amplifier can not only generate a continuous sinusoidal signal, the frequency and amplitude of which is continuous and adjustable, but can also generate the tone-burst signal, whose frequency, amplitude, and cycle number are adjustable. In the experiment, the oscilloscope records the wave data used to measure the voltages of the coil and the 1 Ω sample resistance, which is connected in series with the coil. Thus, the current of the coil can be obtained according to the voltage of the sample resistance. The resistance and inductance of the coil can be calculated in terms of the measured continuous sine voltage and current signal.

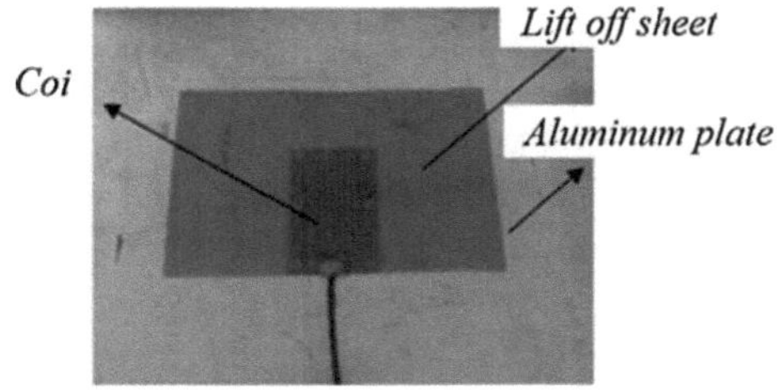

Fig. 2.34 The setup of the experiment

In the calculation of the coil's pulsed magnetic field, the pulsed excitation should be applied on the coil. The tone-burst voltage and current signals used for generating the acoustic waves can also be measured. When the liftoff value of the coil is 1 mm, a tone-burst signal with 500 kHz frequency is applied. The cycle number is 4. The voltage signal in the sample resistance (i.e., the coil's current signal) is shown in Fig. 2.35.

In order to compare with the numerical computation methods, the TSFEM method is used for the calculation of the above model. Since the calculation of the 3D model is too complex, the influence of the coil's end is neglected and it is assumed that the length of the coil and conductor are infinite in their length direction. The 2D model of the coil is calculated in this way. In the 2D model, the calculation can only obtain the y component of the eddy current, not the x component or z component of the magnetic flux density. The calculation is done using the finite element analysis software *Comsol Multiphysics*.

The current signal shown in Fig. 2.35 is used as the excitation in the simulation. Since the frequency of the excitation signal is very high, the skin depth is very small. Meanwhile, the dimensions of the coil conductor are also very small. In order to guarantee the accuracy of the results, it is necessary to draw a dense grid within the conductor's skin depth and the coil conductors. The total number of elements in the TSFEM calculation model is 15,348, as shown in Fig. 2.36.

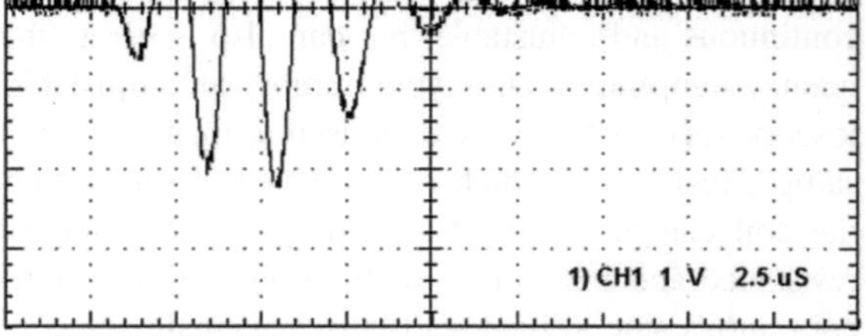

Fig. 2.35 Measured pulsed voltage and current signals

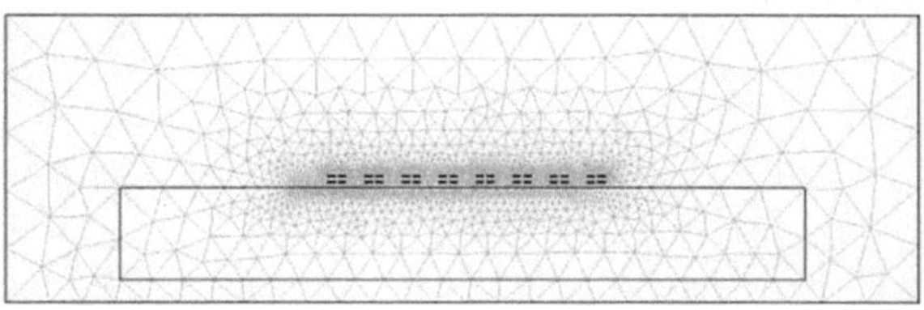

Fig. 2.36 The simulation model of TSFEM

Firstly, the coil's impedance under different liftoff values was calculated. According to (2.141), (2.142), and (2.143), a high-accuracy adaptive recursive *Simpson* method is used to calculate the numerical integration. Since the expression only contains the trigonometric function, but the *Bessel function* is included in the spiral coil's expression, the expression has a very good convergence, so the calculation is much simpler.

The FFT-IFFT method was used to calculate the coil's pulsed magnetic field and the inductive coil in the specimen. For the current signal in Fig. 2.36, 1024 points were selected to make the FFT process, substituting the amplitude corresponding to each frequency into (2.130), (2.131), and (2.132), and calculating the eddy current and magnetic induction intensity corresponding to each frequency using the recursive adaptive method. Finally, we obtain the time-domain signal of the eddy current and magnetic induction intensity by IFFT.

The analytical model's calculated value of the y component of eddy current density and the x and z components of magnetic induction intensity at the coordinates (1.1725 mm, 11.665 mm, −0.1 mm) is shown in Figs. 2.37, 2.38, and 2.39, respectively, as well as the calculated values by the TSFEM method. Figures 2.40 and 2.41 show the calculated values of the x and z components of the magnetic

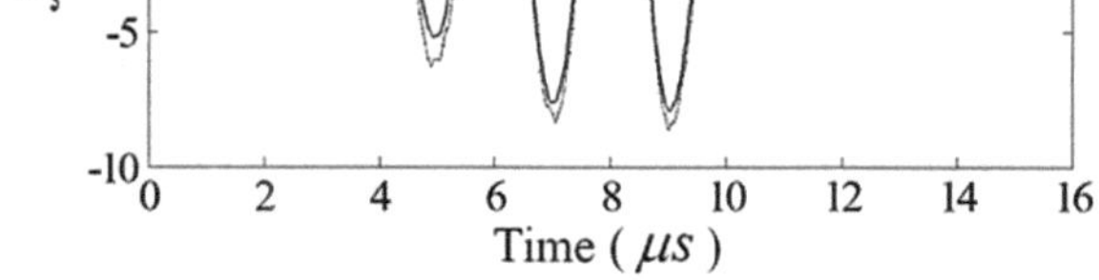

Fig. 2.37 The y component of the eddy current density at the coordinates (1.1725 mm, 11.665 mm, −0.1 mm)

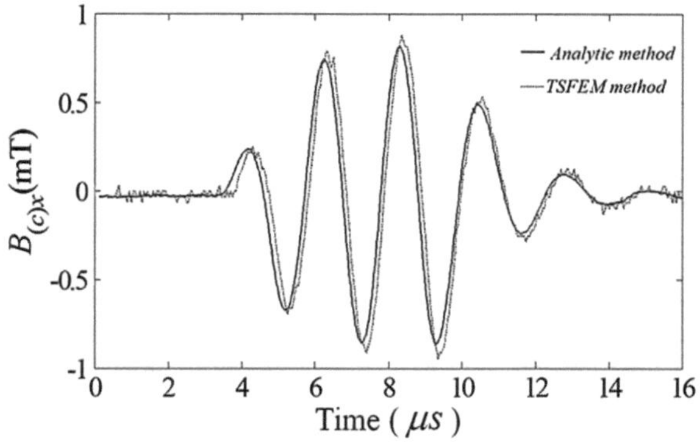

Fig. 2.38 The x component of the magnetic induction intensity at the coordinates (1.1725 mm, 11.665 mm, −0.1 mm)

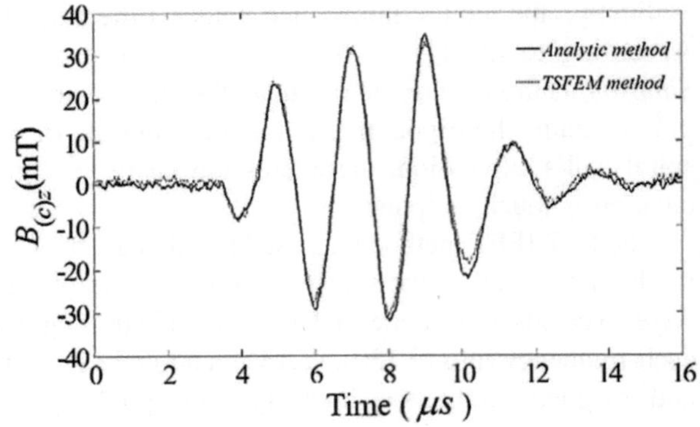

Fig. 2.39 The z component of the magnetic induction intensity at the coordinates (1.1725 mm, 11.665 mm, −0.1 mm)

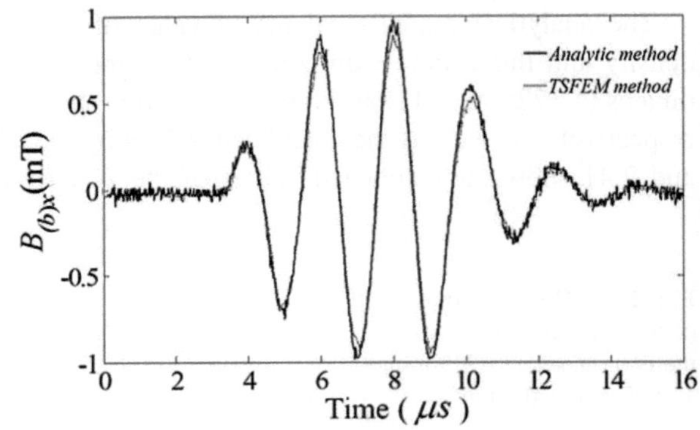

Fig. 2.40 The x component of the magnetic induction intensity at the coordinates (1.1725 mm, 11.665 mm, −0.1 mm)

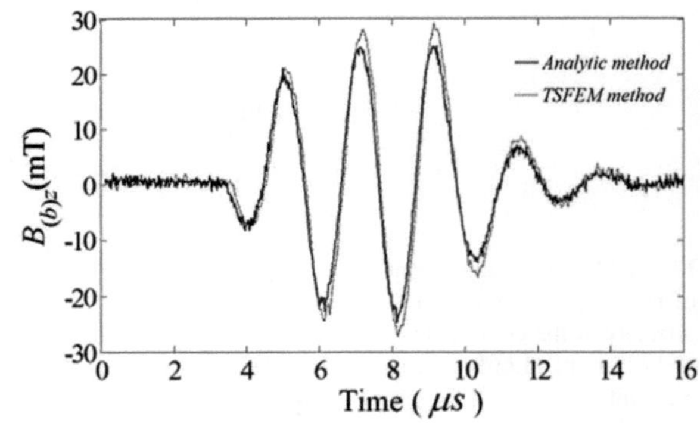

Fig. 2.41 The z component of the magnetic induction intensity at the coordinates (1.1725 mm, 11.665 mm, −0.1 mm)

induction intensity at the coordinates (1.1725 mm, 11.665 mm, −0.1 mm) by the analytical model and the TSFEM method. The horizontal axis in Figs. 2.37 and 2.39 represents the time. It can be seen from these pictures that there is basic agreement of time and amplitude of the calculated values between the analytical model and the TSFEM method, even if a certain error does exist.

Because the coil's skin effect and proximity effect are neglected in the derivation process of the analytical equation, inaccurate simulation results will definitely result between the analytical equation and the actual physical model with the assumption of equality of the current density in the conductor. Since it is difficult to evaluate the skin effect and proximity effect in the coil's analytical model, the neglecting of their effect is a general practice for the derivation of the coil's analytical model. The calculated error caused by this issue is, in general, acceptable in practical engineering applications. In addition, the concept of taking the meander coil equivalent to the superposition of the single fold closing coil itself can also bring about some calculation errors.

Compared with the numerical calculation, the advantages of the analytical calculation are the clarity of the analytic equation's physical meaning and the rapid calculation process.

2.6 The Analytical Method of EMAT Under Impulse Voltage Excitation [5]

In practical applications, most of the excitation power supply of EMAT is a voltage source or a power source. If the *Lorentz force* in EMAT can be calculated when the excitation of the pulsed voltage or power is given, it will be beneficial to the analysis and design of EMAT. Based on the analytic calculation of the EMAT coil impedance, the EMAT's analytical analysis method under voltage excitation can be obtained.

The EMAT analytic method of several coil structure's EMAT can be achieved when the pulsed current is given in advance. Actually, if the coil's terminal voltage or input power is already known, the pulsed voltage or power signal can be transformed into a pulsed current signal through the coil's impedance. Meanwhile, it can enable the analysis of EMAT by using the aforementioned method. Thus, the analytical problem of EMAT under the pulsed voltage can be transformed into the current excitation problem if the EMAT coil's voltage is given.

2.6.1 Calculating the Pulsed Current Using the Analytical Equation

In the sinusoidal steady state, if the excited voltage of the coil is given, the excited current of the coil can be expressed by (2.144).

$$I(\omega) = \frac{U(\omega)}{Z(\omega)} \tag{2.144}$$

For the pulse voltage excitation, the excitation of pulsed current can be calculated using the method of FFT-IFFT, specifically. Firstly, the pulsed voltage signal is transformed into frequency domain using FFT, according to (2.144), and the value of the current's frequency domain under a specific frequency is calculated. Then, IFFT is used to obtain the pulsed current of the time domain.

2.6.2 Calculating the Pulsed Current Using the Field-Circuit Coupling Finite Element Method

Using the field-circuit coupling method, the pulsed current is calculated. No matter what kind of voltage excitation is applied on the coil, narrow pulses, general tone-burst signals, or windowed tone-burst signals, the excitation signal can be expressed as the superposition of a series trigonometric function, as shown in (2.145).

$$u(t) = \sum_{k=1}^{N} U_k \cos(\omega_k t + \varphi_k) \quad 0 \leq t \leq T \tag{2.145}$$

In the above equation, $k = 1, 2, \ldots, N$; U_k is the amplitude of the harmonic component of order k, ω_k is the angular frequency of the kth-order component; φ_k is the initial phase angle of the kth-order component, and T represents the duration of the excitation signal.

In (2.145), U_k, ω_k, and φ_k can be obtained using the FFT process. If the equivalent circuit parameters of the coil and specimen system can be calculated according to *Ohm's law*, the current excitation signal can be obtained through the IFFT process.

Since the frequency of the excitation signal is very high, the eddy current, skin, and proximity effects should be considered. Therefore, the alternating current (AC) resistance and inductance of the coil will change with the excitation frequency, the geometric configuration, and the electrical characteristics of the coils and the specimen.

For a specific frequency, the resistance and inductance of the coil can be calculated through sinusoidal steady-state analysis of the circuit-field coupling. For the system of the coil and specimen, the capacitance value is very small and can be neglected. Then, the system of the coil and specimen can be viewed as a two-port network, as in Fig. 2.42. The excitation voltage of EMATs is the port voltage, and the excitation current is the port current. Thus, the complex power can be expressed by (2.146).

$$\tilde{S}_k = \dot{\mathbf{U}}_k \dot{\mathbf{I}}_k^* = (R_k + j\omega_k L_k) \dot{\mathbf{I}}_k \dot{\mathbf{I}}_k^* = R_k I_k^2 + j\omega_k L_k I_k^2 \tag{2.146}$$

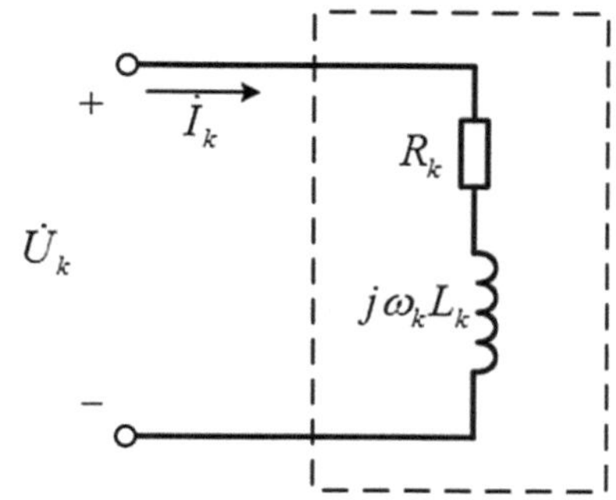

Fig. 2.42 Equivalent two-port network of coil

In the above equation, $\tilde{S}_k$ represents the complex power, $\dot{\mathbf{U}}_k$ represents the port voltage, $\dot{\mathbf{I}}_k$ is the port current, $\dot{\mathbf{I}}_k^*$ is the conjugate of $\dot{\mathbf{I}}_k$, and $\mathbf{I}_k$ is the effective value of $\dot{\mathbf{I}}_k$. R_k and L_k are the equivalent resistance and inductance of the two-port network—the *AC* resistance and inductance of the coil.

Neglecting the displacement current in the coil and specimen field, we obtain the phasor notation of *Poynting theorem*, as shown in (2.147).

$$-\iint\limits_S \left(\dot{E}_k \times \dot{H}_k^*\right) \cdot \mathrm{d}s = \iiint\limits_V \frac{1}{\sigma} J_k^2 \mathrm{d}V + j \iiint\limits_V \omega\mu H_k^2 \mathrm{d}V \tag{2.147}$$

In the above equation, $\dot{E}_k$ is the electrical field strength phasor quantity, $\dot{H}_k^*$ is the conjugate of the magnetic field intensity phasor quantity $\dot{H}_k$, J_k is the current density, and H_k is the magnetic field intensity. In (2.147), the left part of the equation represents the electromagnetic complex power flowing into the field close to the coil and specimen, the first term of the right part represents the dissipative active power and the second term of the right part represents dissipative reactive power.

Comparing (2.146) with (2.147), we obtain (2.148) and (2.149).

$$R_k = \frac{1}{I_k^2} \iiint\limits_V \frac{1}{\sigma} J_k^2 \mathrm{d}V \tag{2.148}$$

$$L_k = \frac{1}{I_k^2} \iiint\limits_V \mu H_k^2 \mathrm{d}V \tag{2.149}$$

In the above equations, ω is the angular frequency.

Therefore, when we calculate the impedance of the coil, we shall use the time-harmonic field equation, considering the eddy current effect, skin effect, and the proximity effect. By replacing the partial differential operator $\partial/\partial t$ with $j\omega$, the expression of the time-harmonic field equation can be converted into (2.150) and (2.151).

$$j\omega\sigma A - \frac{1}{\mu}\nabla^2 A = J_s \tag{2.150}$$

$$I = \iint_{\Omega_\tau} (J_s + J_e)\mathrm{d}s \tag{2.151}$$

Correspondingly, we get (2.152), (2.153), and (2.154).

$$J_e = -j\omega\sigma A \tag{2.152}$$

$$J = J_s + J_e \tag{2.153}$$

$$H = \frac{1}{\mu}\nabla \times A \tag{2.154}$$

When J and H are determined, the resistance R_k and inductance L_k of the coil can be calculated according to (2.148) and (2.149).

When the pulsed excitation of the coil is calculated using the FFT-IFFT method, the characteristics of EMAT can be analyzed using the pulsed current, which was given in the preceding part of the text.

Using the circuit-field coupling FEM method, the simulation methods and calculated steps of the EMAT under the impulse voltage excitation are shown in Fig. 2.43.

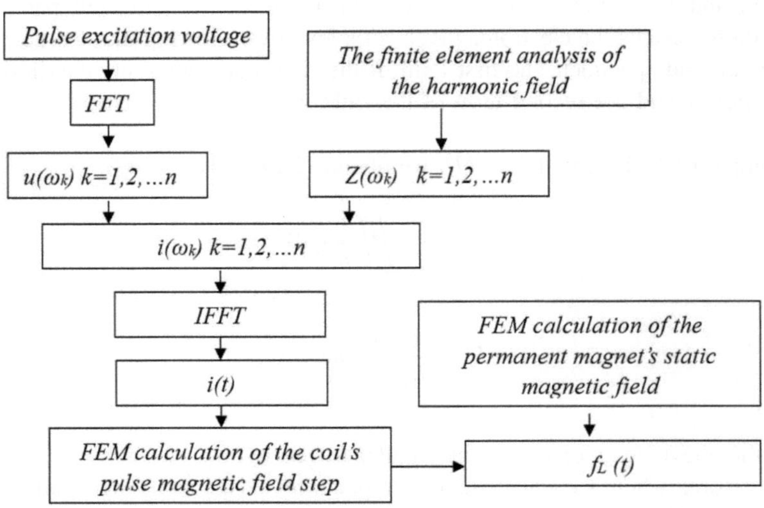

Fig. 2.43 The analytical procedure of EMAT under the pulsed voltage excitation

The given theoretical analysis method under the pulsed voltage of Lorentz force-based EMAT is also suited to the analysis of the EMAT based on the magnetostrictive mechanism

2.6.3 The Coil's Current Calculation Examples Realized Using the Circuit-Field Coupled Finite Element Method

Following the setup shown in Fig. 2.34, calculations and experimental work are performed. The parameters of the coil and the specimen are shown in Tables 2.1 and 2.2. In this work, the liftoff value between coil and specimen is 1 mm. In the experiments, the tone-burst signal was provided by RF power amplifier. An oscilloscope, which can record the wave data, was used to measure the voltages of the coil and 1 Ω sampling resistance, which was connected in series with the coil. The tone-burst signals with 500 kHz frequencies and 3 period number were supplied to the meander coil. The voltage and current signals used for generating the acoustic waves can also be measured, as shown in Fig. 2.44.

In the calculation process, the measured voltage signal in Fig. 2.44 was used as the pulsed voltage excitation. The signal was transformed into the frequency domain by FFT. After that, the coil inductance and resistance of each point can be calculated by the field-circuit coupled method. Then, the amplitude and phase position of the electric current of each frequency point were calculated. Finally, the time-domain waveform of the pulsed current was obtained by IFFT. The time-domain pulsed current can be taken as the excitation source of the simulated analysis for EMAT.

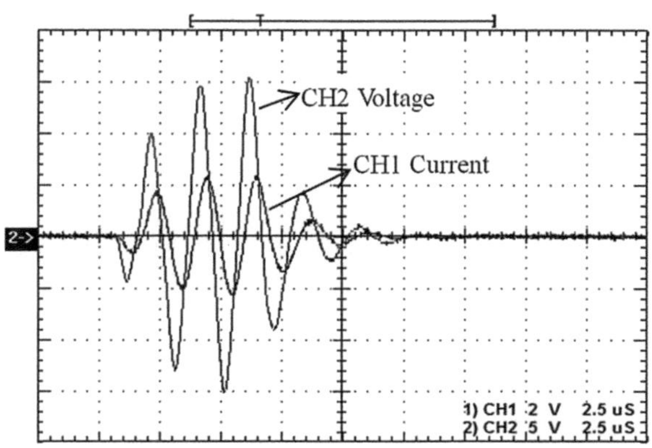

Fig. 2.44 The measured pulsed voltage and current signals of the coil

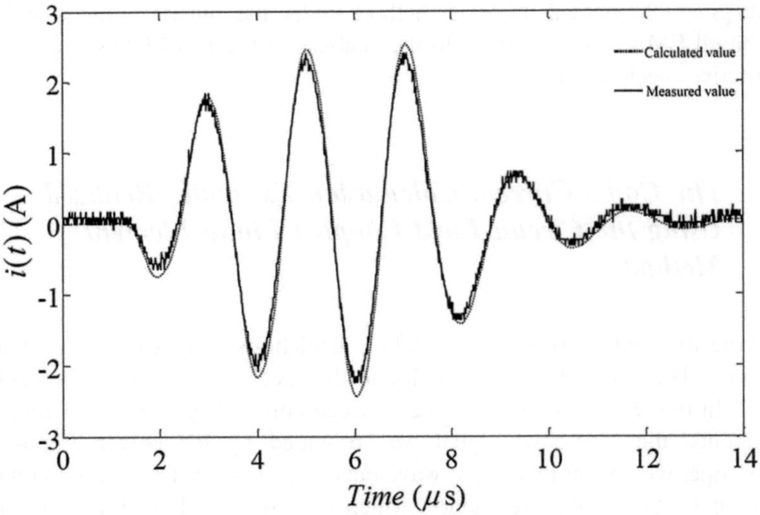

Fig. 2.45 The calculated and measured current signals

The calculated and experimental measured current signals at the liftoff distance of 1 mm are shown in Fig. 2.45. Comparing these two waveforms, we find that both the amplitudes and phase positions are in agreement, which verifies the validity of the proposed field-circuit coupled and FFT-IFFT method, and impedance calculation equations.

References

1. Hao, K., Huang, S., Zhao, W., et al.: Analytical calculation and analysis for meander-coil electromagnetic acoustic transducers. In: 2011 IEEE International Instrumentation and Measurement Technology Conference, I2MTC 2011—Proceedings, pp. 231–234
2. Hao, K., Huang, S., Zhao, W., et al.: Modeling and finite element analysis of transduction process of electromagnetic acoustic transducers for nonferromagnetic material testing. J. Cent. S. Univ. Technol. **18**(3), 749–754 (2011)
3. Hao, K., Huang, S., Zhao, W., et al.: Analytical modeling and calculation of pulsed magnetic field and input impedance for EMATs with planar spiral coils. NDT & E Int. **44**(3), 274–280 (2011)
4. Hao, K., Huang, S., Zhao, W., et al.: Analytical modelling and calculation of impedance and pulsed magnetic field for rectangular meander coil based on second order potential. Acta Phys. Sin. **60**(7), 078103 (2011) (in Chinese)
5. Hao, K., Huang, S., Zhao, W., et al.: Circuit-field coupled finite element analysis method for an electromagnetic acoustic transducer under pulsed voltage excitation. Chin. Phys. B **20**(6), 791–800 (2011)

Chapter 3
Analytical Method of EMAT Based on Magnetostrictive Mechanism

If the tested specimen is ferromagnetic, the material will have magnetostrictive properties. Generally, the ultrasonic waves will be excited and received by EMAT in the specimen so that both the mechanisms of *Lorentz force* and magnetostriction are involved at the same time. In this chapter, the EMAT based on a magnetostrictive mechanism is studied, and its analytical method is provided. Compared with the EMAT based on *Lorentz force*, the energy conversion process of EMAT based on magnetostrictive mechanism has a close relationship with the magnetization characteristics and magnetostrictive properties of the tested specimen. Since the magnetization of the ferromagnetic material is nonlinear and is difficult to obtain, the analysis of EMAT based on a magnetostrictive mechanism is more complex. If we want to get the ultrasonic wave in the specimen by calculation, we should first calculate the stress condition inside the specimen, which has a relationship with the specimen's magnetostrictive property, called the magnetostrictive force. From the point of view of receiving, the magnetic effect caused by particle motion in the specimen also has a relationship with the magnetostrictive property. The electric current density generated by the magnetic effect inside the specimen is called the magnetostrictive current density. Deriving the equations of magnetostrictive force and magnetostrictive current density is key stage to achieving the simulation of EMAT based on a magnetostrictive mechanism. When the various equations and parameters are given, the numerical simulation and the analysis of the whole energy conversion process of EMAT based on magnetostrictive mechanism can be obtained using the finite element method. For two different types of EMAT based on the magnetostrictive mechanism, both are special and have important applications, which are the SH guided wave EMAT and the axisymmetric *T*-mode guided wave EMAT in the pipeline. Their directions of the bias magnetic field are parallel to that of the coil's conductor. Considering the transverse isotropic characteristics of the specimen under a magnetization state, the analytical modeling and calculation can be achieved easily. This approach provides an alternative way for the theoretical analysis of those two kinds of EMAT.

S. Huang et al., *Electromagnetic Ultrasonic Guided Waves*, Springer Series in Measurement Science and Technology, DOI 10.1007/978-981-10-0564-0_3

3.1 Magnetic and Magnetostrictive Property of Ferromagnetic Materials

The energy conversion principle of EMAT based on a magnetostrictive mechanism has a close relationship with the specimen's material property. In order to achieve the simulated analysis of EMAT based on a magnetostrictive mechanism, it is necessary to have a deep understanding and characterization of the magnetization characteristics and magnetostrictive properties of ferromagnetic materials.

3.1.1 Magnetic Characteristics and Magnetic Permeability of Ferromagnetic Materials

The magnetization characteristics of ferromagnetic material can be characterized using the magnetizing curve; the magnetizing curve of a typical ferromagnetic material is shown in Fig. 3.1. It can be seen from the curve that magnetic permeability $\mu = B/H$ is the nonlinear function of the magnetic field H. According to various applications, the magnetic permeability can be defined differently.

When the value of the magnetic field H is very small, the magnetization process is reversible. The slope of the B-H curve at the origin can be taken as the magnetic permeability in the low-value zone of the magnetic field, known as initial permeability, as shown in (3.1).

$$\mu_{\text{in}} = \lim_{\Delta H \to 0} \left(\frac{\Delta B}{\Delta H} \right) \Bigg|_{H=0,B=0} \tag{3.1}$$

The slope at each point of the magnetizing curve at the initial stage, known as differential permeability, is shown in (3.2).

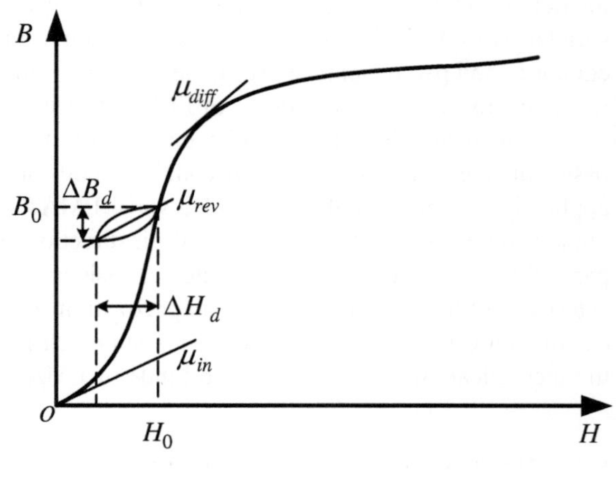

Fig. 3.1 The magnetizing curve of a typical ferromagnetic material

$$\mu_{\text{diff}} = \lim_{\Delta H \to 0} \left(\frac{\Delta B}{\Delta H} \right) \Big|_{H,B} = \frac{\text{d}B}{\text{d}H} \tag{3.2}$$

In some application fields, a small periodic dynamic magnetic field H_d is coupled with a bigger static magnetic field, which leads to a small hysteresis on the *B-H* curve. The working point on the *B-H* curve is determined by the big static magnetic field. At this condition, the slope of the small hysteresis at the operating point can be defined as the reversible permeability, as shown in (3.3).

$$\mu_{\text{rev}} = \lim_{\Delta H_\text{d} \to 0} \left(\frac{\Delta B_\text{d}}{\Delta H_\text{d}} \right) \Big| H = 0, B = 0 \tag{3.3}$$

The magnitude of the reversible permeability is related to the magnitude of the static magnetic field and the frequency of the dynamic magnetic field.

In EMAT, the magnetization problem of ferromagnetic material is about a small periodic dynamic magnetic field H_d coupled with a big static magnetic field. Thus, it is necessary to use the reversible permeability for the analysis and calculation. For the isotropic ferromagnetic material, under the effect of the bias magnetic field, its mechanical property and magnetic permeability will show a transverse isotropy. This situation is similar to the magnetostrictive property of the polycrystalline ferromagnetic material under the effect of the bias magnetic field and the piezoelectric property of the piezoelectric material of a 6/m series. In this way, on the interface perpendicular to the bias magnetic field, the value of the reversible permeability remains the same. The direction of the bias magnetic field is assumed to be the same as x_3. The matrix of permeability is shown in (3.4).

$$[\mu] = \begin{bmatrix} \mu_{11} & 0 & 0 \\ 0 & \mu_{11} & 0 \\ 0 & 0 & \mu_{33} \end{bmatrix} \tag{3.4}$$

3.1.2 Magnetostrictive Property of the Ferromagnetic Material

The microstructure of ferromagnetic materials is the same as crystals, the iron's anion being surrounded by a cloud of electrons in the center of the iron's positive ion. In the ferromagnetic material, a unit magnetic torque will be generated due to the electrons spinning between adjacent atoms. There is an interaction force between the unit magnetic torques, which drives the adjacent unit magnetic torques in a parallel arrangement in the same direction as a magnetic domain, the interaction between those magnetic domains being very small. If there is no external magnetic field, the magnetic domains are kept in equilibrium with each other; the total

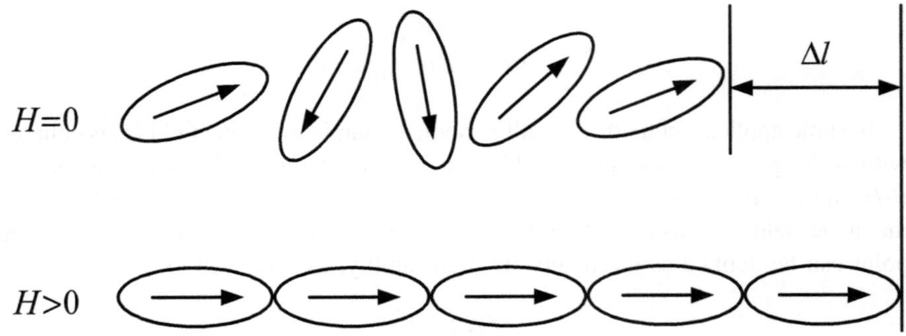

Fig. 3.2 The magnetostriction of ferromagnetic material

magnetization of the material is equal to zero. If there is an external magnetic field, the equilibrium will be disrupted. The magnetization vectors of the magnetic domain will turn toward the external magnetic field direction. When the magnetization state of ferromagnetic materials is changed, the magnetic domain will rotate, which will cause a small change in its length or volume. This phenomenon is called a "magnetostrictive effect." As shown in Fig. 3.2, the magnetostrictive effect has two kinds: line magnetostriction and volume magnetostriction.

When the ferromagnetic materials are magnetized, it causes the spontaneous deformation of the crystal lattice, including extending or shortening along the magnetization direction, known as "line magnetostriction." When the line magnetostriction occurs in the magnet body, its volume almost remains the same. Before the unsaturated magnetic state, the change mainly lies in the magnet's length. The line magnetostrictions can be divided into two kinds: longitudinal magnetostriction and transverse magnetostriction. The magnetostriction whose dimension changes along the external magnetic field direction is called "longitudinal magnetic field extension"; the magnetostriction whose dimension changes vertical to the direction of the external magnetic field is called "transverse magnetic field extension." The line magnetostriction coefficient is defined as $\gamma = \Delta l/l$. Here, Δl is the variation in material length, and l is the material length. A positive value of γ, which indicates that the extending of the material strain is enhancing the external magnetic field, is called "positive magnetostriction"; otherwise, it indicates the shortening of the material strain is enhancing the external magnetic field, which is called "negative magnetostriction." The relationship between the material magnetostriction and the magnetic field can be illustrated by the magnetostrictive curve. The magnetostrictive curve of the single crystal ferromagnetic is provided in Fig. 3.3.

Volume magnetostriction refers to the expansion and contraction of volume through changes to the magnet's magnetized states. After the saturation of magnetization, the magnetostriction mainly refers to volume magnetostriction. In general ferromagnetic materials, the volume magnetostriction is very small and so rarely considered in measurements and investigations. Therefore, the general magnetostrictions are related to line magnetostriction.

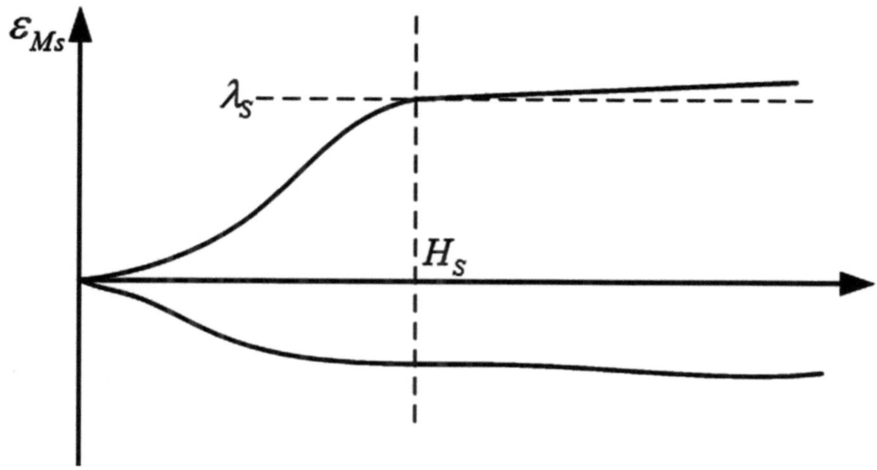

Fig. 3.3 The magnetostrictive curve of the single crystal ferromagnetic material

The magnetostrictive property of the ferromagnetic material is similar to the linear piezoelectricity of piezocrystals; thereby, it is also called the piezomagnetic property of the ferromagnetic material. In contrast to the piezomagnetic property, the strain generated in the magnetostrictive material has no relationship with the polarity of the applied magnetic field. There is a quadratic relationship between magnetostriction and magnetic field, which is similar to the quadratic effect of the electromagnetic expansion. When a small dynamic magnetic field is superimposed upon a big static bias magnetic field, the relationship between magnetostriction and magnetic field shows a linear characteristic.

In contrast to the magnetostrictive property, when the ferromagnetic material dimension changes, it will cause the movement and rotation of the magnetic domain, and thus, the magnetic effect will be generated in the material. This phenomenon is called the "inverse magnetostrictive effect." The ferromagnetic material's magnetostrictive effect and inverse magnetostrictive effect are the main principles of the excitation and reception of ultrasonic waves in the ferromagnetic materials.

The relationship between the ferromagnetic material mechanical properties and magnetic properties can be expressed using the magnetoelastic constitutive equation, as shown in (3.5), (3.6), and (3.7).

$$S_i = s_{ij}^H \sigma_j + d_{ki} H_k \tag{3.5}$$

$$\sigma_i = c_{ij}^H S_j - e_{ki} H_k \tag{3.6}$$

$$B_m = d_{mj}\sigma_j + \mu_0\mu_{mk}^{\mathrm{T}} H_k \tag{3.7}$$

In the above equation, $i, j = 1, \ldots, 6$, $m, k = 1, 2, 3$; (3.5) and (3.6) describe the ferromagnetic material magnetostrictive effect, and (3.7) describes the material inverse magnetostrictive effect. In the equation, σ_i, S_{ij}^H, and C_{ij}^H represent the stress matrix, flexible coefficient matrix, reversible matrix permeability, and stiffness matrix, respectively, under a specific magnetic field; B_m and μ_{mk}^{T} are the total magnetic induction intensity and anisotropic reversible permeability, respectively, under the constant static stress; and d_{ki} and e_{ki} are the piezomagnetic matrix and inverse piezomagnetic matrix, respectively, as defined in (3.8) and (3.9).

$$d_{ki} = \left.\frac{\partial S_i}{\partial H_k}\right|_{\sigma} \tag{3.8}$$

$$e_{ki} = \left.\frac{\partial \sigma_i}{\partial H_k}\right|_{\varepsilon} \tag{3.9}$$

In the above equations, H_k is the component of the dynamic magnetic field.

From (3.5) to (3.7), in the ferromagnetic material, the change of magnetic field will generate elastic strain in the material, while the elastic strain will also generate the magnetic field in the material. This is the key point of the energy conversion of EMAT based on the magnetostrictive mechanism. Equations (3.8) and (3.9) indicate that the piezomagnetic coefficient d_{ki} and inverse piezomagnetic matrix e_{ki} can be calculated according to the material magnetostrictive curve.

For the theoretical analysis of EMAT based on the magnetostrictive mechanism, the bias magnetic field is assumed to have unidirection and uniformity. The reason is if the distribution of the bias magnetic field is non-uniform, the magnetization direction and magnetization intensity of each point inside the ferromagnetic material will be different. It is very difficult to give a mathematical description for the magnetization and magnetostriction properties for the analytical calculation of EMAT in a microscopic manner. In addition, it is also possible to approximately obtain a uniformally distributed bias magnetic field through a reasonable design of the magnetic body.

3.2 Finite Element Method of EMAT Based on the Magnetostrictive Mechanism [1]

3.2.1 Basic Physical Equations

The calculation process of EMAT based on the magnetostrictive mechanism is similar to the analysis process of EMAT based on *Lorentz force*, while the differences are as follows: (1) since the assumption of the bias magnetic field is that it is of unidirectional uniform distribution, there is no necessity to build a magnet model to calculate the bias magnetic field; and (2) the changes of the material's properties will cause changes to the calculations of the coil's dynamic magnetic field. In the EMAT based on *Lorentz force*, the specimen's magnetic property does not change with the external magnetic field, but for the EMAT based on the magnetostrictive mechanism, the ferromagnetic material has magnetization characteristics and magnetostrictive properties. For this case, the calculations of magnetostriction force and magnetostrictive current density are critical to achieve the multifield coupling of EMAT based on the magnetostrictive mechanism.

From the magnetic property of ferromagnetic material and the working principle of EMAT, under the effect of the strong bias magnetic field, the dynamic magnetic field generated by the excitation coil is relatively small. In this situation, the fluctuation of the stacked synthetic dynamic magnetic field is around the bias magnetic field. So, corresponding to the reversible permeability, the material permeability is a constant in the bias magnetic field. The specific value can be measured through experimental tests.

In the excitation process of ultrasonic waves, the current density generated by the inverse magnetostrictive effect is much less than the density of the eddy current in the specimen. Thus, its effects are often neglected. Therefore, the dynamic magnetic field generated by the coil satisfies (3.10).

$$\frac{1}{\mu}\nabla^2 A - \sigma\frac{\partial A}{\partial t} + \frac{1}{S}\iint\limits_{S}\sigma\frac{\partial A}{\partial t}\mathrm{d}s = -\frac{i}{S} \tag{3.10}$$

The ferromagnetic material shows a transverse magnetic isotropic characteristic under the magnetization state. The magnetic conductivity matrix should be determined by the direction of the bias magnetic field. In addition, it can be expressed by a general matrix of permeability, as shown in (3.11).

$$\mu = \begin{bmatrix} \mu_{11} & 0 & 0 \\ 0 & \mu_{22} & 0 \\ 0 & 0 & \mu_{33} \end{bmatrix} \tag{3.11}$$

The forces that cause the vibrations in the material include the *Lorentz force*, magnetic force, and magnetostrictive force. Compared with the other two, the magnetic force is relatively small and so can be neglected. The magnitude of magnetostrictive force has a relationship with the strength of dynamic magnetic fields H and the inverse piezomagnetic matrix e. In the case of free stress, according to (3.7), the magnetostriction force can be expressed by (3.12), in the above equation, $e = [e_{kj}]$.

$$f_{\mathrm{Ms}} = -\nabla_t(e^{\mathrm{T}}H) \tag{3.12}$$

The specimen's *Lorentz force* is expressed in (2.17).

Therefore, the total stresses of the ferromagnetic specimen are given by the following equation.

$$f = f_{\mathrm{L}} + f_{\mathrm{M}}$$

Under the combined action of *Lorentz force* and magnetostrictive force, the equilibrium equation of the particle displacement in the specimen can be calculated using (3.13).

$$(\nabla \cdot \mathrm{c}\nabla \mathrm{u}) + f = \rho \frac{\partial^2 u}{\partial t^2} \tag{3.13}$$

In the above equation, we have the following relationship.

$$c = \begin{bmatrix} c_{11} & c_{12} & c_{13} & c_{14} & c_{15} \\ c_{21} & c_{22} & c_{23} & c_{24} & c_{25} \\ c_{31} & c_{32} & c_{33} & c_{34} & c_{35} \\ c_{41} & c_{42} & c_{43} & c_{44} & c_{45} \\ c_{51} & c_{52} & c_{53} & c_{54} & c_{55} \end{bmatrix} u = \begin{bmatrix} u & v & w \end{bmatrix}^{\mathrm{T}}$$

When receiving the ultrasonic signals, the dynamic magnetic field will be generated by the particle motion in the specimen under the effect of the static bias magnetic field. The voltage will be induced by the dynamic magnetic field within the receiving coil; thereby, the receiving of the ultrasonic wave signal can be realized.

From the material inverse magnetostrictive effect, we have (3.14).

$$B_{\mathrm{Ms}} = eS \tag{3.14}$$

In the above equation, B_{Ms} is the magnetic flux density generated by the material particle displacement. Thus, the electric current density in the specimen is expressed in (3.15).

$$J_{\mathrm{Ms}} = \nabla \times B_{\mathrm{M}} \tag{3.15}$$

In the above equation, J_{Ms} is known as the magnetostrictive current density.

The equation of *Lorentz* current density caused by particle motion is expressed in (2.22).

Thereby, the total current densities in the specimen are expressed in (3.16).

$$J = J_{\mathrm{L}} + J_{\mathrm{Ms}} \tag{3.16}$$

Thus, the governing equation satisfied by the receiving coil and the specimen is expressed in (3.17).

$$-\frac{1}{\mu}\nabla^2 A+\sigma\frac{\partial A}{\partial t}-\frac{\sigma}{S}\frac{\partial}{\partial t}\iint\limits_{\Omega_\tau} A\,\mathrm{d}s=J \tag{3.17}$$

After the calculations of magnetic vector potential in various regions, the induced electromotive force of the receiving coil can be calculated. The induced electric field within the coil conductor can be expressed by (3.18).

$$E=-\frac{\partial A}{\partial t} \tag{3.18}$$

The electromotive force of the conductor at a certain point in the coil can be obtained by performing a line integral of the electric field intensity, so that we have (3.19).

$$V_{\text{pout}}=\int\limits_{l}-\frac{\partial A}{\partial t}\cdot \mathrm{d}l \tag{3.19}$$

Then, the coil's output voltage can be obtained by averaging the electromotive forces of the point conductor included in the coil, as shown in (3.20).

$$V_{\text{out}}=\frac{\int_\Omega V_{\text{pout}}\mathrm{d}\Omega}{\int_\Omega \mathrm{d}\Omega} \tag{3.20}$$

3.2.2 *Calculations of Magnetostrictive Force and Magnetostrictive Current Density in the Two-Dimensional Cartesian Coordinate System*

A two-dimensional finite element method can be used to do the calculation for EMAT. The rectangular coordinate system is used to establish the mathematical model for the EMAT that is used in the flat specimen. In the two-dimensional *Cartesian coordinate system*, only the x and y components of the magnetic field and mechanical field are considered, the magnetic vector potential and eddy current only having the z component. In order to achieve the coupling between different fields, the calculated equations of magnetostriction force and magnetostrictive current density should be derived.

To solve the electromagnetic field by the finite element method, the results are magnetic vector potential A_z. The magnetostriction force should be expressed by A_z.

According to (3.13), the inverse piezomagnetic matrix is substituted into it and the magnetostrictive force in the two-dimensional *Cartesian coordinate system* can be expressed by (3.21).

$$f_{\mathrm{Ms}} = \begin{bmatrix} f_{\mathrm{Ms}x} \\ f_{\mathrm{Ms}y} \\ f_{\mathrm{Ms}z} \end{bmatrix} = -\begin{bmatrix} \frac{\partial}{\partial x} & 0 & 0 & 0 & 0 & \frac{\partial}{\partial y} \\ 0 & \frac{\partial}{\partial y} & 0 & 0 & 0 & \frac{\partial}{\partial x} \\ 0 & 0 & 0 & \frac{\partial}{\partial y} & \frac{\partial}{\partial x} & 0 \end{bmatrix} \begin{bmatrix} e_{11} & e_{21} & e_{31} \\ e_{12} & e_{22} & e_{32} \\ e_{13} & e_{23} & e_{33} \\ e_{14} & e_{24} & e_{34} \\ e_{15} & e_{25} & e_{35} \\ e_{16} & e_{26} & e_{36} \end{bmatrix} \begin{bmatrix} H_x \\ H_y \\ 0 \end{bmatrix} \quad (3.21)$$

In the above equation, H_x and H_y are the x and y components of the dynamic magnetic field intensity, respectively.

The matrix can be expressed by (3.22).

$$f_{\mathrm{Ms}} = \begin{bmatrix} f_{\mathrm{Ms}x} \\ f_{\mathrm{Ms}y} \\ f_{\mathrm{Ms}z} \end{bmatrix} = -\begin{bmatrix} e_{11} & e_{21} & e_{16} & e_{26} \\ e_{16} & e_{26} & e_{12} & e_{22} \\ e_{15} & e_{25} & e_{14} & e_{24} \end{bmatrix} \begin{bmatrix} \frac{\partial H_x}{\partial x} \\ \frac{\partial H_y}{\partial x} \\ \frac{\partial H_x}{\partial y} \\ \frac{\partial H_y}{\partial y} \end{bmatrix} \quad (3.22)$$

The dynamic magnetic field intensity H, magnetic induction intensity B, and A_z satisfy the relationship, as shown in (3.23).

$$\begin{bmatrix} H_x \\ H_y \end{bmatrix} = \begin{bmatrix} \frac{1}{\mu_{xx}} & 0 \\ 0 & \frac{1}{\mu_{yy}} \end{bmatrix} \begin{bmatrix} B_x \\ B_y \end{bmatrix} = \begin{bmatrix} \frac{1}{\mu_{xx}} & 0 \\ 0 & \frac{1}{\mu_{yy}} \end{bmatrix} \begin{bmatrix} \frac{\partial A_z}{\partial y} \\ -\frac{\partial A}{\partial x} \end{bmatrix} \quad (3.23)$$

Taking the derivative of the magnetic field intensity H_x and $H_{y,}$ with respect to x and y, respectively, we obtain (3.24).

$$\begin{bmatrix} \frac{\partial H_x}{\partial x} \\ \frac{\partial H_y}{\partial x} \\ \frac{\partial H_x}{\partial y} \\ \frac{\partial H_y}{\partial y} \end{bmatrix} = \begin{bmatrix} \frac{1}{\mu_{xx}} & 0 & 0 & 0 \\ 0 & -\frac{1}{\mu_{yy}} & 0 & 0 \\ 0 & 0 & \frac{1}{\mu_{xx}} & 0 \\ 0 & 0 & 0 & -\frac{1}{\mu_{yy}} \end{bmatrix} \begin{bmatrix} \frac{\partial^2 A_z}{\partial x \partial y} \\ \frac{\partial^2 A_z}{\partial x^2} \\ \frac{\partial^2 A_z}{\partial y^2} \\ \frac{\partial^2 A_z}{\partial y \partial x} \end{bmatrix} \quad (3.24)$$

And then, we have (3.25).

$$f_{\mathrm{Ms}} = \begin{bmatrix} f_{\mathrm{Ms}x} \\ f_{\mathrm{Ms}y} \\ f_{\mathrm{Ms}z} \end{bmatrix} = -\begin{bmatrix} e_{11} & e_{21} & e_{16} & e_{26} \\ e_{16} & e_{26} & e_{12} & e_{22} \\ e_{15} & e_{25} & e_{14} & e_{24} \end{bmatrix} \begin{bmatrix} \frac{1}{\mu_{xx}} & 0 & 0 & 0 \\ 0 & -\frac{1}{\mu_{yy}} & 0 & 0 \\ 0 & 0 & \frac{1}{\mu_{xx}} & 0 \\ 0 & 0 & 0 & -\frac{1}{\mu_{yy}} \end{bmatrix} \begin{bmatrix} \frac{\partial^2 A_z}{\partial x \partial y} \\ \frac{\partial^2 A_z}{\partial x^2} \\ \frac{\partial^2 A_z}{\partial y^2} \\ \frac{\partial^2 A_z}{\partial y \partial x} \end{bmatrix} \quad (3.25)$$

Under the two-dimensional *Cartesian coordinate system*, (3.15) is expanded. Considering the relationship between dynamic magnetic field intensity H and magnetic induction intensity B, we can obtain (3.26).

$$\begin{bmatrix} H_{\mathrm{Ms}x} \\ H_{\mathrm{Ms}y} \end{bmatrix} = \begin{bmatrix} \frac{1}{\mu_{xx}} & 0 \\ 0 & \frac{1}{\mu_{yy}} \end{bmatrix} \begin{bmatrix} e_{11} & e_{12} & e_{13} & e_{14} & e_{15} & e_{16} \\ e_{21} & e_{22} & e_{23} & e_{24} & e_{25} & e_{26} \end{bmatrix} \begin{bmatrix} S_1 \\ S_2 \\ S_3 \\ S_4 \\ S_5 \\ S_6 \end{bmatrix} \tag{3.26}$$

In the above equation, $S = \begin{bmatrix} S_1 \\ S_2 \\ S_3 \\ S_4 \\ S_5 \\ S_6 \end{bmatrix}$ is the strain tensor matrix and $H_{\mathrm{M}x}$ and $H_{\mathrm{M}y}$ are the magnetostrictive magnetic field intensity.

The relationship between strain and displacement satisfies the flowing geometric equation, as shown in (3.27).

$$\begin{bmatrix} S_1 \\ S_2 \\ S_3 \\ S_4 \\ S_5 \\ S_6 \end{bmatrix} = \begin{bmatrix} \frac{\partial}{\partial x} & 0 & 0 \\ 0 & \frac{\partial}{\partial y} & 0 \\ 0 & 0 & 0 \\ 0 & 0 & \frac{\partial}{\partial y} \\ 0 & 0 & \frac{\partial}{\partial x} \\ \frac{\partial}{\partial y} & \frac{\partial}{\partial x} & 0 \end{bmatrix} \begin{bmatrix} \mu \\ v \\ w \end{bmatrix} \tag{3.27}$$

Substituting S into (3.26), and taking derivatives of it with respect to $H_{\mathrm{M}x}$ and $H_{\mathrm{M}y}$, we can get (3.28).

$$\begin{bmatrix} \frac{\partial H_{\mathrm{Ms}y}}{\partial x} \\ \frac{\partial H_{\mathrm{Ms}x}}{\partial y} \end{bmatrix} = \begin{bmatrix} \frac{1}{\mu_{xx}} \left[e_{11} \frac{\partial^2 \mu}{\partial y \partial x} + e_{12} \frac{\partial^2 v}{\partial y^2} + e_{14} \frac{\partial^2 w}{\partial y^2} + e_{15} \frac{\partial^2 w}{\partial y \partial x} + e_{16} \left(\frac{\partial^2 u}{\partial y^2} + \frac{\partial^2 v}{\partial y \partial x} \right) \right] \\ \frac{1}{\mu_{yy}} \left[e_{21} \frac{\partial^2 u}{\partial x^2} + e_{22} \frac{\partial^2 v}{\partial x \partial y} + e_{24} \frac{\partial^2 w}{\partial x \partial y} + e_{25} \frac{\partial^2 w}{\partial x^2} + e_{26} \left(\frac{\partial^2 v}{\partial x^2} + \frac{\partial^2 u}{\partial x \partial y} \right) \right] \end{bmatrix} \tag{3.28}$$

According to (3.16), the magnetostrictive current density is expressed in (3.29).

$$J_{\mathrm{Ms}} = \nabla \times H_{\mathrm{Ms}} = \begin{vmatrix} \hat{x} & \hat{y} & \hat{z} \\ \frac{\partial}{\partial x} & \frac{\partial}{\partial y} & 0 \\ H_{\mathrm{Ms}x} & H_{\mathrm{Ms}y} & 0 \end{vmatrix} = \left(\frac{\partial H_{\mathrm{Ms}y}}{\partial x} - \frac{\partial H_{\mathrm{Ms}x}}{\partial y} \right) \hat{z} \tag{3.29}$$

In the above equation, $\hat{x}$, $\hat{y}$, and $\hat{z}$ make up the direction vector.

3.2.3 Calculation of the Magnetostrictive Force and Magnetostrictive Current Density in the Axisymmetric Coordinates

It is more appropriate to do the analysis in the cylindrical coordinate system for the thickness measurement of ferromagnetic material and, for the excitation, propagation and receiving of the guided wave propagating along the axial direction of the pipeline. Considering both the spiral coil and the pipeline with a circular cross section have axisymmetric characteristics, the problem can be simplified as a two-dimensional axisymmetric problem. Under the two-dimensional axisymmetric coordinates, the azimuth of the vector magnetic potential A_θ is the only variable to be considered.

According to the magnetostrictive property of the ferromagnetic material, the stress caused by the magnetic field in the specimen can be expressed by (3.30).

$$\begin{bmatrix} \sigma_1 \\ \sigma_2 \\ \sigma_3 \\ \sigma_4 \\ \sigma_5 \\ \sigma_6 \end{bmatrix} = -\begin{bmatrix} e_{11} & e_{21} & e_{31} \\ e_{12} & e_{22} & e_{32} \\ e_{13} & e_{23} & e_{33} \\ e_{14} & e_{24} & e_{34} \\ e_{15} & e_{25} & e_{35} \\ e_{16} & e_{26} & e_{36} \end{bmatrix} \begin{bmatrix} H_r \\ 0 \\ H_z \end{bmatrix} \quad (3.30)$$

In the above equation, $\sigma = \begin{bmatrix} \sigma_1 \\ \sigma_2 \\ \sigma_3 \\ \sigma_4 \\ \sigma_5 \\ \sigma_6 \end{bmatrix}$ is the stress tensor matrix and H_r and H_z are the magnetic field intensity in the r and z direction, respectively.

According to (3.13), the magnetostrictive force under the two-dimensional axisymmetric coordinates can be expressed using (3.31).

$$f_{\mathrm{Ms}} = \begin{bmatrix} f_{\mathrm{Ms}r} \\ f_{\mathrm{Ms}\theta} \\ f_{\mathrm{Ms}z} \end{bmatrix}$$
$$= -\begin{bmatrix} e_{11} & e_{31} & e_{16} & e_{36} \\ e_{14} & e_{34} & e_{15} & e_{35} \\ e_{16} & e_{36} & e_{13} & e_{33} \end{bmatrix} \begin{bmatrix} \frac{\partial H_r}{\partial r} \\ \frac{\partial H_z}{\partial r} \\ \frac{\partial H_r}{\partial z} \\ \frac{\partial H_z}{\partial z} \end{bmatrix} - \frac{1}{r}\begin{bmatrix} e_{11} - e_{12} & e_{31} - e_{32} \\ 2e_{14} & 2e_{34} \\ e_{16} & e_{36} \end{bmatrix} \begin{bmatrix} H_r \\ H_z \end{bmatrix} \quad (3.31)$$

According to the relationship between the magnetic field intensity H and magnetic induction intensity B, we can obtain (3.32).

$$\begin{bmatrix} H_r \\ H_z \end{bmatrix} = \begin{bmatrix} \frac{1}{\mu_{rr}} & 0 \\ 0 & \frac{1}{\mu_{zz}} \end{bmatrix} \begin{bmatrix} B_r \\ B_z \end{bmatrix} = \begin{bmatrix} \frac{1}{\mu_{rr}} & 0 \\ 0 & \frac{1}{\mu_{zz}} \end{bmatrix} \begin{bmatrix} -\frac{\partial A_\theta}{\partial z} \\ \frac{\partial A_\theta}{\partial r} \end{bmatrix} \tag{3.32}$$

In the above equation, μ_{rr} and μ_{zz} are the permeability of the r and z direction, respectively.

The partial differential is calculated for the above equation with respect to r and z.

$$\begin{bmatrix} \frac{\partial H_r}{\partial r} \\ \frac{\partial H_z}{\partial r} \\ \frac{\partial H_r}{\partial z} \\ \frac{\partial H_z}{\partial z} \end{bmatrix} = \begin{bmatrix} 0 & 0 & -\frac{1}{\mu_{rr}} \\ \frac{1}{\mu_{zz}} & 0 & 0 \\ 0 & -\frac{1}{\mu_{xx}} & 0 \\ 0 & 0 & \frac{1}{\mu_{zz}} \end{bmatrix} \begin{bmatrix} \frac{\partial^2 A_\theta}{\partial r^2} \\ \frac{\partial^2 A_\theta}{\partial z^2} \\ \frac{\partial^2 A_\theta}{\partial r \partial z} \end{bmatrix} \tag{3.33}$$

Substituting (3.33) into (3.31), we obtain the magnetostrictive force represented by A_θ, as shown in (3.34).

$$f_{\mathrm{Ms}} = \begin{bmatrix} f_{\mathrm{Ms}r} \\ f_{\mathrm{Ms}\theta} \\ f_{\mathrm{Ms}z} \end{bmatrix} = -\begin{bmatrix} e_{11} & e_{31} & e_{16} & e_{36} \\ e_{14} & e_{34} & e_{15} & e_{35} \\ e_{16} & e_{36} & e_{13} & e_{33} \end{bmatrix} \begin{bmatrix} 0 & 0 & -\frac{1}{\mu_{rr}} \\ \frac{1}{\mu_{zz}} & 0 & 0 \\ 0 & -\frac{1}{\mu_{xx}} & 0 \\ 0 & 0 & \frac{1}{\mu_{zz}} \end{bmatrix} \begin{bmatrix} \frac{\partial^2 A_\theta}{\partial r^2} \\ \frac{\partial^2 A_\theta}{\partial z^2} \\ \frac{\partial^2 A_\theta}{\partial r \partial z} \end{bmatrix} - \frac{1}{r} \begin{bmatrix} e_{11} - e_{12} & e_{31} - e_{32} \\ 2e_{14} & 2e_{34} \\ e_{16} & e_{36} \end{bmatrix} \begin{bmatrix} -\frac{1}{\mu_{rr}} & 0 \\ 0 & \frac{1}{\mu_{zz}} \end{bmatrix} \begin{bmatrix} \frac{\partial A_\theta}{\partial z} \\ \frac{\partial A_\theta}{\partial r} \end{bmatrix} \tag{3.34}$$

In the two-dimensional axisymmetric coordinates, (3.14) is expanded and the relationship between the dynamic magnetic field intensity H and the magnetic induction intensity B is introduced to get (3.25).

$$\begin{bmatrix} H_{\mathrm{Ms}r} \\ H_{\mathrm{Ms}z} \end{bmatrix} = \begin{bmatrix} \frac{1}{\mu_{rr}} & 0 \\ 0 & \frac{1}{\mu_{zz}} \end{bmatrix} \begin{bmatrix} e_{11} & e_{12} & e_{13} & e_{14} & e_{15} & e_{16} \\ e_{31} & e_{32} & e_{33} & e_{34} & e_{35} & e_{36} \end{bmatrix} \begin{bmatrix} S_1 \\ S_2 \\ S_3 \\ S_4 \\ S_5 \\ S_6 \end{bmatrix} \tag{3.35}$$

In the above equation, $H_{\mathrm{Ms}r}$ and $H_{\mathrm{Ms}z}$ are the magnetostrictive magnetic field intensity in the r and z direction, respectively.

The relationship between strain and displacement satisfies (3.36).

$$\begin{bmatrix} S_1 \\ S_2 \\ S_3 \\ S_4 \\ S_5 \\ S_6 \end{bmatrix} = \begin{bmatrix} \frac{\partial}{\partial r} & 0 & 0 \\ 0 & 0 & 0 \\ 0 & 0 & \frac{\partial}{\partial z} \\ 0 & \frac{\partial}{\partial z} & 0 \\ \frac{\partial}{\partial z} & 0 & \frac{\partial}{\partial r} \\ 0 & \frac{\partial}{\partial r} - \frac{1}{r} & 0 \end{bmatrix} \begin{bmatrix} u \\ v \\ w \end{bmatrix} \tag{3.36}$$

Substituting (3.36) into (3.35), and calculating the differential for the magnetic field intensity, we can obtain (3.37) and (3.38).

$$\frac{\partial H_{\mathrm{Msr}}}{\partial z} = \frac{1}{\mu_{rr}} \left[e_{11} \frac{\partial^2 u}{\partial r \partial z} + e_{13} \frac{\partial^2 w}{\partial z^2} + e_{14} \frac{\partial^2 v}{\partial z^2} + e_{15} \left(\frac{\partial^2 u}{\partial z^2} + \frac{\partial^2 w}{\partial r \partial z} \right) + e_{16} \left(\frac{\partial^2 v}{\partial r \partial z} - \frac{1}{r} \frac{\partial w}{\partial z} \right) \right] \tag{3.37}$$

$$\frac{\partial H_{\mathrm{Msz}}}{\partial r} = \frac{1}{\mu_{ZZ}} \left[e_{31} \frac{\partial^2 u}{\partial r^2} + e_{33} \frac{\partial^2 w}{\partial z \partial r} + e_{34} \frac{\partial^2 v}{\partial z \partial r} + e_{35} \left(\frac{\partial^2 u}{\partial z \partial r} + \frac{\partial^2 w}{\partial r^2} \right) + e_{36} \left(\frac{\partial^2 v}{\partial r^2} - \frac{1}{r} \frac{\partial w}{\partial r} \right) \right] \tag{3.38}$$

According to (3.16), the magnetostrictive current density in the two-dimensional axisymmetric coordinates can be expressed by (3.39).

$$J_{\mathrm{Ms}} = \nabla \times H_{\mathrm{Ms}} = \begin{vmatrix} \frac{1}{r}\hat{r} & \hat{\theta} & \frac{1}{r}\hat{z} \\ \frac{\partial}{\partial r} & 0 & \frac{\partial}{\partial z} \\ H_{\mathrm{Msr}} & 0 & H_{\mathrm{Mz}} \end{vmatrix} = \frac{1}{r} \left(\frac{\partial H_{\mathrm{Msz}}}{\partial r} - \frac{\partial H_{\mathrm{Msr}}}{\partial z} \right) \hat{\theta} \tag{3.39}$$

In the above equation, $\hat{r}$, $\hat{\theta}$, and $\hat{z}$ make up the direction vector.

3.2.4 Determination of the Piezomagnetic Coefficient

It can be seen from the expressions of magnetostrictive force and magnetostrictive current that if we want to obtain the magnetostrictive force and magnetostrictive current density by calculations, it is necessary to know the inverse piezomagnetic matrix e. According to (3.9), if each strain caused by the external magnetic field is already given in the specimen, each element in the inverse magnetostrictive matrix can be obtained by taking derivatives with respect to the magnetic field intensity components. But, in a real situation, it is very difficult to measure each strain component caused by the external magnetic field in the specimen. Currently, it is not necessary to know exactly each of the components of the inverse piezomagnetic matrix for the simulation of EMAT based on the magnetostrictive mechanism. According to the relationship of the numerical value and direction between the static bias magnetic field and dynamic magnetic field, the component of the inverse piezomagnetic matrix in the calculation of EMAT can be obtained approximately. This makes it possible to realize the simulated analysis of EMAT based on the magnetostrictive mechanism.

$$S_i^{(\mathrm{MS})} = d_{ki}^{(\mathrm{MS})} H_k \tag{3.40}$$

$$\sigma_i^{(\mathrm{MS})} = -e_{ki}^{(\mathrm{MS})} H_k \tag{3.41}$$

From the stress–strain relationship $\sigma_i^{(\mathrm{MS})} = -c_{ij}^H S_j^{(\mathrm{MS})}$, we can obtain (3.42).

$$\sigma_i^{(\mathrm{MS})} = -c_{ij}^H d_{ki}^{(\mathrm{MS})} H_k = -e_{ik}^{(\mathrm{MS})} H_k \tag{3.42}$$

Thus, the inverse piezomagnetic matrix can be calculated by the piezomagnetic matrix, as shown in (3.43).

$$e_{ik}^{(\mathrm{MS})} = d_{ki}^{(MS)} c_{ij}^H \tag{3.43}$$

The calculation of the inverse magnetostrictive coefficient should be processed according to the relationship between the direction of the bias magnetic field and the dynamic magnetic field. Following analysis is conducted on the platelike specimen in the rectangular coordinate system, as shown in Fig. 3.4. The inverse magnetostrictive coefficient in the cylindrical coordinate system can be obtained by the comparisons between r, θ, z and x, y, z.

If the static bias magnetic field is along the y direction, meaning the bias magnetic field is perpendicular to the specimen's surface, then there are two kinds of EMAT: body wave EMAT and *Lamb wave* EMAT. The magnetostriction of the ferromagnetic specimen is assumed to be in the state of free stress, the direction of which is the same as that of the static bias magnetic field $\varepsilon(H_{0y})$. Since the change of magnetostriction is an isovolumetric deformation, the quantity of deformation that is perpendicular to the direction of the static magnetic field is $-\varepsilon(H_{0y})/2$ (Fig. 3.5).

Strain generated by the static magnetic field in the specimen can be expressed by (3.44), (3.45), and (3.46).

$$S_2^0 = \varepsilon(H_{0y}) \tag{3.44}$$

$$S_1^0 = S_3^0 = -\varepsilon(H_{0y})/2 \tag{3.45}$$

$$S_4^0 = S_5^0 = S_6^0 = 0 \tag{3.46}$$

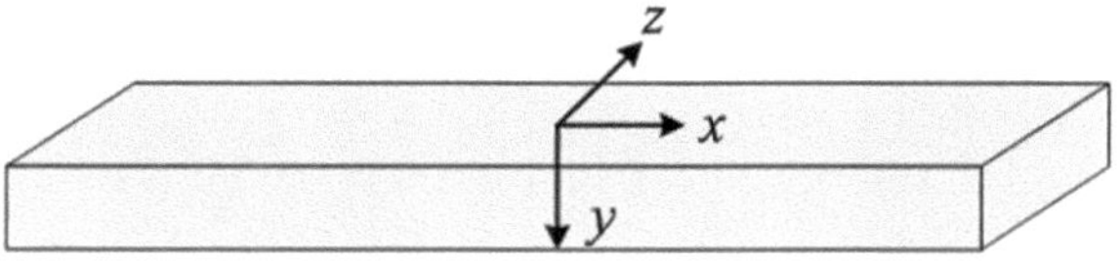

Fig. 3.4 The rectangular coordinate system of the platelike specimen

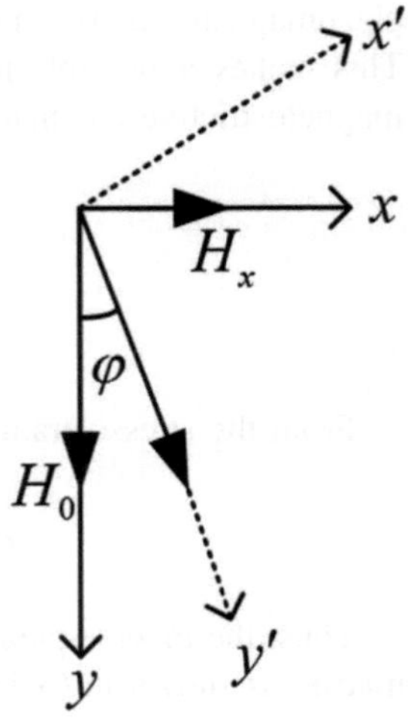

Fig. 3.5 The field direction

In the two-dimensional coordinate system, the dynamic magnetic field only has H_x and H_y components. Firstly, let us consider the effect of H_x; in this case, the ferromagnetic material's magnetostriction will be along the direction of the resultant total magnetic field H_t of the static magnetic field and dynamic magnetic field. The included angle of H_t and H_0 is referred to by φ. The value of φ will change with the dynamic magnetic field. The resultant magnetic field will cause the expansion of the ferromagnetic material. S'_1, S'_2 and S'_3 are used to express the principle strain corresponding to x', y', and z' in the rectangular coordinate system. The new coordinate system can be obtained by rotating the original coordinate system anticlockwise θ degrees. The deformation caused by magnetostriction is assumed to be isometric. Based on the knowledge of elastic mechanics, there is no shear strain component, and the principle strain of S'_1, S'_2 and S'_3 is along the principle direction x', y', and z' as expressed by (3.47), (3.48), and (3.49).

$$S'_2 = \varepsilon_{\mathrm{M}}(H_t) \tag{3.47}$$

$$S'_1 = S'_2 = -\frac{1}{2}\varepsilon_{\mathrm{M}}(H_t) \tag{3.48}$$

$$S'_4 = S'_5 = S'_6 = 0 \tag{3.49}$$

According to the coordinate transformation equation of strain, we have (3.50) and (3.51).

$$\varepsilon'_{ij} = Q_{i'k}Q_{j'l}\varepsilon_{kl} \tag{3.50}$$

$$Q_{i'k} = \vec{e_i}' \cdot \vec{e_k}' = Q_{ki'} \tag{3.51}$$

The expressions for each strain under the original coordinal system are listed as follows:

$$S_1 = S_1' \cos^2 \varphi + S_2' \sin^2 \varphi \tag{3.52}$$

$$S_2 = S_1' \sin^2 \varphi + S_2' \cos^2 \varphi \tag{3.53}$$

$$S_3 = S_3' \tag{3.54}$$

$$S_4 = S_5 = 0 \tag{3.55}$$

$$S_6 = (S_2' - S_1')\sin\varphi \tag{3.56}$$

Thus, from (3.8), we can obtain each element of the piezomagnetic matrix, as shown in the following equations, in which λ represents the slope of the magnetostrictive curve of the specimen.

$$d_{3z}^{(MS)} = d_{1x}^{(MS)} = \left.\frac{\partial S_1}{\partial H_x}\right|_\sigma = \frac{3\varepsilon_M}{H_0 \cos\varphi} \cos^3 \varphi \sin\varphi + \lambda \sin\varphi\left(-\frac{1}{2}\cos^2\varphi + sin^2\varphi\right) \tag{3.57}$$

$$d_{2z}^{(MS)} = d_{2x}^{(MS)} = \left.\frac{\partial S_2}{\partial H_x}\right|_\sigma = -\frac{3\varepsilon_M}{H_0 \cos\varphi} \cos^3 \varphi \sin\varphi + \lambda \sin\theta\left(-\frac{1}{2}\sin^2\varphi + \cos^2\varphi\right) \tag{3.58}$$

$$d_{1z}^{(MS)} = d_{3x}^{(MS)} = \left.\frac{\partial S_3}{\partial H_x}\right|_\sigma = -\frac{1}{2}\lambda \sin\varphi \tag{3.59}$$

$$d_{4z}^{(MS)} = d_{5x}^{(MS)} = 0 \tag{3.60}$$

$$d_{4z}^{(MS)} = d_{6x}^{(MS)} = \left.\frac{\partial S_6}{\partial H_x}\right|_\sigma = \frac{3\varepsilon_M}{H_0 \cos\varphi} \cos^2 \varphi \cos 2\varphi + \frac{3}{2}\lambda \sin 2\varphi \sin\varphi \tag{3.61}$$

$$d_{5z}^{(MS)} = d_{6x}^{(MS)} = 0 \tag{3.62}$$

In addition, for the effect of H_y, if the value of the dynamic magnetic field is much less than that of the static magnetic field, the following equations can be obtained approximately:

$$S_2 = \left(\frac{\partial S_2}{\partial H_y}\right) H_y \quad (3.63)$$

$$S_1 = S_3 = -\frac{1}{2}\left(\frac{\partial S_2}{\partial H_y}\right) H_y \quad (3.64)$$

$$S_4 = S_5 = S_6 = 0 \quad (3.65)$$

Thus, the piezomagnetic coefficients in the y direction are provided in (3.66), (3.67), and (3.68).

$$d_{1y}^{(\mathrm{MS})} = d_{3Y}^{(\mathrm{MS})} = -\frac{1}{2}\lambda \quad (3.66)$$

$$d_{2Y}^{(\mathrm{MS})} = \lambda \quad (3.67)$$

$$d_{4y}^{(\mathrm{MS})} = d_{5y}^{(\mathrm{MS})} = d_{6y}^{(\mathrm{MS})} = 0 \quad (3.68)$$

In this way, the piezomagnetic matrix of this magnetic material can be expressed by (3.69).

$$\left[d_{ik}^{(\mathrm{MS})}\right] = \begin{bmatrix} d_{1x}^{(\mathrm{MS})} & d_{2x}^{(\mathrm{MS})} & d_{3x}^{(\mathrm{MS})} & 0 & 0 & d_{6x}^{(\mathrm{MS})} \\ d_{1y}^{(\mathrm{MS})} & d_{2y}^{(\mathrm{MS})} & d_{1y}^{(\mathrm{MS})} & 0 & 0 & 0 \\ d_{3x}^{(\mathrm{MS})} & d_{2x}^{(\mathrm{MS})} & d_{1x}^{(MS)} & d_{6x}^{(\mathrm{MS})} & 0 & 0 \end{bmatrix}^{\mathrm{T}} \quad (3.69)$$

When the static bias magnetic field is considerably larger than the dynamic magnetic field $\varphi \approx 0$, the piezomagnetic matrix can be simplified, as shown in (3.70).

$$\left[d_{ik}^{(\mathrm{MS})}\right] = \begin{bmatrix} 0 & 0 & 0 & 0 & 0 & \frac{3\varepsilon_{\mathrm{M}}}{H_0} \\ -\frac{\lambda}{2} & \lambda & -\frac{\lambda}{2} & 0 & 0 & 0 \\ 0 & 0 & 0 & \frac{3\varepsilon_{\mathrm{M}}}{H_0} & 0 & 0 \end{bmatrix}^{\mathrm{T}} \quad (3.70)$$

If the strong bias magnetic field is applied on the isotropic ferromagnetic material, the mechanical property and magnetic conductivity will have a transverse isotropic property. For this reason, the magnetostriction of polycrystalline ferromagnetic materials is similar to the piezoelectric property of the 6/m hexagonal piezoelectric material where the bias magnetic field is applied. If the direction of the bias magnetic field is along x_2, based on the symmetry of the crystal structure, the stiffness matrix of the ferromagnetic material can be expressed by (3.71).

$$[c_{ij}] = \begin{bmatrix} c_{11} & c_{13} & c_{12} & 0 & 0 & 0 \\ c_{13} & c_{11} & c_{13} & 0 & 0 & 0 \\ c_{12} & c_{13} & c_{11} & 0 & 0 & 0 \\ 0 & 0 & 0 & c_{44} & 0 & 0 \\ 0 & 0 & 0 & 0 & c_{66} & 0 \\ 0 & 0 & 0 & 0 & 0 & c_{44} \end{bmatrix} \tag{3.71}$$

Then, the inverse piezomagnetic matrix is shown in (3.72).

$$e_{ki}^{(\mathrm{MS})} = \begin{bmatrix} 0 & 0 & 0 & 0 & 0 & c_{44}\frac{3\varepsilon_M}{H_0} \\ [c_{13} - \frac{1}{2}(c_{11} + c_{22})]\lambda & (c_{11} - c_{13})\lambda & [c_{13} - \frac{1}{2}(c_{11} + c_{22})]\lambda & 0 & 0 & 0 \\ 0 & 0 & 0 & c_{44}\frac{3\varepsilon_M}{H_0} & 0 & 0 \end{bmatrix} \tag{3.72}$$

If the static bias magnetic field is along the z direction, meaning the bias magnetic field is parallel to the surface of the specimen, the corresponding EMAT is the SH guided wave EMAT.

Considering the symmetric relationship of the coordinate system, if we exchange the components of the z-axis and y-axis in the piezomagnetic matrix, the piezomagnetic matrix in this situation can be expressed by (3.73).

$$\left[d_{ik}^{(\mathrm{MS})}\right] = \begin{bmatrix} d_{1x}^{(\mathrm{MS})} & d_{2x}^{(\mathrm{MS})} & d_{3x}^{(\mathrm{MS})} & 0 & d_{6x}^{(\mathrm{MS})} & 0 \\ d_{3x}^{(\mathrm{MS})} & d_{2x}^{(\mathrm{MS})} & d_{1x}^{(\mathrm{MS})} & d_{6x}^{(\mathrm{MS})} & 0 & 0 \\ d_{1z}^{(\mathrm{MS})} & d_{2z}^{(\mathrm{MS})} & d_{1z}^{(\mathrm{MS})} & 0 & 0 & 0 \end{bmatrix}^{\mathrm{T}} \tag{3.73}$$

If the direction of the bias magnetic field is along x_3, the ferromagnetic material shows an isotropic characteristic along x_3. The stiffness matrix in this case is expressed in (3.74).

$$[c_{ij}] = \begin{bmatrix} c_{11} & c_{13} & c_{12} & 0 & 0 & 0 \\ c_{13} & c_{11} & c_{13} & 0 & 0 & 0 \\ c_{12} & c_{13} & c_{11} & 0 & 0 & 0 \\ 0 & 0 & 0 & c_{44} & 0 & 0 \\ 0 & 0 & 0 & 0 & c_{44} & 0 \\ 0 & 0 & 0 & 0 & 0 & c_{66} \end{bmatrix} \tag{3.74}$$

Then, the corresponding inverse piezomagnetic matrix can be expressed by (3.75).

$$e_{ki}^{(\mathrm{MS})} = \begin{bmatrix} 0 & 0 & 0 & 0 & c_{44}\frac{3\varepsilon_M}{H_0} & 0 \\ 0 & 0 & 0 & c_{44}\frac{3\varepsilon_M}{H_0} & 0 & 0 \\ [c_{13} - \frac{1}{2}(c_{11} + c_{22})]\lambda & [c_{13} - \frac{1}{2}(c_{11} + c_{22})]\lambda & (c_{11} - c_{13})\lambda & 0 & 0 & 0 \end{bmatrix} \tag{3.75}$$

If the bias magnetic field is along the x direction and parallel to the surface of the specimen, the corresponding EMAT is a *Lamb wave* or surface wave EMAT.

The inverse piezomagnetic matrix can be obtained by a similar principle to the former two cases and can be expressed by (3.76).

$$e_{ki}^{(MS)} = \begin{bmatrix} (c_{11}-c_{13})\lambda & [c_{13}-\frac{1}{2}(c_{11}+c_{22})]\lambda & [c_{13}-\frac{1}{2}(c_{11}+c_{22})]\lambda & 0 & 0 & 0 \\ 0 & 0 & 0 & 0 & 0 & c_{44}\frac{3\varepsilon_M}{H_0} \\ 0 & 0 & 0 & 0 & c_{44}\frac{3\varepsilon_M}{H_0} & 0 \end{bmatrix} \tag{3.76}$$

3.2.5 *Numerical Simulation of EMAT Based on Magnetostrictive Mechanism*

The weak form of the finite element software COMSOL Multiphysics is used to achieve the multifield coupled numerical simulation of the EMAT based on the magnetostrictive mechanism.

Firstly, the virtual displacement method is used to derive the weak form of (3.11) and (3.14) in the two-dimensional *Cartesian coordinate system*. Through derivation, the coil's pulsed eddy current field and the weak form of the equilibrium equation of the specimen can be provided by (3.77).

$$\int\limits_{\Omega}\left(\frac{1}{\mu}\frac{\partial A_z}{\partial x}\frac{\partial \delta A_z}{\partial x}+\frac{1}{\mu}\frac{\partial A_z}{\partial y}\frac{\partial \delta A_z}{\partial y}\right)\mathrm{d}A = -\int\limits_{\Omega}\sigma\frac{\partial A_z}{\partial t}\delta A\mathrm{d}A + \int\limits_{\Omega}\frac{1}{S}\iint\limits_{S}\sigma\frac{\partial A_z}{\partial t}\mathrm{d}s\delta A_z\mathrm{d}A + \int\limits_{\Omega}J_{sz}\delta A_z\mathrm{d}A + \int\limits_{\Gamma}\frac{1}{\mu}\frac{\partial A_z}{\partial n}\delta A_z\mathrm{d}l \tag{3.77}$$

$$\int\limits_{\Omega}(\sigma_x\delta\varepsilon_x+\sigma_y\delta\varepsilon_y+\sigma_{xy}\delta\varepsilon_{xy})\mathrm{d}V = +\int\limits_{V}\left(\rho\frac{\partial^2 u}{\partial t^2}\delta u+\rho\frac{\partial^2 v}{\partial t^2}\delta v\right)\mathrm{d}A + \int\limits_{\Omega}(f_x\delta u+f_y\delta v)\mathrm{d}A + \int\limits_{\Gamma}(T_x\delta u+T_y\delta v)\mathrm{d}l \tag{3.78}$$

In the above equation, μ represents the magnetic permeability matrix.

In the axisymmetric coordinates, the weak form of the coil's pulsed eddy current field and the specimen's equilibrium equation are provided by (3.79) and (3.80).

$$\int_{\Omega} \frac{1}{\mu}\left(r\frac{\partial A_\theta}{\partial r}\frac{\partial \delta A_\theta}{\partial r} + r\frac{\partial A_\theta}{\partial z}\frac{\partial \delta A_\theta}{\partial z}\right)\mathrm{d}A = -\int_{\Omega} r\sigma\frac{\partial A_\theta}{\partial t}\delta A_\theta \mathrm{d}A + \int_{\Omega} rJ_{s\theta}\delta A_\theta \mathrm{d}A + \int_{\Omega} r\frac{1}{S}\iint_{S}\sigma\frac{\partial A_\theta}{\partial t}\mathrm{d}s\delta A_\theta \mathrm{d}A + \int_{\Gamma} r\frac{1}{\mu}\frac{\partial A_\theta}{\partial n}\delta A_\theta \mathrm{d}l \tag{3.79}$$

$$\begin{aligned}&\int_{\Omega}(\sigma_{rr}\delta\varepsilon_{rr} + \sigma_{zz}\delta\varepsilon_{zz} + \sigma_{rz}\delta\varepsilon_{rz})\mathrm{d}A + \int_{\Omega}\left(\rho\frac{\partial^2 u}{\partial t^2}\delta u + \rho\frac{\partial^2 v}{\partial t^2}\delta v\right)\mathrm{d}A \\ &= \int_{\Omega}\left[\left(f_r + \frac{\sigma_{rr} - \sigma_{\theta\theta}}{r}\right)\delta u + \left(\frac{\sigma_{rz}}{r} + f_z\right)\delta v\right]\mathrm{d}A + \int_{\Gamma}(T_r\delta u + T_z\delta v)\mathrm{d}l\end{aligned} \tag{3.80}$$

Since the specimen satisfies the traction-free boundary condition, the internal force boundary integral term in (3.78) and (3.80) is 0.

If the computational equations of the multifield coupling equation, magnetostrictive force and magnetostrictive current density of the EMAT based on the magnetostrictive mechanism are given, the finite element software COMSOL Multiphysics can be used to achieve the numerical simulation of the whole process of the ultrasonic wave's excitation, propagation and reception. In the process of the numerical simulation, the magnetostrictive force and magnetostrictive current density have been realized in the coupling of the electromagnetic field to the mechanical field and of the mechanical field to the electromagnetic field, respectively.

Using COMSOL Multiphysics, the calculated procedures of the numerical simulation analysis of EMAT based on the magnetostrictive mechanism are listed as follows:

(1) According to the magnetizing curve and magnetostrictive curve of ferromagnetic material, calculating the value of ε_M and λ of the ferromagnetic material in the operating condition; determining the specimen's permeability of the different direction and the value of the stiffness matrix element; and then obtaining the element's value of each element.
(2) According to the concrete model of EMAT, the 2D rectangular coordinate system or axisymmetric coordinate system is selected.
(3) The model of each field is selected as the weak form. Three models should be chosen.
(4) Setting the geometric model of each component of EMAT. In the calculation of the electromagnetic field, a solving region should be given; thus, a solving region is determined outside the model.
(5) According to (3.77) and (3.80) and the boundary condition, respectively, the solving area and boundary are set. Setting the coupling variables in the various fields, the *Lorentz force* and magnetostrictive force are set as the driving forces of the particle vibration in the specimen. The Lorentz current density and

magnetostrictive current density are set as the source of current density in the specimen when receiving the ultrasonic waves. Since the induced electromotive force inside the coil conductor should be calculated when the coil is receiving the ultrasonic wave signal, the integral coupling variable should be set to calculate the result of the induced electromotive force of each solution step.

(6) Meshing the solving area. Two points should be noted to improve the accuracy of the calculation: One is setting more than two elements within the skin depth of the upper surface, and the other is guaranteeing that there are more than 7 elements within the wavelength of the excited ultrasonic wave in the specimen.

(7) Obtaining the transient solution of the model. The step size, relative error, and absolute error should be set in the solving process. Since the used pulse signal is a high-frequency tone burst signal, a relative small time step size should be set, so that the smoothness and steadiness of solutions can be guaranteed. When solving the particle displacement in the specimen, the displacement's value is relatively small; the given absolute error also needs to be small enough, so that the correctness of the solution can be guaranteed.

(8) Postprocessing of the result, including generation of surface plots of each field and the transient waveform.

In order to verify the effectiveness and correctness of the equation derived previously and the given analytical method, the results from the numerical simulation examples and physical experimental method are used to verify the correctness of the calculation results.

The example of the model is shown in Fig. 3.6; the *Lamb wave* is excitated and received on a thin nickel strap by the meander coil. The coil is a single-layer non-split meander coil, and the nickel strap is a pure nickel belt. The dimensions of coil and nickel strap and the material parameters are listed in Tables 3.1 and 3.2, respectively. The dispersion curves of the phase velocity and group velocity of the *Lamb waves* in the nickel strap are shown in Fig. 3.7.

According to the dispersion curves of the nickel strap, the excited S_0 *Lamb wave* mode in the nickel strap is selected. In order to satisfy the matching relationship between the meander coil's interval and the wave length of the *Lamb wave*, the selected exciting frequency is 783.3 kHz. Under this frequency, the phase velocity and group velocity of the corresponding S_0 mode *Lamb wave* are 5074 and 5067 m/s, respectively.

The pulse echo method is used in this work. The ultrasonic wave excited by the coil in the nickel strap propagates along two directions. The ultrasonic waves, after reflecting from the boundary, propagate to the coil to transform into the received signal. Since the distances from the coil to both sides of the ends are the same, the reflection signals from the two boundaries will arrive at the coil at the same time.

In the experiment, the RITEC RPR-4000 pulser and receiver were chosen for the excitation source of the EMAT probe to generate a RF tone burst signal. The amplitude and frequency of the signal can be adjusted according to requirements.

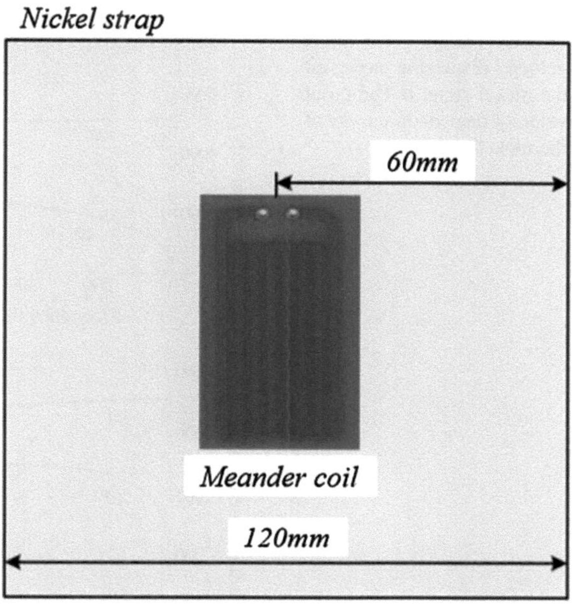

Fig. 3.6 The model of calculation and experiment

Table 3.1 The coil's dimension and material parameters

Parameter name	Value
Substrate thickness	0.500 mm
Copper wire width	1.625 mm
Copper wire thickness	0.05 mm
Fold spacing	6.50 mm
Copper wire electric conductivity	2.667×10^7 S/m

Table 3.2 The dimension and parameter of thin nickel plate

Parameter name	Value
Length	150 mm
Width	120 mm
Thickness	0.3 mm
Electric conductivity	$1.43 \times e^7$ S/m
Elastic modulus	206.9 GPa
Density	8880 kg/m^3
Poisson's ratio	0.31

Meanwhile, the RPR-4000 can be used to filter and amplify the received signal, and the bandwidth of the filtering and magnification can also be adjusted. For the RPR-4000 and the coil of EMAT, the matching between the coil impedance and excitation source output impedance can be achieved using the impedance matching device, so a larger power output can be realized. The receiving signal of the RPR-4000 can be shown in the oscilloscope, and the data acquisition software

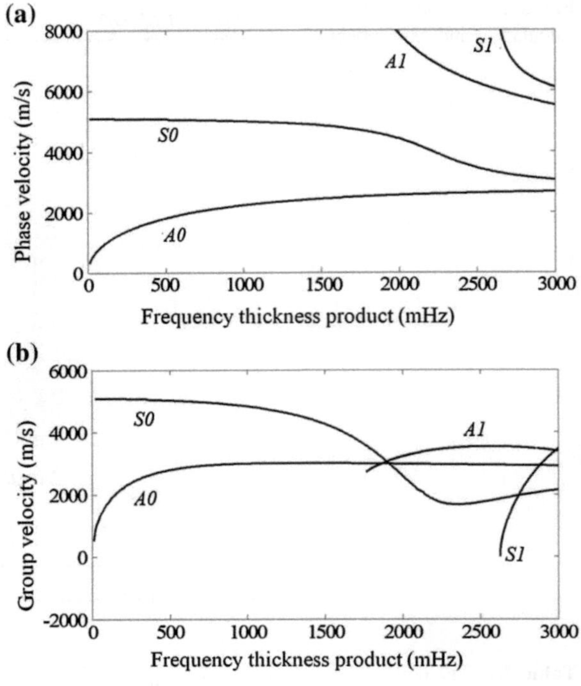

Fig. 3.7 Dispersion curves of the nickel strap. **a** The phase velocity dispersion curves of the nickel strap. **b** The group velocity dispersion curves of the nickel strap

Wave Star is used with the oscilloscopes together to achieve the data acquisition. The waveform is displayed on the PC. The adopted experimental equipment and its connection are shown in Fig. 3.8.

In this example, the direction of the static bias magnetic field is perpendicular to the surface of the nickel strap. The static bias magnetic field is provided by the residual magnetism after the magnetization of the nickel belt, and the residual magnetic flux density of the nickel strap is 0.335 T. The magnetizing curve and magnetostrictive curve of pure nickel are taken from the references as shown in Figs. 3.9 and 3.10. In order to express the numerical relationship among magnetostriction, magnetic induction intensity, and magnetic field intensity, the exponential fitting method is used to do the fitting for the two groups of curves; the specific fitting equations are shown in (3.81) and (3.82).

$$\varepsilon_t = -32.31\mathrm{e}^{0.0005659H} + 33.33\mathrm{e}^{-0.0898H} \tag{3.81}$$

$$B = 7.229\mathrm{e}^{-0.0001943H} - 4.465\mathrm{e}^{-0.1129H} \tag{3.82}$$

According the (3.82), if the magnetic induction intensity is 0.3325 T, which corresponds to magnetic field intensity H_0 = 2.1 kA/m, then after submitting the value of magnetic field intensity into (3.81), the magnetostriction of the thin nickel strap will be obtained, ε_M = −4.7467 ppm; the slope of the thin nickel strap's

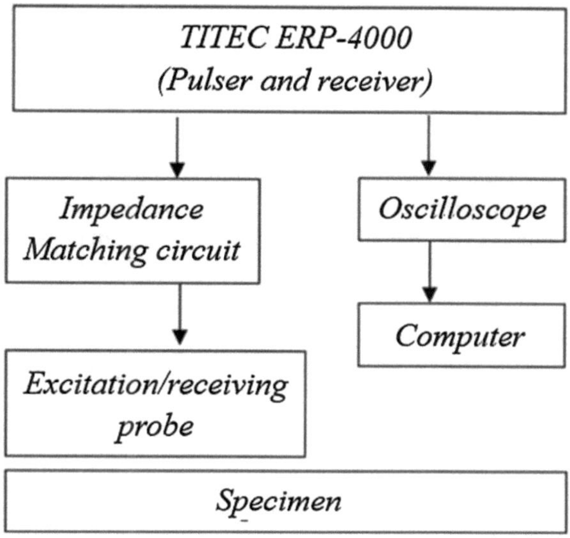

Fig. 3.8 Experimental setup of the *Lamb wave* test by EMAT

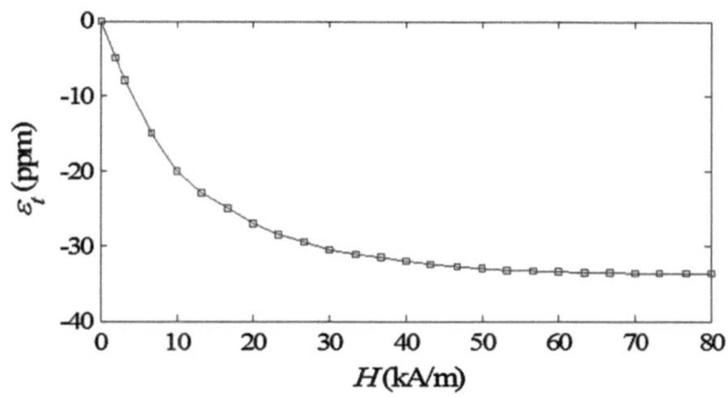

Fig. 3.9 The magnetostrictive curve of pure nickel

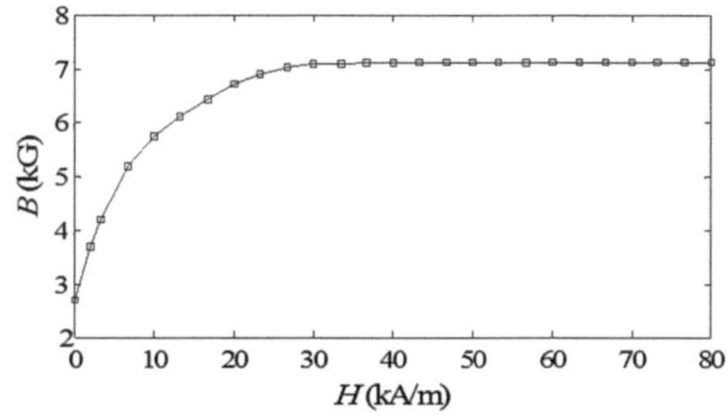

Fig. 3.10 The magnetizing curve of pure nickel

magnetostrictive curve *is* $\lambda = -2.5171$ ppm. The nickel strap's dynamic permeability is $1.5833 \times e^{-4}$ H/m (the relative permeability is 126). If the direction of the dynamic magnetic field and the bias magnetic field is the same, only the amplitude of the total magnetic field changes; thus, the nickel strap's magnetic permeability along the bias magnetic field direction is the differential permeability. According to the magnetizing curve in the figure, when $H_0 = 2.1$ kA/m, the nickel strap's differential permeability is $3.963 \times e^{-4}$ H/m (the relative permeability is 315.4). The elements of the nickel belt's stiffness matrix are $c_{11} = 3.906 \times 10^{12}$ Pa, $c_{13} = 6.554 \times 10^{11}$ Pa, $c_{12} = 0$, and $c_{44} = 1.97 \times 10^{11}$ Pa.

In this situation, the nickel belt acts under the combined effects of *Lorentz force* and magnetostrictive force. The calculated equations of the *Lorentz force* and magnetostrictive force are provided by (3.83).

$$f_{\mathrm{L}} = -\sigma B_0 \frac{\partial A_z}{\partial t} \tag{3.83}$$

$$f_{\mathrm{Ms}x} = c_{44} \frac{1}{\mu_{xx}} \frac{3\varepsilon_{\mathrm{M}}}{H_0} \frac{\partial^2 A_z}{\partial y^2} - \left[c_{13} - \frac{1}{2}(c_{11} + c_{12})\right] \lambda \frac{1}{\mu_{yy}} \frac{\partial^2 A_z}{\partial x^2} \tag{3.84}$$

When receiving the ultrasonic signal, the electric current density of the *Lorentz force* and magnetostriction in the specimen are shown in (3.85) and (3.86).

$$J_{\mathrm{L}} = B_0 \sigma \frac{\partial u_x}{\partial t} \tag{3.85}$$

$$\begin{aligned} J_{\mathrm{Ms}} = & \left[\frac{1}{\mu_{yy}}(c_{11} - c_{13})\lambda - \frac{1}{\mu_{xx}} c_{44} \frac{3\varepsilon_{\mathrm{M}}}{H_0}\right] \frac{\partial^2 u_y}{\partial x \partial y} \\ & + \frac{1}{\mu_{yy}}\left(c_{13} - \frac{1}{2}c_{11} - \frac{1}{2}c_{12}\right)\lambda \frac{\partial^2 u_x}{\partial x^2} - \frac{1}{\mu_{xx}} c_{44} \frac{3\varepsilon_{\mathrm{M}}}{H_0} \frac{\partial^2 u_x}{\partial y^2} \end{aligned} \tag{3.86}$$

The frequency of the excited ultrasonic wave is 783.3 kHz, and the cycle number of the generated signal is 3. Its waveform is shown in Fig. 3.11.

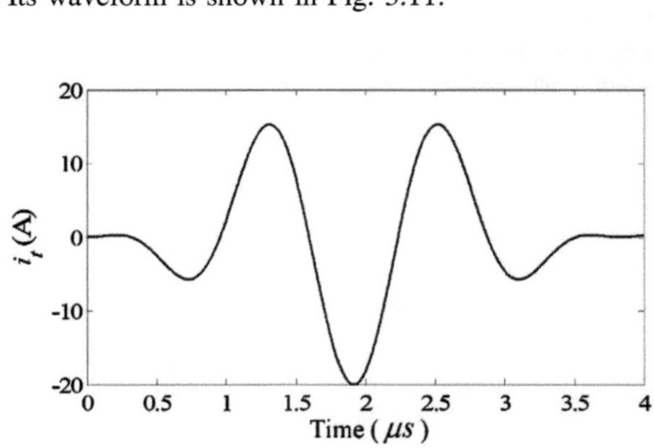

Fig. 3.11 The waveform of the excitation current

Entering the excitation current shown in Fig. 3.11 into the software COMSOL Multiphysics, we can establish the model and calculate the value for EMAT according to procedure 2–8. In the simulation, the element number within each ultrasonic wavelength in the tested specimen is assumed to be 8 and the size of time step is 2 ns. In order to present the calculated results conveniently, the observation point P is selected in the specimen, whose distance from the left end of the specimen is 60 mm and from the upper surface is 0.01 mm.

Figure 3.12 provides the equipotential line of the magnetic vector potential generated by the excitation current inside the coil at 10 μs. It clearly shows that the dynamic magnetic field is generated by the coil around and in the specimen; within the specimen's skin depth, the eddy current will generate *Lorentz force* under the effects of the residual magnetism of the nickel belt. Under the effects of those two forces, the vibrations of particle motion generate the ultrasonic waves in the specimen. Figure 3.13 shows the distribution of the equipotential line of the vector magnetic potential generated by *Lorentz force* current density and magnetostrictive current density near the tested specimen. The pulsed magnetic field generated by the electric current density of the *Lorentz force* and magnetostriction in the specimen induces the voltage within the coil to realize the receiving of the signal.

From Figs. 3.14 and 3.15, the *Lorentz force* density and magnetostrictive force density can be calculated by (3.83) and (3.84) at point P on the specimen. The comparison of these two figures shows that the *Lorentz force* used to excite the *Lamb wave* in the nickel strap specimen is much less than the magnetostriction force. It can be used to explain why the magnetostrictive effect plays a key function in the process of *Lamb wave* excitation.

From Figs. 3.16 and 3.17, the time-dependent waveforms of the x and y components of the particle displacement at point P are given. From these two figures, it is found that the x component is bigger than the y component, but the group velocities of both components are the same. This shows that there is no existence of other ultrasonic wave modes.

The pictures of the particle displacement of the x component at different times are provided in Fig. 3.18. Figure 3.18 clearly shows the ultrasonic wave propagation in the specimen with the change of time, its reflection from the end, and its second reflection in the tested specimen, and so on. It can be found that the time from the ultrasonic excitation to the reception of the reflected wave is approximately 25 μs.

Figure 3.19 shows the waveform of the signal received by the receiving probe. Figure 3.20 shows the received signal using the same number of periods, frequency, and current excitation as the experimental method shown in Fig. 3.8. In order to make the comparison between them, these two waveforms were normalized and the amplitude of the first waveform was used as the reference. Through the comparisons, it is found that there is not much difference in the waveforms corresponding to the time between the two signals. In the two waveforms, the first wave packet is the receiving signal when the coil is exciting and propagating the ultrasonic wave, while the second, third, and fourth wave packets are the first, second, and third reflections of the received signal from the end.

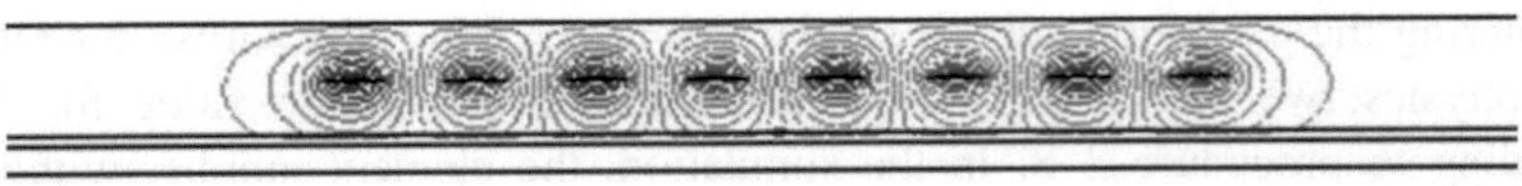

Fig. 3.12 The equipotential line of the vector magnetic potential generated by the coil's electricity at the time of 10 μs

Fig. 3.13 The equipotential line of the vector magnetic potential generated by the specimen's electric current density at the time of 25 μs

Fig. 3.14 The density of the *Lorentz force* at the point P

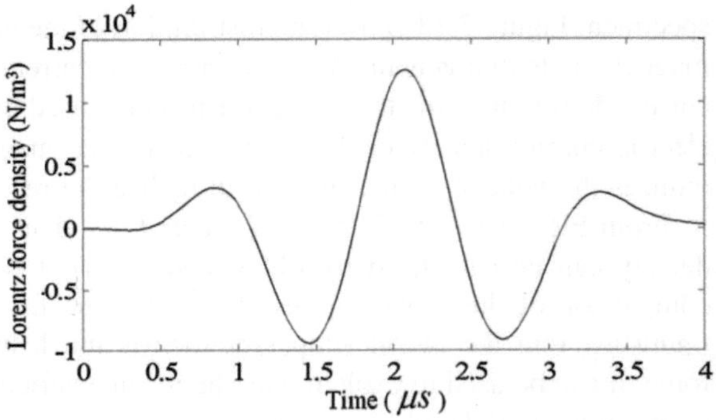

Fig. 3.15 The *x* component of the magnetostrictive force density at the point P

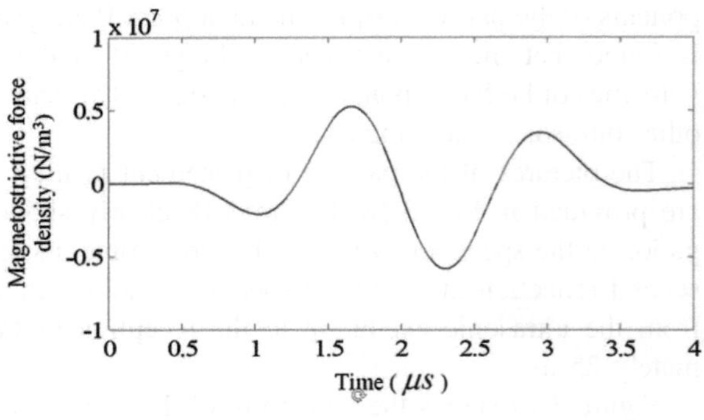

The group velocity of the ultrasonic wave was calculated based on the propagation distance and time. For the time difference between the second and third wave packet's peak value in Figs. 3.19 and 3.20, the propagation distance of the ultrasonic wave is 10 mm. The comparisons of the values of the theoretical velocity, modeling velocity, and experimental velocity are provided in Table 3.3.

From Table 3.3, it can be seen that the velocities obtained by the modeling and theoretical calculations are quite similar; there are certain errors among the velocities obtained by the experiment, modeling, and simulation, but they are not that

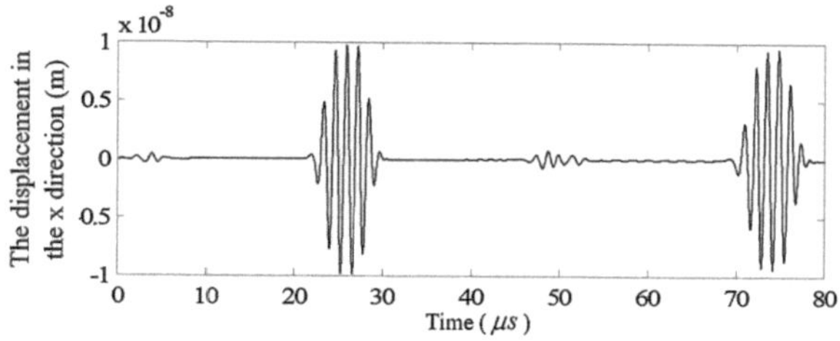

Fig. 3.16 The time-dependent waveform of the *x* component of the particle displacement at point P

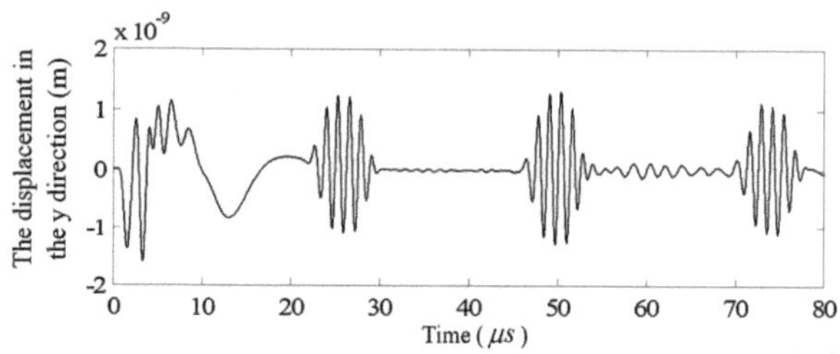

Fig. 3.17 The time-dependent waveform of the *y* component of the particle displacement at point P

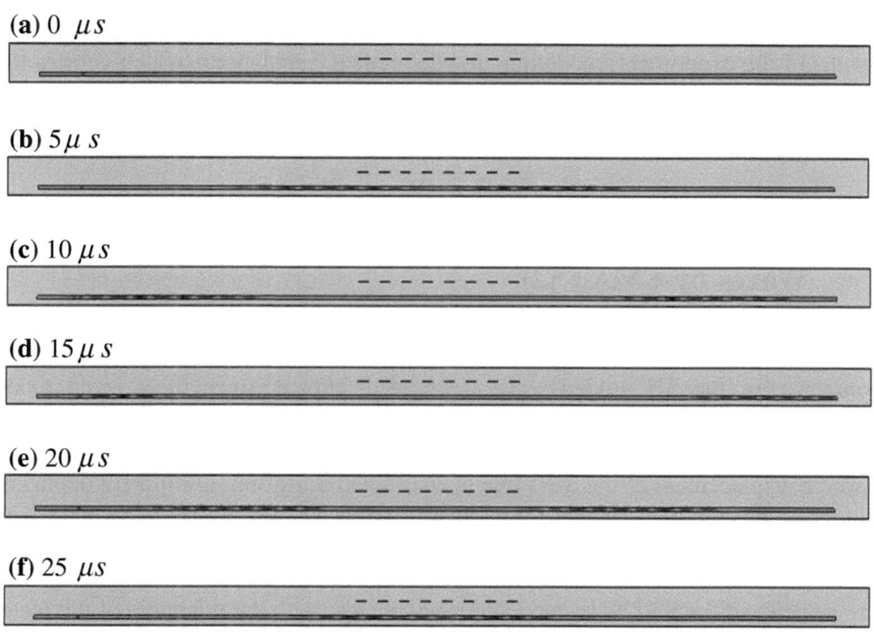

Fig. 3.18 The picture of the particle displacement of the *x* component at different times

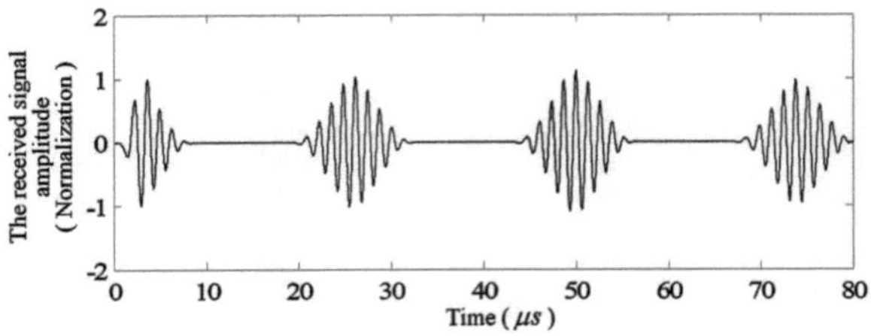

Fig. 3.19 The modeling of the receiving signal

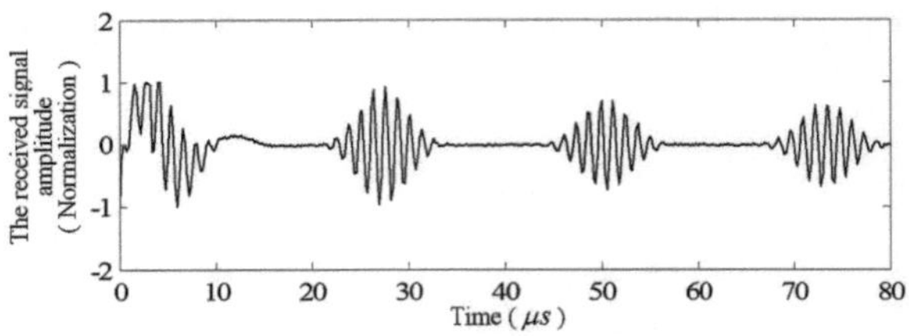

Fig. 3.20 The experimental signal received by the coil

Table 3.3 The comparison of the wave velocities obtained by the three methods

The theoretical velocity (m/s)	The modeling velocity (m/s)	The experimental velocity (m/s)
5067	5038	5217

significant. The reason for the errors could be the differences in the material parameters between the experimental nickel strap and theoretical calculation. Table 3.3 also explains that both the simulation and experiment have obtained the S_0 *Lamb wave* mode at the designed frequency and velocity. There is general agreement among the results of the experiment, modeling, and theoretical calculations, which verifies the correctness and effectiveness of the presented modeling method of the whole energy conversion process of EMAT based on the magnetostrictive mechanism.

3.3 Analytical Modeling and Calculation of SH Guided Waves by EMAT [2]

For the nondestructive testing of ferromagnetic material by electromagnetic ultrasonic waves, the SH guided wave is an ideal choice, since there is no mode conversion for SH wave propagation and reflection from the end, so it is very convenient for the signal receiving and processing of the SH guided wave. The particle displacement of the SH wave is distributed uniformly in different depths of the ferromagnetic plate, which means that it will travel uniformly in the plate. The reflection will happen at any position where the crack is located; thus, it can be used to detect the crack at any position in the specimen. For the SH wave modes, since the velocity of the SH_0 wave mode does not change with the thickness of the plate, it is suitable to be used widely to test plates with different thicknesses.

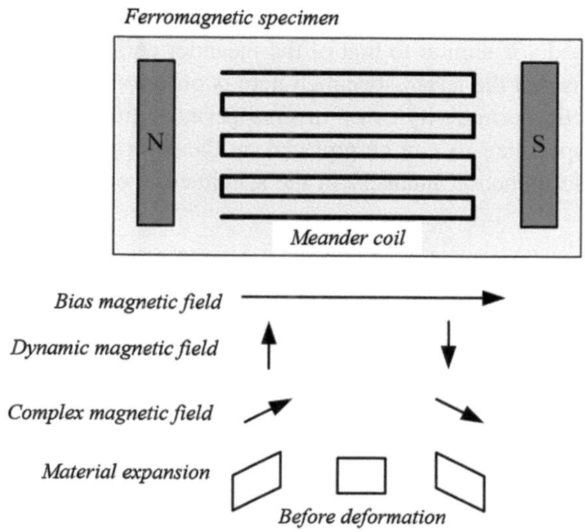

Fig. 3.21 The structure and working principle of the SH guided wave EMAT

The sketch and working principle of the SH guided wave EMAT is shown in Fig. 3.21. The direction of the bias magnetic field and coil conductor are parallel to each other, and the direction of the bias magnetic field is perpendicular to that of the dynamic magnetic field. After coupling them and making the ferromagnetic specimen generate periodic deformation, the ultrasonic wave can be excited. In the SH guided wave mode generated by EMAT, the direction of the bias magnetic field is parallel to that of the eddy current generated by the coil in the specimen. There is no effect of *Lorentz force* on the specimen, and the electromagnetic ultrasonic waves will only be generated by the magnetostrictive mechanism (Fig. 3.22).

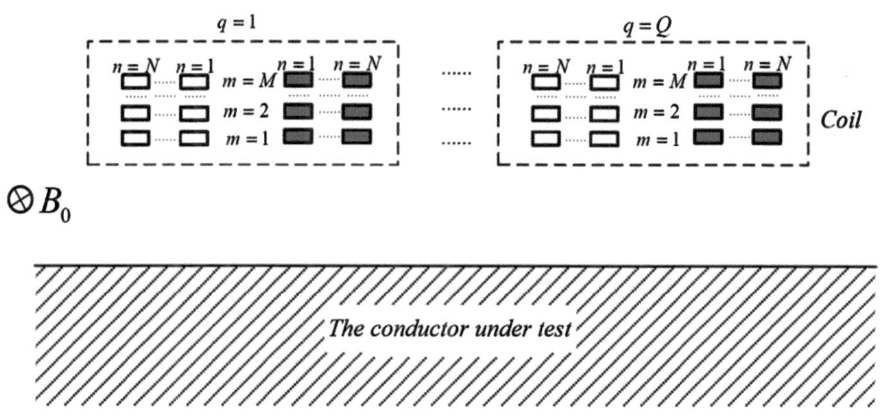

Fig. 3.22 The model of the multilayer and multisplit coil

If the multilayer and multisplit coil is placed on the specimen, considering this situation, the model is similar to that of the meander coil, as described in Chap. 2. The difference is that the tested specimen here is of ferromagnetic material. So, the material magnetic permittivity is of inverse permeability, which means that the ferromagnetic specimen μ_0 can be replaced by the reversible permeability μ_{rev}.

The magnetic induction intensity in the specimen is expressed in (3.87).

$$B_{(c)} = \frac{\mu_0 I}{2\pi^2 (h_2 - h_1) w} \int_{-\infty}^{+\infty} \int_{-\infty}^{+\infty} \left[\frac{e^{\lambda z}}{-k^2} \times \frac{2k\mu_r \lambda}{k\mu_r + \lambda} \times (j\alpha \hat{x} + j\beta \hat{y} + \lambda \hat{z}) \right] \sum_{m=1}^{M} (e^{-kh_{2m}} - e^{-kh_{1m}}) \times \frac{1}{\alpha\beta} \sum_{n=1}^{N} P_n \sum_{q=1}^{Q} e^{-j\alpha x'_q} e^{j\alpha y} e^{j\beta y} \mathrm{d}\alpha \mathrm{d}\beta \quad (3.87)$$

In the above equation, $\hat{x}$ and $\hat{y}$ are the directional vector.

The impedance of the coil is listed in (3.88).

$$Z = Z_d + Z_0 + \Delta Z \quad (3.88)$$

In the above equation, Z_d is the DC impedance, Z_0 is the impedance of free space, and ΔZ represents the variation in impedance caused by the eddy current.

The expression of each impedance shown in Eq. (3.88) is expressed in the following equations, where L is the total length of the meander coil and σ_c is the electrical conductivity of the coil conductor.

$$\begin{aligned} Z_0 = {} & \frac{2j\omega\mu_0}{\pi^2 (h_2 - h_1)^2 w^2} \sum_{m'=1}^{M} \int_{-\infty}^{+\infty} \int_{-\infty}^{+\infty} \Bigg\{ 2(h_{2m'} - h_{1m'}) \\ & + \frac{2}{k} [e^{-k(h_{2m'} - h_{1m'})} - 1] + \sum_{m=1}^{m'-1} \frac{1}{k} (e^{kh_{2m}} - e^{kh_{1m}})(e^{-kh_{1m'}} - e^{-kh_{2m'}}) \\ & + \sum_{m=m'+1}^{M} \frac{1}{k} (e^{-kh_{2m}} - e^{-kh_{1m}})(e^{kh_{1m'}} - e^{kh_{2m'}}) \Bigg\} \\ & \frac{1}{x(\alpha\beta)^2} \left(\sum_{n=1}^{N} P_n \right)^2 \left(\sum_{q=1}^{Q} e^{-j\alpha x'_q} + \sum_{q=1}^{Q} e^{j\alpha x'_q} \right) \mathrm{d}\alpha \mathrm{d}\beta \end{aligned} \quad (3.89)$$

$$\begin{aligned} \Delta Z = {} & \frac{j2\omega\mu_0}{\pi^2 (h_2 - h_1)^2 w^2} \int_{-\infty}^{+\infty} \int_{-\infty}^{+\infty} \frac{1}{k(\alpha\beta)^2} \left[\sum_{m=1}^{M} (e^{-kh_{2m}} - e^{-kh_{1m}}) \right]^2 \\ & \left(\sum_{n=1}^{N} P_n \right)^2 \left(\sum_{q=1}^{Q} e^{-j\alpha x'_q} \times \sum_{q=1}^{Q} e^{j\alpha x'_q} \right) \times \frac{k\mu_r - \lambda}{k\mu_r + \lambda} \mathrm{d}\alpha \mathrm{d}\beta \end{aligned} \quad (3.90)$$

$$Z_{\mathrm{d}} = \frac{L}{\sigma_c(h_2 - h_1)w} \tag{3.91}$$

When the direction of the static bias magnetic field is parallel to that of the coil conductor and the surface of the specimen, the SH guided wave can be excited and received by EMAT. The excitation of the SH guided wave is determined by the dynamic magnetic field component H_x and the shear strain S_6 generated by the static magnetic field H_{0z}. This means that for the SH guided wave EMAT, the specimen satisfies the flowing constitutive relations like elastic media, as shown in (3.92) and (3.93).

$$S_6 = s_{66}\sigma_{66} + d_{16}H_x \tag{3.92}$$

$$B_x = d_{16}\sigma_6 + \mu_{\mathrm{rev}}H_x \tag{3.93}$$

In (3.92), the superposition of the dynamic magnetic field and the static bias magnetic field results in the dynamic strain in the ferromagnetic specimen. The dynamic strain is the source of the ultrasonic wave, while the inverse magnetostrictive effect given by (3.93) shows that the dynamic stress generates the dynamic magnetic field in the specimen. The equation can be used for the reception of the ultrasonic wave signal.

According to (3.75) and the constitutive relationship between magnetostriction and inverse magnetostriction, the magnetostrictive strain $\mathrm{S}_{\mathrm{Ms6}}$ in the specimen is expressed by (3.94).

$$S_{\mathrm{Ms6}} = \frac{3\varepsilon_{\mathrm{M}}}{H_0}H_x \tag{3.94}$$

The corresponding shear strength is expressed in (3.95).

$$T_{\mathrm{Ms6}} = -c_{44}\frac{3\varepsilon_{\mathrm{M}}}{H_0}H_x \tag{3.95}$$

The value of the above two equations determines the magnitude of the generated SH guided wave, so the magnitude of the SH guided wave is proportional to the material piezomagnetic coefficient. The magnitude of the SH guided wave can be quantified using the above two equations.

According to (3.87), substituting H_x into the above two equations, the expressions of the frequency domain of magnetostrictive strain and magnetostrictive stress can be obtained, as shown in (3.96) and (3.97).

$$S_{\mathrm{Ms6}} = \frac{3\varepsilon_{\mathrm{M}}}{H_0}\frac{I}{2\pi^2\mu_r(h_2-h_1)w}\int_{-\infty}^{+\infty}\int_{-\infty}^{+\infty}\left[\frac{e^{\lambda z}}{-k^2}\times\frac{2k\mu_r\lambda}{k\mu_r+\lambda}\times j\beta\right] \sum_{m=1}^{M}\left(e^{-kh_{2m}}-e^{-kh_{1m}}\right)\times\frac{1}{\alpha\beta}\sum_{n=1}^{N}P_n\sum_{q=1}^{Q}e^{-j\alpha x'_q}e^{j\alpha x}e^{j\beta y}\mathrm{d}\alpha\mathrm{d}\beta \tag{3.96}$$

$$T_{\mathrm{Ms6}} = -c_{44}\frac{3\varepsilon_t}{H_{0z}}\frac{I}{2\pi^2\mu_r(h_2-h_1)w}\int_{-\infty}^{+\infty}\int_{-\infty}^{+\infty}\left[\frac{e^{\lambda z}}{-k^2}\times\frac{2k\mu_r\lambda}{k\mu_r+\lambda}\times j\alpha\right] \sum_{m=1}^{M}\left(e^{-kh_{2m}}-e^{-kh_{1m}}\right)\times\frac{1}{\alpha\beta}\sum_{n=1}^{N}P_n\sum_{q=1}^{Q}e^{-j\alpha x'_q}e^{j\alpha x}e^{j\beta y}\mathrm{d}\alpha\mathrm{d}\beta \tag{3.97}$$

As shown in Fig. 3.6, the single-layer non-split meander coil and thin nickel strap are used as the coil and specimen for the SH guided wave EMAT, the direction of the static bias magnetic field being along the coil conductor.

In the simulation model, the meander coil is put on the thin nickel strap. The liftoff value between the coil and the thin nickel strap is 1 mm, as shown in Fig. 3.23.

In the analytical calculation, the calculation of S_{Ms6} is used by the FFT-IFFT method. The excitation frequency of the coil is 500 kHz, and the waveform of the tone burst signal has number of periods of 3. Its waveform is shown in Fig. 3.24.

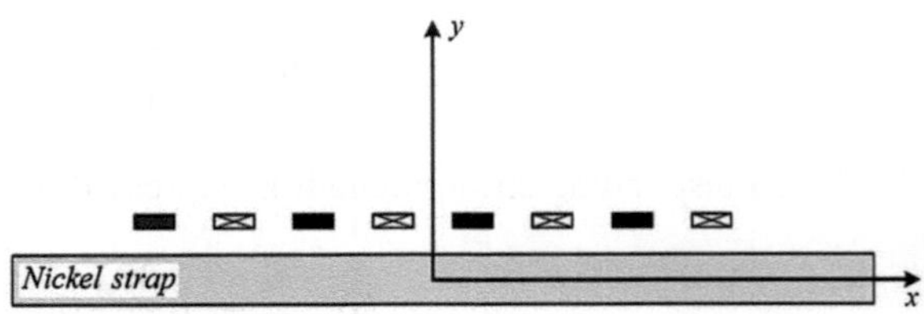

Fig. 3.23 The meander coil is placed on the nickel strap

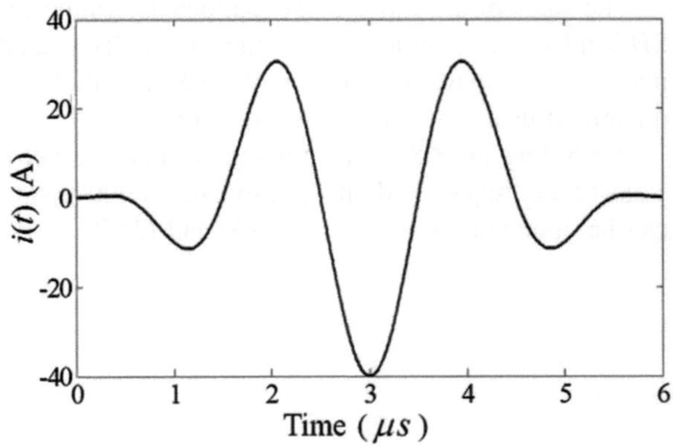

Fig. 3.24 The waveform of the excitation current

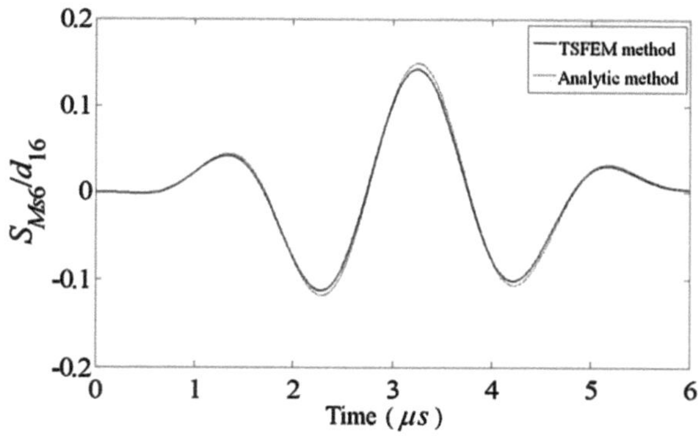

Fig. 3.25 The result of the analytical and finite element calculation

The result of S_{Ms6}/d_{16} at a depth of 0.0001 mm in the specimen is shown in Fig. 3.25. The calculation is specific to the lower part of the first conductor at the left side of the coil, and the analytical calculation and time-stepping finite element method are used.

From the calculated results in Fig. 3.25, the results of the analytical calculation are consistent with that of the finite element simulation; this verifies the correctness of the calculation. In this work, we only calculated the value of S_{Ms6}/d_{16}. Under the specific bias voltage and excitation frequency, d_{16} is a constant, proportional to the value of the shear strain S_{Ms6}.

From (3.94) and (3.95), the amplitude of the SH ultrasonic guided wave excited by the SH guided wave EMAT relates to two key factors: one is the piezomagnetic coefficient determined by the material stiffness matrix, magnetostrictive property, and static bias magnetic field, and the other is the amplitude value of the dynamic magnetic field generated by the coil.

3.4 Analytical Modeling and Calculation of an Axial Guided Wave in a Pipe by EMAT

The axisymmetric guided waves generated by the magnetostrictive transducer are an effective choice for defect detection in a ferromagnetic pipe. For the structure of the *L*-mode EMAT, M.J. Sablik provided the method of analytical calculation. The analytical modeling and calculation of *T*-mode EMAT will be discussed in this section [3].

Figure 3.26 shows the picture of a cross section of a *T*-mode guided wave EMAT. The idea of analytical modeling and calculation of the *T*-mode guided wave EMAT is as follows: First, the coil's excitation is assumed to be in a sinusoidal steady state of excitation; second, calculating the expression of the frequency of magnetic field and eddy current of the EMAT and then transforming the expression of the frequency domain into that of the time domain by an inverse *Fourier transformation*. The method of FFT-IFFT is used in this realization.

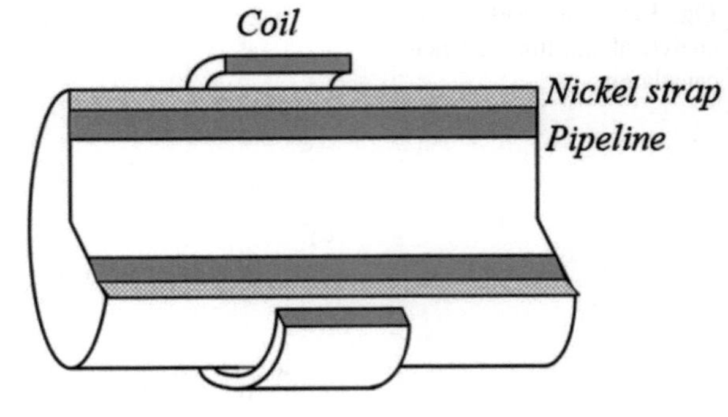

Fig. 3.26 The *T*-mode guided wave EMAT

When the axisymmetric guided wave is used for the inspection in the pipe, generally, the frequency of the excitated signal is less than 1 MHz for the satisfaction of quasi-static approximation conditions. Since there is no free charge existing in the solution domain, and the velocity effect of the particle motion is neglected, the flowing equation is satisfied containing the magnetic vector potential, as shown in (3.98).

$$\nabla^2 A - j\omega\mu\sigma A = 0 \tag{3.98}$$

For the calculation of EMAT, the axisymmetric model is used. The magnetic vector potential generated by the coil is symmetrical along the axial direction in the pipe. There is only the azimuthal component, as shown in (3.99), in which $\hat{\theta}$ is the direction vector along the circumferential direction in the pipe.

$$A = A(r, z)\hat{\theta} \tag{3.99}$$

3.4.1 The Magnetic Vector Potential of the δ Coil

As shown in Fig. 3.27, the δ coil is considered first. In this situation, the solution area is divided into 5 parts, that is, the fluid region 1 within the pipeline, the area of

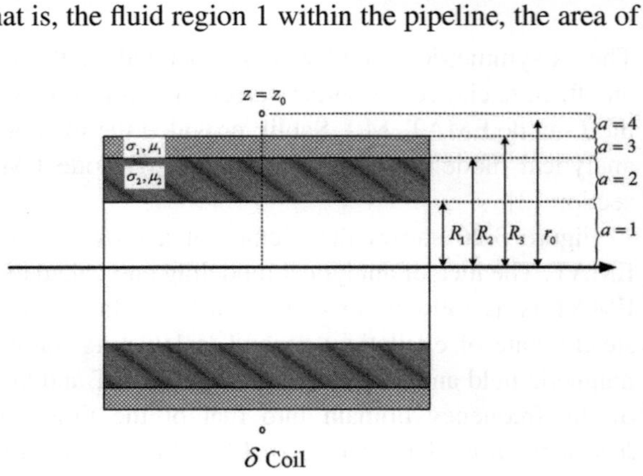

Fig. 3.27 The model of the δ coil

pipeline's wall 2, the area of nickel belt wall 3, and the area below the δ coil 4 and above the δ coil 5.

Under the excitation of the δ coil, the magnetic vector potential satisfies (3.100), where $A_a(\omega, r, z)$ is the magnetic vector potential; $i(\omega)$ is the density of the excitation current; μ_a and σ_a are the medium's magnetic permeability and electrical conductivity, respectively, and the subscript a represents the area variable.

$$\begin{aligned}&\left(\frac{\partial^2}{\partial r^2}+\frac{1}{r}\frac{\partial}{\partial r}+\frac{\partial^2}{\partial z^2}-\frac{1}{r^2}-j\omega\mu_\alpha\sigma_a\right)A_a(\omega,r,z)\\&=-\mu_a i(\omega)\delta(r-r_0)\delta(z-z_0)\end{aligned} \tag{3.100}$$

Considering the symmetry of the model, the general solution of (3.100) can be expressed by the method of variable separation, as shown in (3.101).

$$A_a(\omega,r,z)=\int_0^\infty[B_a(\alpha)I_1(\alpha_a r)+C_a(\alpha)_1K_1(\alpha_a r)]\cos\alpha(z-z_0) \tag{3.101}$$

In the above equation, I_1 and K_1 are the revised *Bessel* functions; α_a can be defined as follows:

$$\alpha_a=\sqrt{\alpha^2+j\omega\mu_a\sigma_a}$$

In the above equation, $B_a(\alpha)$ and $C_a(\alpha)$ can be determined based on the boundary condition of the interface since there is no interface current existing at the coil's position. Thus, at the boundary position outside of the δ coil's interface, we get the relationship shown in (3.102) and (3.103), where a = 1, 2, 3.

$$A_a(\omega,R_a,z)=A_{a+1}(\omega,R_a,z) \tag{3.102}$$

$$\frac{\partial A_a(\omega,R_a,z)}{\partial r}=\frac{\partial A_{a+1}(\omega,R_a,z)}{\partial r} \tag{3.103}$$

While at the interface of the δ coil, we get the relationship shown in (3.104).

$$\frac{\partial A_4(\omega,R_m,z)}{\partial r}=\frac{\partial A_5(\omega,R_m,z)}{\partial r}+\mu_0 I\delta(z-z_0) \tag{3.104}$$

In addition, when r = 0, A is a finite value and when $r = \infty$, A = 0.

$$\begin{aligned}C_1(\alpha)&=0\\B_5(\alpha)&=0\end{aligned}$$

Substituting (3.101) into (3.102) and (3.104), and using Fourier theorem, we get (3.105).

$$\frac{1}{\pi}\int_{0}^{\infty} f(\alpha)\left[\int_{0}^{\infty}\cos(\alpha x)\cos(\beta x)\mathrm{d}x\right]\mathrm{d}\alpha = f(\beta) \tag{3.105}$$

At the outside of the coil boundary, we can obtain (3.106).

$$\begin{bmatrix} I_1(\alpha_a R_a) & K_1(\alpha_a R_a) \\ I_1'(\alpha_a R_a) & K_1'(\alpha_a R_a) \end{bmatrix}\begin{bmatrix} B_a(\alpha) \\ C_a(\alpha) \end{bmatrix} = \begin{bmatrix} I_1(\alpha_{a+1} R_a) & K_1(\alpha_{a+1} R_a) \\ \frac{\alpha_{a+1}}{\alpha_a} I_1'(\alpha_{a+1} R_a) & \frac{\alpha_{a+1}}{\alpha_a} K_1'(\alpha_{a+1} R_a) \end{bmatrix}\begin{bmatrix} B_{a+1}(\alpha) \\ C_{a+1}(\alpha) \end{bmatrix} \tag{3.106}$$

After the compilation of the above equation, we can get (3.107).

$$\begin{bmatrix} B_{a+1}(\alpha) \\ C_{a+1}(\alpha) \end{bmatrix} = D_a \begin{bmatrix} B_a(\alpha) \\ C_a(\alpha) \end{bmatrix} \tag{3.107}$$

In the above equation, we have the following relationship, as shown in (3.108).

$$\begin{aligned} D_a &= \begin{bmatrix} I_1(\alpha_{a+1} R_a) & K_1(\alpha_{a+1} R_a) \\ \frac{\alpha_{a+1}}{\alpha_a} I_1'(\alpha_{a+1} R_a) & \frac{\alpha_{a+1}}{\alpha_a} K_1'(\alpha_{a+1} R_a) \end{bmatrix}^{-1} \times \begin{bmatrix} I_1(\alpha_a R_a) & K_1(\alpha_a R_a) \\ I_1'(\alpha_a R_a) & K_1'(\alpha_a R_a) \end{bmatrix} \\ &= R_a \begin{bmatrix} \begin{matrix} \alpha_{a+1} K_0(\alpha_{a+1} R_a) I_1(\alpha_a R_a) \\ +\alpha_a K_1(\alpha_{a+1} R_a) I_0(\alpha_a R_a) \end{matrix} & \begin{matrix} \alpha_a K_1(\alpha_{a+1} R_a) K_0(\alpha_a R_a) \\ -\alpha_{a+1} K_0(\alpha_{a+1} R_a) K_1(\alpha_a R_a) \end{matrix} \\ \begin{matrix} \alpha_a I_1(\alpha_{a+1} R_a) I_0(\alpha_a R_a) \\ -\alpha_{a+1} I_0(\alpha_{a+1} R_a) I_1(\alpha_a R_a) \end{matrix} & \begin{matrix} \alpha_a I_1(\alpha_{a+1} R_a) K_0(\alpha_a R_a) \\ +\alpha_{a+1} I_0(\alpha_{a+1} R_a) K_1(\alpha_a R_a) \end{matrix} \end{bmatrix} \end{aligned} \tag{3.108}$$

Thus, we get (3.109).

$$\begin{bmatrix} B_4(\alpha) \\ C_4(\alpha) \end{bmatrix} = D_3 D_2 D_1 \begin{bmatrix} B_1(\alpha) \\ 0 \end{bmatrix} = \begin{bmatrix} d_{11} & d_{12} \\ d_{21} & d_{22} \end{bmatrix}\begin{bmatrix} B_1(\alpha) \\ 0 \end{bmatrix} \tag{3.109}$$

While at the boundary of coil, we obtain (3.110).

$$\begin{bmatrix} I_1(\alpha_4 r_0) & K_1(\alpha_4 r_0) \\ I_1'(\alpha_4 r_0) & K_1'(\alpha_4 r_0) \end{bmatrix}\begin{bmatrix} B_4(\alpha) \\ C_4(\alpha) \end{bmatrix} = \begin{bmatrix} K_1(\alpha_5 r_0) \\ K_1'(\alpha_5 r_0) \end{bmatrix} C_5(\alpha) + \begin{bmatrix} 0 \\ \frac{\mu_0 I}{\pi \alpha} \end{bmatrix} \tag{3.110}$$

Solving (3.109) and (3.110), we can get (3.111), (3.112), and (3.113).

$$C_4(\alpha) = \frac{\mu_0 I r_0 K_1(\alpha r_0)}{\pi Q} \tag{3.111}$$

$$B_4(\alpha) = \frac{\mu_0 I r_0 K_1(\alpha r_0)}{\pi} \tag{3.112}$$

$$C_5(\alpha) = \frac{\mu_0 I r_0 \left[I_1(\alpha r_0) + \frac{K_1(\alpha r_0)}{Q(\alpha)} \right]}{\pi} \tag{3.113}$$

In the equations mentioned above, we have the following relationship:

$$Q(\alpha) = \frac{d_{11}}{d_{12}}$$

From (3.107), we can obtain (3.114).

$$\begin{bmatrix} B_3(\alpha) \\ C_3(\alpha) \end{bmatrix} = D_3^{-1} \begin{bmatrix} B_4(\alpha) \\ C_4(\alpha) \end{bmatrix} = \begin{bmatrix} m_{11} & m_{12} \\ m_{21} & m_{22} \end{bmatrix} \begin{bmatrix} B(\alpha_4) \\ C(\alpha_4) \end{bmatrix} \tag{3.114}$$

The required magnetic vector potential can be expressed as the follows.

$$\begin{aligned} A_3 = \frac{\mu_0 I}{\pi} r_0 \int_0^\infty \frac{1}{Q(\alpha)} [&m_{11} Q(\alpha) K_1(\alpha r_0) K_1(\alpha_3 r) + m_{12} K_1(\alpha r_0) K_1(\alpha_3 r) \\ &+ m_{21} Q(\alpha) K_1(\alpha r_0) I_1(\alpha_3 r) + m_{22} K_1(\alpha r_0) I_1(\alpha_3 r)] \cos(z - z_0) \mathrm{d}\alpha \end{aligned} \tag{3.115}$$

$$A_4 = \frac{\mu_0 I}{\pi} r_0 \int_0^\infty \frac{1}{Q(\alpha)} [Q(\alpha) K_1(\alpha r_0) K_1(\alpha r) + K_1(\alpha r_0) I_1(\alpha r)] \cos(z - z_0) \mathrm{d}\alpha \tag{3.116}$$

$$A_5 = \frac{\mu_0 I}{\pi} r_0 \int_0^\infty \frac{1}{Q(\alpha)} [Q(\alpha) I_1(\alpha r_0) + K_1(\alpha r)] K_1(\alpha r) \cos(z - z_0) \mathrm{d}\alpha \tag{3.117}$$

3.4.2 *Magnetic Vector Potential of the Coil with the Rectangular Cross Section*

When it comes to the magnetic vector potential of the coil with a rectangular cross section composed of the n-turns coil, it can be expressed as the superposition of the δ coil's magnetic vector potential (Fig. 3.28).

$$A_{n\text{-turn}} = \int_{r_1}^{r_2} \int_{l_1}^{l_2} A(r, z, r_0, z_0) \mathrm{d}z_0 \mathrm{d}r_0 \tag{3.118}$$

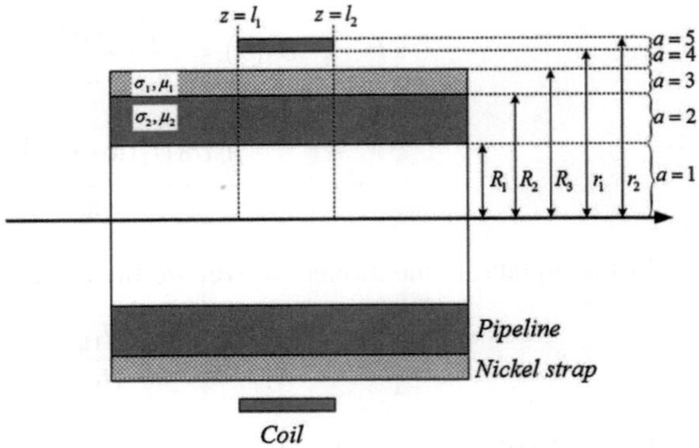

Fig. 3.28 The coil with a rectangular cross section

The electric current density of the coil with a rectangular cross section is assumed to have a uniform distribution, as shown in (3.119).

$$J = \frac{nI}{(l_2 - l_1)(R_2 - R_1)} \tag{3.119}$$

In areas 4 and 6, we get (3.120) and (3.121).

$$A_{4n\text{-turn}} = \frac{\mu_0 J}{\pi} \int_0^{\infty} \frac{1}{\alpha}[\sin \alpha(z - l_1) - \sin \alpha(z - l_2)] S_4(\alpha, r) \mathrm{d}\alpha \tag{3.120}$$

$$A_{6n\text{-turn}} = \frac{\mu_0 J}{\pi} \int_0^{\infty} \frac{1}{\alpha}[\sin \alpha(z - l_1) - \sin \alpha(z - l_2)] S_6(\alpha, r) \mathrm{d}\alpha \tag{3.121}$$

In the above equations, we have the following relationship:

$$S_4(\alpha, r) = \int_{r_1}^{r_2} \frac{1}{Q(\alpha)} r_0 [Q(\alpha) K_1(\alpha r_0) K_1(\alpha r) + K_1(\alpha r_0) I_1(\alpha r)] \mathrm{d}r_0$$

$$S_6(\alpha, r) = \int_{r_1}^{r_2} \frac{1}{Q(\alpha)} r_0 [Q(\alpha) I_1(\alpha r_0) + K_1(\alpha r)] K_1(\alpha r) \mathrm{d}r_0$$

The magnetic vector potential of the coil's cross-sectional area can be obtained from the above two equations, and substituting $l_2 = z$ into (3.121), substituting $l_1 = z$ into (3.120), and adding the above two equations together, we can get (3.122).

$$A_{5n-\text{turn}} = \frac{\mu_0 J}{\pi}\int_0^\infty \frac{1}{\alpha}[\sin\alpha(z-l_1)S_6(\alpha,r) - \sin\alpha(z-l_2)S_4(\alpha,r)]\mathrm{d}\alpha \tag{3.122}$$

In the conductor area 3, its vector magnetic potential is listed in (3.123), where $P(\alpha) = \int_{r_1}^{r_2} xK_1(\alpha x)\mathrm{d}x$.

$$A_{3n-\text{turn}} = \frac{2\mu_0 nI}{\pi(l_2-l_1)(r_2-r_1)}\int_0^\infty \frac{1}{Q(\alpha)}\{[Q(\alpha)m_{11}+m_{12}]K_1(\alpha_3 r) + [Q(\alpha)m_{21}+m_{22}]I_1(\alpha_3 r)\}\frac{P(\alpha)}{\alpha}\sin\left[\frac{\alpha(l_2-l_1)}{2}\right]\cos\alpha(z-z_0)\mathrm{d}\alpha \tag{3.123}$$

3.4.3 The Impedance, Eddy Current, and Magnetic Induction Intensity of the Coil

For a single-turn coil, the induced voltage can be calculated by the flowing line integral listed in (3.124).

$$U = j\omega\oint_{\text{coil}} A\cdot \mathrm{d}\boldsymbol{r} = j\omega 2\pi r_0 A(r_0, z_0) \tag{3.124}$$

The induced voltage of the n-turns coil can be calculated by the superposition of the single-turn coil's induced voltage, as shown in (3.125).

$$U = \frac{j\omega 2\pi n}{(l_2-l_1)(r_2-r_1)}\int_{l_1}^{l_2}\int_{l_1}^{l_2} rA_{5n-\text{turn}}(r,z)\mathrm{d}r\mathrm{d}z \tag{3.125}$$

With the turns of the coil assumed to be uniformly distributed, the coil's impedance is expressed in (3.126).

$$Z = \frac{U}{I} = \frac{16j\omega\mu_0 n^2}{(l_2-l_1)^2(r_2-r_1)^2}\int_0^\infty \frac{1}{\alpha^2}\sin^2\left[\frac{\alpha(l_2-l_1)}{2}\right] \times\left[\int_{r_1}^{r_2}\int_{r_1}^{r_2} xyK_1(\alpha x)I_1(\alpha y)\mathrm{d}x\mathrm{d}y + \frac{1}{Q(\alpha)}\left(\int_{r_1}^{r_2} xK_1(\alpha x)\mathrm{d}x\right)^2\right]\mathrm{d}\alpha \tag{3.126}$$

The eddy current within the conductor's area 3 is expressed in (3.127).

$$J_e = -j\omega A_{3n-\text{turn}} = -\frac{2j\omega\mu_0 nI}{\pi(l_2 - l_1)(r_2 - r_1)} \int_0^\infty \frac{1}{Q(\alpha)} \{[Q(\alpha)m_{11} + m_{12}]K_1(\alpha_3 r) + [Q(\alpha)m_{21} + m_{22}]I_1(\alpha_3 r)\} \frac{P(\alpha)}{\alpha} \sin\left[\frac{\alpha(l_2 - l_1)}{2}\right] \cos\alpha(z - z_0)\mathrm{d}\alpha \tag{3.127}$$

The coil's magnetic induction intensity within area 3 is expressed in (3.128).

$$B = \nabla \times A_{3n-\text{turn}} \tag{3.128}$$

Thus, we can obtain (3.129) and (3.130).

$$B_r = -\frac{2\mu_0 nI}{\pi(l_2 - l_1)(r_2 - r_1)} \int_0^\infty \frac{1}{Q(\alpha)} \{[Q(\alpha)m_{11} + m_{12}]K_1(\alpha_3 r) + [Q(\alpha)m_{21} + m_{22}]I_1(\alpha_3 r)\} P(\alpha) \sin\left[\frac{\alpha(l_2 - l_1)}{2}\right] \sin\alpha(z - z_0)\mathrm{d}\alpha \tag{3.129}$$

$$B_z = -\frac{2\mu_0 nI}{\pi(l_2 - l_1)(r_2 - r_1)} \left\{ \int_0^\infty \frac{1}{Q(\alpha)r} \{[Q(\alpha)m_{11} + m_{12}]K_1(\alpha_3 r). + [Q(\alpha)m_{21} + m_{22}]I_1(\alpha_3 r)\} + r\alpha_3[Q(\alpha)m_{11} + m_{12}]K_1'(\alpha_3 r) + r\alpha_3[Q(\alpha)m_{21} + m_{22}]I_1'(\alpha_3 r) \right\} \frac{P(\alpha)}{\alpha} \sin\left[\frac{\alpha(l_2 - l_1)}{2}\right] \cos\alpha(z - z_0)\mathrm{d}\alpha \tag{3.130}$$

3.4.4 One-Layer Conductor

If the pipe can be magnetized in its circumferential direction, the *T*-mode guided wave will be generated in the pipe. So, only one-layer conductor in the pipe should be considered for the calculated model, as shown in Fig. 3.29. This condition can be taken as the special situation of the *T*-mode guided wave EMAT, which means that the parameters of area 1 and 2 are set the same.

In this condition, the expressions of magnetic vector potential in all regions are expressed in the following equations.

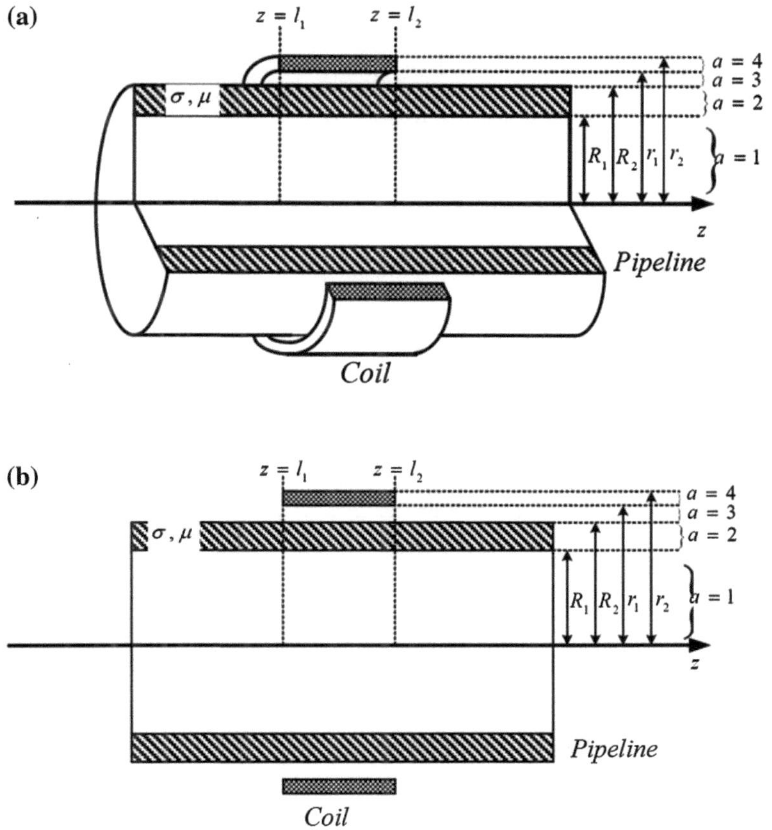

Fig. 3.29 The structure of EMAT for the excitation of the *T*-mode guided wave in the pipe. **a** The cross-sectional drawing. **b** The expanded view

$$A_{1n-\text{turn}} = \frac{\mu_0 I}{\pi}\int_0^\infty \frac{1}{R_1R_2\alpha^3D}K(r_2,r_1)I_1(\alpha_1 r)[\sin\alpha(z-l_1)-\sin\alpha(z-l_2)]\mathrm{d}\alpha \tag{3.131}$$

$$\begin{aligned}A_{2n-\text{turn}} = \frac{\mu_0 I}{\pi}\int_0^\infty \frac{1}{R_2\alpha^3D}K(r_2,r_1)\{[\alpha_2 I_1(\alpha_1R_1)I_0(\alpha_2R_1)\\ -\alpha_1I_1(\alpha_2R_1)I_0(\alpha_1R_1)]K_1(\alpha_2 r)+[\alpha_2K_0(\alpha_2R_1)I_1(\alpha_1R_1)\\ +\alpha_1K_1(\alpha_2R_1)I_0(\alpha_1R_1)]I_1(\alpha_2 r)\}[\sin\alpha(z-l_1)-\sin\alpha(z-l_2)]\mathrm{d}\alpha\end{aligned} \tag{3.132}$$

$$\begin{aligned} A_{3n\text{-turn}} = & \frac{\mu_0 I}{\pi}\int_0^\infty \frac{1}{\alpha^3}K(r_2,r_1)\Big\{I_1(\alpha r) - \Big\{\frac{K_1(\alpha_2 R_2)}{R_2 D K_1(R_1 R_2)}[\alpha_1 I_1(\alpha_2 R_1)I_0(\alpha_1 R_1) \\ & - \alpha_2 I_1(\alpha_1 R_1)I_0(\alpha_2 R_1)] - \frac{I_1(\alpha_2 R_2)}{R_2 D I_1(R_1 R_2)}[\alpha_2 K_0(\alpha_2 R_1)I_1(\alpha_1 R_1) \\ & + \alpha_1 K_1(\alpha_2 R_1)I_0(\alpha_1 R_1)] + \frac{I_1(R_1 R_2)}{K_1(R_1 R_2)}\Big\}K_1(\alpha r)\Big\}[\sin\alpha(z-l_1) - \sin\alpha(z-l_2)]\mathrm{d}\alpha \end{aligned} \tag{3.133}$$

$$\begin{aligned} A_{5n\text{-turn}} = & \frac{\mu_0 I}{\pi}\int_0^\infty \Big\{\frac{1}{\alpha^3}K(r_2,r_1)K_1(\alpha r) \\ & \Big\{\frac{K_1(\alpha_2 R_2)[\alpha_2 I_1(\alpha_1 R_1)I_0(\alpha_2 R_1) - \alpha_1 I_1(\alpha_2 R_1)I_0(\alpha_1 R_1)]}{K_1(R_1 R_2)R_2 D} \\ & + \frac{I_1(\alpha_2 R_2)[\alpha_2 K_0(\alpha_2 R_1)I_1(\alpha_1 R_1) + \alpha_1 K_1(\alpha_2 R_1)I_0(\alpha_1 R_1)]}{K_1(R_1 R_2)R_2 D} \\ & - \frac{I_1(R_1 R_2)}{K_1(R_1 R_2)}\Big\} + \frac{1}{\alpha^3}I(r_2,r_1)K_1(\alpha r)\Big\}[\sin\alpha(z-l_1) - \sin\alpha(z-l_2)]\mathrm{d}\alpha \end{aligned} \tag{3.134}$$

The expression of the magnetic vector potential in the cross-sectional area of the coil is listed in (3.135).

$$\begin{aligned} A_{4n\text{-turn}} = & \frac{\mu_0 I}{\pi}\int_0^\infty \frac{1}{\alpha^3}\Big\{[K(r,r_1)I_1(\alpha r). + I(r_2,r)K_1(\alpha r)] \\ & + K_1(\alpha r)[K(r,r_1) + K(r_2,r)] \\ & \Big\{\frac{K_1(\alpha_2 R_2)[\alpha_2 I_1(\alpha_1 R_1)I_0(\alpha_2 R_1) - \alpha_1 I_1(\alpha_2 R_1)I_0(\alpha_1 R_1)]}{K_1(R_1 R_2)R_2 D} \\ & + \frac{I_1(\alpha_2 R_2)[\alpha_2 K_0(\alpha_2 R_1)I_1(\alpha_1 R_1) + \alpha_1 K_1(\alpha_2 R_1)I_0(\alpha_1 R_1)]}{K_1(R_1 R_2)R_2 D} \\ & - \frac{I_1(R_1 R_2)}{K_1(R_1 R_2)}\Big\}\Big\}[\sin\alpha(z-l_1) - \sin\alpha(z-l_2)]\mathrm{d}\alpha \end{aligned} \tag{3.135}$$

In the above equation, there are some relationships as follows:

$$\begin{aligned} D = & [\alpha_2 K_0(\alpha_2 R_2) K_1(\alpha R_2) - \alpha K_0(\alpha R_2) K_1(\alpha_2 R_2)] \\ & [\alpha_1 I_1(\alpha_2 R_1) I_0(\alpha_1 R_1) - \alpha_2 I_1(\alpha_1 R_1) I_0(\alpha_2 R_1)] \\ & + [\alpha_2 K_0(\alpha_2 R_1) I_1(\alpha_1 R_1) + \alpha_1 K_1(\alpha_2 R_1) I_0(\alpha_1 R_1)] \\ & [\alpha I_1(\alpha_2 R_2) K_0(\alpha R_2) + \alpha_2 I_0(\alpha_2 R_2) K_1(\alpha R_2)] \end{aligned}$$

$$I(r_2, r_1) = \int_{\alpha r_1}^{\alpha r_2} x I_1(x) \mathrm{d}x$$

$$K(r_2, r_1) = \int_{\alpha r_1}^{\alpha r_2} x K_1(x) \mathrm{d}x$$

The number of the coil's turn is assumed to have a uniform distribution. The expression of the coil's impedance is expressed in (3.136).

$$\begin{aligned} Z = & \frac{4j\omega\mu_0 n^2}{(l_2 - l_1)^2 (r_2 - r_1)^2} \int_0^{\infty} \frac{1}{\alpha^4} \\ & \left\{ M + N \left\{ \frac{K_1(\alpha_2 R_2)[\alpha_2 I_1(\alpha_1 R_1) I_0(\alpha_2 R_1) - \alpha_1 I_1(\alpha_2 R_1) I_0(\alpha_1 R_1)]}{K_1(R_1 R_2) R_2 D} \right.\right. \\ & + \frac{I_1(\alpha_2 R_2)[\alpha_2 K_0(\alpha_2 R_1) I_1(\alpha_1 R_1) + \alpha_1 K_1(\alpha_2 R_1) I_0(\alpha_1 R_1)]}{K_1(R_1 R_2) R_2 D} \\ & \left.\left. - \frac{I_1(R_1 R_2)}{K_1(R_1 R_2)} \right\} \right\} [1 - \cos \alpha(l_1 - l_2)] \mathrm{d}\alpha \end{aligned} \tag{3.136}$$

In the above equation, we have the following relationship.

$$M = \int_{r_1}^{r_2} r[K(r, r_1) I_1(\alpha r) + I(r_2, r) K_1(\alpha r)] \mathrm{d}r$$

$$N = \int_{r_1}^{r_2} r K_1(\alpha r)[K(r, r_1) + K(r_2, r)] \mathrm{d}r$$

The magnetic induction intensity within the wall of the pipe is listed in (3.137).

$$B = \nabla \times A_{2n-\mathrm{turn}} \tag{3.137}$$

Substituting (3.132) into the above equation, the expressions of the frequency domain of the magnetic induction intensity within the wall of the pipeline are listed in (3.138) and (3.139).

$$\begin{aligned} B_r = \frac{\mu_0 I}{\pi} \int_0^{\infty} \frac{1}{R_2 \alpha^2 D} K(r_2, r_1)\{[\alpha_2 I_1(\alpha_1 R_1) I_0(\alpha_2 R_1) \\ - \alpha_1 I_1(\alpha_2 R_1) I_0(\alpha_1 R_1)] K_1(\alpha_2 r) + [\alpha_2 K_0(\alpha_2 R_1) I_1(\alpha_1 R_1) \\ + \alpha_1 K_1(\alpha_2 R_1) I_0(\alpha_1 R_1)] I_1(\alpha_2 r)\}[\cos\alpha(z - l_1) - \cos\alpha(z - l_2)]\mathrm{d}\alpha \end{aligned} \tag{3.138}$$

$$\begin{aligned} B_z = \frac{\mu_0 I}{\pi} \int_0^{\infty} \frac{1}{R_2 \alpha^3 D} K(r_2, r_1)\{[\alpha_2^2 I_1(\alpha_1 R_1) I_0(\alpha_2 R_1) \\ - \alpha_1 \alpha_2 I_1(\alpha_2 R_1) I_0(\alpha_1 R_1)] K_0(\alpha_2 r) + [\alpha_2^2 K_0(\alpha_2 R_1) I_1(\alpha_1 R_1) \\ + \alpha_1 \alpha_2 K_1(\alpha_2 R_1) I_0(\alpha_1 R_1)] I_0(\alpha_2 r)\}[\sin\alpha(z - l_1) - \sin\alpha(z - l_2)]\mathrm{d}\alpha \end{aligned} \tag{3.139}$$

3.4.5 *Magnetic Elasticity of the Axial Guided Wave EMAT in Pipe*

For the *T*-mode axial guided wave EMAT in the pipe, the directions of the bias magnetic field and coil are the same, which means both of them are along the circumferential direction of the pipe. So, the piezomagnetic matrix and inverse piezomagnetic matrix of the specimen are expressed in (3.75) and (3.77). The *T*-mode guided wave is generated by the static magnetic field $H_{0\theta}$ and shear strain S_5, caused by the horizontal dynamic magnetic field H_z. So, the magnetoelastic elasticity relationship for the *T*-mode guided wave EMAT can be expressed by (3.140) and (3.141).

$$S_5 = s_{55}\sigma_5 + d_{15}H_z \tag{3.140}$$

$$B_z = d_{15}\sigma_5 + \mu_{\mathrm{rev}} H_z \tag{3.141}$$

In (3.140), the superposition of the dynamic magnetic field and static bias magnetic field will cause dynamic strain in the specimen as the source of the ultrasonic waves. But the inverse magnetostrictive effect given by (3.141) gives the dynamic magnetic field generated by the dynamic stress in the specimen. The equation can be used for the reception of the ultrasonic signal.

According to the constitutive relation between magnetostriction and inverse magnetostriction, the magnetostrictive shear stress S_{Ms5} in the specimen is expressed in (3.142).

$$S_{\mathrm{Ms5}} = \frac{3\varepsilon_{\mathrm{M}}}{H_0} H_z \tag{3.142}$$

The corresponding inverse magnetostrictive shear stress is expressed in (3.143).

$$T_{\mathrm{Ms5}} = -\frac{3\varepsilon_{\mathrm{M}}}{H_0} H_z \tag{3.143}$$

According to (3.142) and (3.143), we can analyze the characteristics of the *T*-mode guided waves.

3.4.6 *Calculation of the Pulsed Magnetic Field of the* T-*Mode Guided Wave*

From (3.142) and (3.143), for the *T*-mode guided wave EMAT, the value of strain in the specimen is determined by the piezomagnetic coefficient of the specimen itself and the magnetic field intensity in the specimen. In other words, the amplitude of the excited ultrasonic waves can be determined by those parameters. If the bias magnetic field and excitation current are given, the value of the piezomagnetic coefficient d_{15} can be determined through (3.75). The calculation of the magnetic induction intensity in the specimen can be done by (3.129) and (3.130).

The magnetic induction intensity in the specimen is calculated to verify the correctness of the given analytical expression in this section. If the pulsed excitation of the coil is given, the calculation of the pulsed magnetic force can be conducted by FFT-IFFT (referring to Chap. 2). The infinite integral shown in (3.129) and (3.130) can be calculated by the self-adaptive numerical integral method. The TEFEM method is used to do the calculation for the same model, and a comparison is made between the calculated results to verify the effectiveness of the derived computational equation.

The *T*-mode axial guided wave EMAT in the pipe consists of enameled wire and a premagnetized thin nickel belt. The calculation model of the coil and nickel belt is shown in Fig. 3.28, while the dimensions of the coil and thin nickel belt are given in Tables 3.4 and 3.5.

For the *T*-mode axial guided wave in the pipe, we make a premagnetization of the nickel strap along the circumferential direction of the pipe. The residual

Table 3.4 The coil's dimension and parameters

Parameter name	Value
Height	5 mm
Width	6.7 mm
Number of turns	40
Electrical conductivity	$5.7 \times e^7$ S/m

Table 3.5 The dimension and parameters of the thin nickel plate

Parameter name	Value
Length	300 mm
Width	55 mm
Thickness	0.3 mm
Electrical conductivity	$1.43 \times e^7$ S/m
Elasticity modulus	206.9 GPa
Poisson's ratio	0.31

magnetic flux density of the nickel strap itself is used to provide the static bias magnetic field for the electromagnetic ultrasonic transducer. The reversible permeability when the nickel strap is generating the ultrasonic guided waves under the effect of residual magnetism is assumed to be $1.5833 \times e^{-4}$ H/m.

The calculated waveform of the tone burst current excitation signal whose frequency is 200 kHz with 4 cycles is shown in Fig. 3.30.

The waveforms of the magnetic induction intensity components r and z at a depth of 0.01 mm inside the nickel strap, just below the coil, are shown in Figs. 3.31 and 3.32, by the analytic method and time-stepping finite element

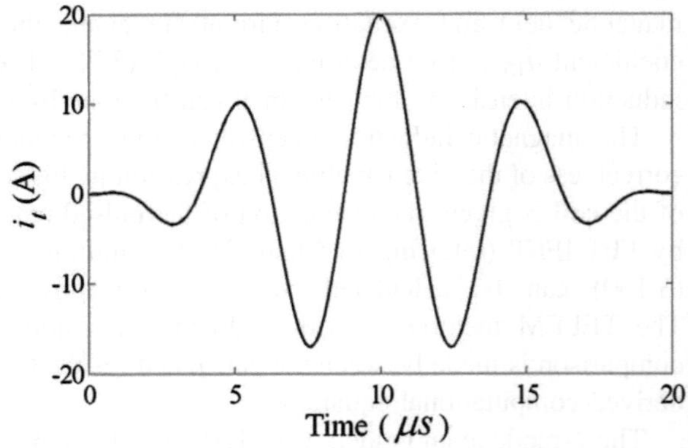

Fig. 3.30 The waveform of the excitation current

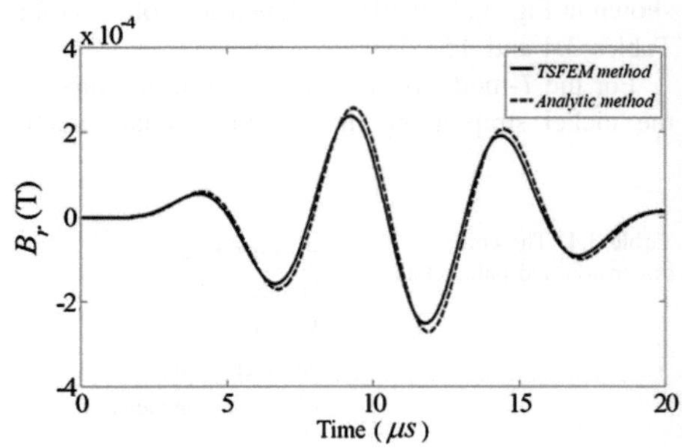

Fig. 3.31 The r component of the magnetic induction intensity

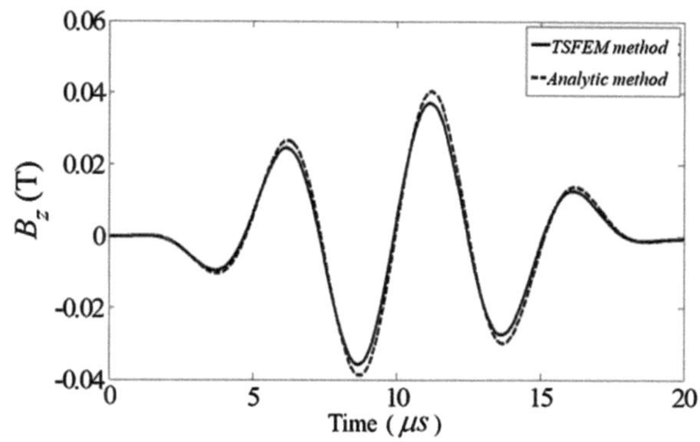

Fig. 3.32 The z component of the magnetic induction intensity

method, respectively. From the comparisons of these waveforms, there is no significant difference between the calculated results of the analytical calculation and the time-stepping finite element analysis, which is also a verification of the accuracy and effectiveness of the analytical equation and modeling method given in this book.

References

1. Hao, K.S., Huang, S.L., Zhao, W., et al.: Multi-belts coil longitudinal guided wave magnetostrictive transducer for ferromagnetic pipes testing. Sci China Technol Sci **54**(2), 502–508 (2011)
2. Hao, K.S.: Analysis and optimal design of electromagnetic ultrasonic transducers. Tsinghua University, Beijing (2008). (In Chinese)
3. Hao, K.S., Huang, S.L., Zhao, W., et al.: A new frequency-tuned longitudinal wave transducer for nondestructive inspection of pipes based on magnetostrictive effect. In: 2010 IEEE Sensors Applications Symposium Proceedings, pp. 64–68 (2010)

Chapter 4
The Propagation Characteristics of Ultrasonic Guided Waves in Plate and Pipe

For the use of ultrasonic guided waves generated by an electromagnetic acoustic transducer for crack inspection in a natural gas pipeline, the guided wave will propagate along the circumferential direction in the pipe. The discussions of dispersion are the basis and fundamentals for the study and application of guided wave theory. For instance, the design of the parameters of transducers and the excitation and verification of the selected mode of guided wave are highly dependent on the dispersion curves. The dispersion of the circumferential guided wave is the main content of this chapter. The circumferential guided wave in the pipe is the ultrasonic guided wave that is used in the detection of cracks in the pipe; however, the investigations of guided waves in plates are also provided as the theoretical basis of this book. In some references, the theoretical analysis and experimental results that are obtained from the plate-like specimen are used approximately in the circumferential direction of the pipe without any verification. It is not reasonable for this approximate application. In this chapter, this issue will be further discussed based on the frequency-dispersive characteristics of circumferential guided waves in the plate and in the pipe. If there is an approximate relationship between the guided waves in plates and circumferential guided waves in pipes, thus the investigations of guided waves in plates can be used for circumferential guided waves in pipes. So, the approximation could make the study of circumferential guided waves in pipes easier and more convenient.

4.1 Dispersion and Wave Structures of the Lamb Waves in the Plate

Compared with the ultrasonic bulk waves in infinite elastic media, ultrasonic guided waves are much more complex. In this work, the explanation of the *Lamb wave* generated in the plate is given first. As shown in Fig. 4.1, the longitudinal wave is

S. Huang et al., *Electromagnetic Ultrasonic Guided Waves*, Springer Series in Measurement Science and Technology, DOI 10.1007/978-981-10-0564-0_4

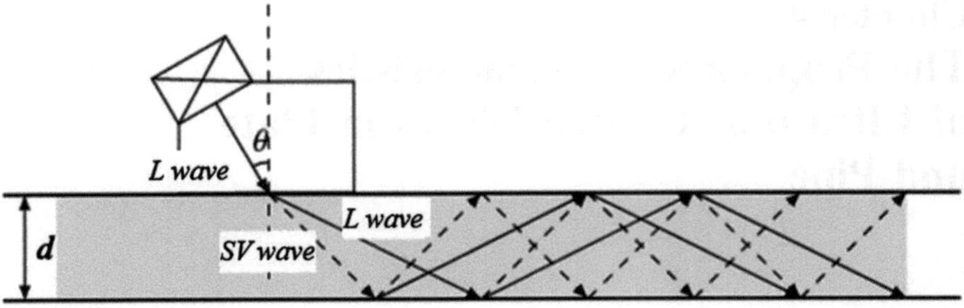

Fig. 4.1 The oblique incidence for the generation principle of the *Lamb* waves in plate

incident obliquely by the piezoelectric transducer in the plate, in accordance with the refractional law of the longitudinal wave (*L* wave). The wave undergoes mode conversion, reflection and refraction, including a vertical shear wave (*SV* wave) and an *L* wave at the interfaces. The reflection of the *SV* wave on the interface of the plate will generate an *SV* wave and an *L* wave, and we can get the same results from the reflection of the *L* wave on the interface of the plate. After a certain propagation distance away from the transducer, the *SV* wave and *L* wave will no longer be individually identifiable but will have been superimposed into a wave packet, constructing the *Lamb waves* propagating steadily in the plate.

Generally, guided wave propagation is associated with dispersion, which means that the propagating velocity of the guided wave is not only dependent on the material properties but also the geometrical shape of the waveguide. The velocity of the guided wave is the function of frequency, while there are curves of multimodes for the guided wave propagation. The relationship between propagating velocity and frequency of the steady-state guided wave is recorded as phase velocity dispersion curves, while the group velocity dispersion curves are used to describe the propagation of a burst or wave packet. Since in practical guided wave testing, a burst with a limited length is generally used as the incident signal, the actual propagation velocity of the guided wave signal is the group velocity.

4.1.1 The Dispersion Characteristics of the Lamb Waves in the Plate

As shown in Fig. 4.2, the partial displacements of the *Lamb waves* in the plate can be decomposed into two displacement components *u* and *v*, which are located in the cross section of the plate. In fact, the problem of the *Lamb waves* in the plate is a plane-strain problem.

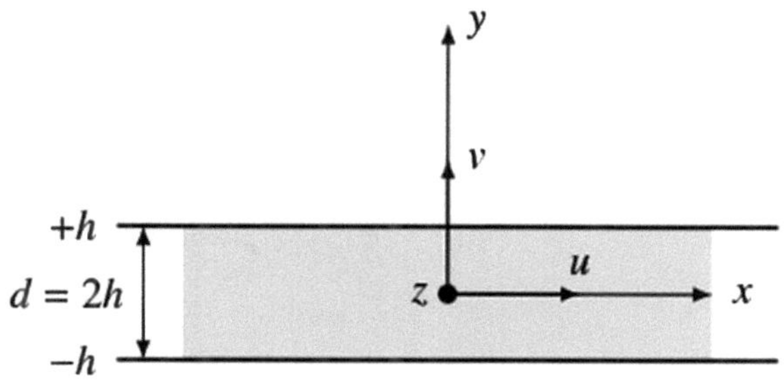

Fig. 4.2 The geometry and partial displacement of the *Lamb waves* in the free plate

The procedures to obtain the dispersion characteristics of the *Lamb wave* in the plate by the decomposition method of displacement potentials are as follows:

1. Deriving the *Navier governing equation* of motion; the displacement vector is decomposed by *Helmholtz decomposition*, expressed as $u = \nabla\varphi + \nabla \times \psi$, where the scalar φ and vector ψ are the displacement potential function.
2. According to the strain–displacement and strain–stress relationships in the elasticity, the stress can be expressed by the potential function φ and ψ where some unknown coefficients are included.
3. As a result of the assumption of free plate and plane strain, the traction-free boundary condition can be obtained, thereby providing the expressions of zero surface stress. As matter of fact, this expression is a system of linear homogeneous equations with some unknown coefficients for the stress equations.
4. In order to get nonzero solutions, the determinant of the coefficient matrix of the system of homogeneous equations should be zero; thereby, the frequency dispersion equations can be obtained. The relationship between the velocity and frequency is obtained from the solutions of the equations, and the dispersion curves of the phase velocity can be obtained by drawing the relationships.

The phase velocity dispersion curves of the *Lamb waves* propagating in the steel plate with 1 mm thickness is shown in Fig. 4.3a, where S_n represents the symmetric mode, A_n represents the antisymmetric mode, and n = 0, 1, 2....

As discussed above, the phase velocity dispersion curves correspond to the situation of the guided waves in steady excitation. But in the real test, the burst signal of a limited length of time is used as the excitation, its propagation velocity being called the group velocity. The group velocity can be obtained from the phase velocity. The dispersion curves of the group velocity dispersion curves corresponding to the phase velocity dispersion curves shown in Fig. 4.3a are given in Fig. 4.3b.

4.1.2 The Wave Structures of the Lamb Waves in the Plate

The wave structure of the *Lamb waves* in the plate will be discussed in the following section. The wave structures of the *Lamb waves* are the variations of displacement and stress across the thickness of the plate.

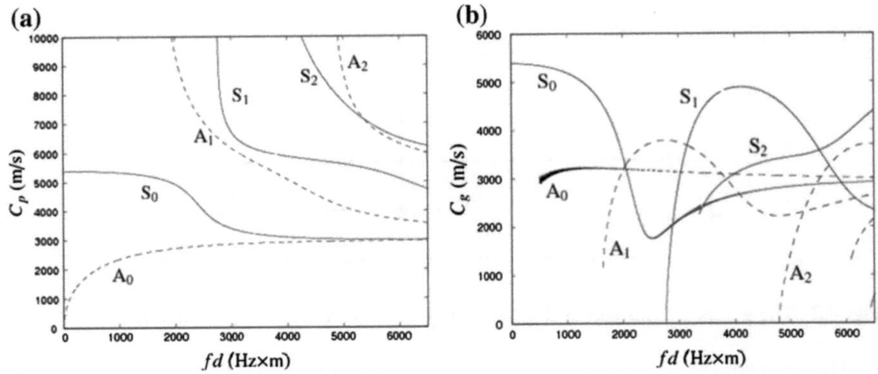

Fig. 4.3 The frequency dispersion curves of the *Lamb waves* in the 1-mm-thick steel plate. **a** Phase velocity dispersion curves. **b** Group velocity dispersion curves

For symmetric modes, it can be expressed as follows:

$$\begin{cases} \frac{u}{A_2} = ik\cos(py) + q\frac{B_1}{A_2}\cos(qy) \\ \frac{v}{A_2} = -p\sin(py) - ik\frac{B_1}{A_2}\sin(qy) \\ \frac{\tau_{xy}}{\mu A_2} = -2ikp\sin(py) + (k^2 - q^2)\frac{B_1}{A_2}\sin(qy) \\ \frac{\sigma_y}{\mu A_2} = \left[-\frac{\lambda}{\mu}(k^2 + p^2) - 2p^2\right]\cos(py) - 2ikq\frac{B_1}{A_2}\cos(qy) \end{cases}$$

In the above equation, k is the wave number; p and q are defined as follows: $p^2 = w^2/c_l^2 - k^2$ and $q^2 = w^2/c_t^2 - k^2$, respectively; λ and μ are the *Lame constants*. Then, we have the following equation, where h is the half thickness of the plate:

$$\frac{B_1}{A_2} = \frac{2ikp\sin(ph)}{(k^2 - q^2)\sin(qh)}$$

We can obtain the following equation:

$$\frac{\lambda}{\mu} = \left(\frac{C_L}{C_T}\right)^2 - 2$$

The expressions of displacement and stress for antisymmetric modes are given as follows:

$$\begin{cases} \frac{u}{A_1} = ik\sin(py) - q\frac{B_2}{A_1}\sin(qy) \\ \frac{v}{A_1} = p\cos(py) - ik\frac{B_2}{A_1}\cos(qy) \\ \frac{\tau_{xy}}{\mu A_1} = 2ikp\cos(py) + (k^2 - q^2)\frac{B_2}{A_1}\cos(qy) \\ \frac{\sigma_y}{\mu A_1} = \left[-\frac{\lambda}{\mu}(k^2 + p^2) - 2p^2\right]\sin(py) + 2ikq\frac{B_2}{A_1}\sin(qy) \end{cases}$$

In the above equation

$$\frac{B_2}{A_1} = -\frac{2ikp\cos(ph)}{(k^2 - q^2)\cos(qh)}$$

Taking the maximum value of the displacement and stress distribution in two directions as the reference and the normalization, respectively, the curves of both the displacement and the stress wave structures for the S_0 and A_0 *Lamb wave* modes are shown in Figs. 4.4 and 4.5, where the examined frequency-thickness products are 600, 1800, 3000, and 4200 Hz × m, respectively.

The wave structure of the guided wave mode has some effect on the guided wave inspection; for example, by avoiding energy leakage from the water loading or insulation, improved sensitivity can be obtained as a result of controlling the off-plane displacement v to make it as small as possible. Paradoxically, there should be a certain magnitude of off-plane displacement v on the plate surface when the

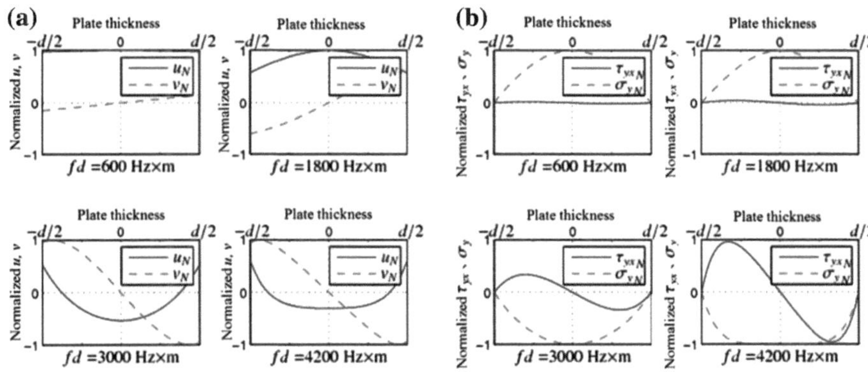

Fig. 4.4 S_0 mode wave structure. **a** The displacement wave structure. **b** The stress wave structure

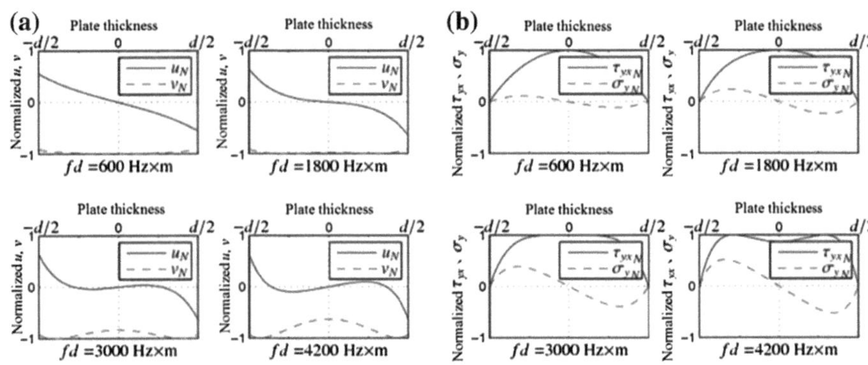

Fig. 4.5 A_0 mode wave structure. **a** The displacement wave structure. **b** The stress wave structure

piezoelectric transducers are used to generate and receive the *Lamb waves*, otherwise the energy conversion cannot happen properly. Each point on the dispersion curves can be used for testing. It will be helpful for the selection of the suitable wave mode, as well as the frequency thickness product *fd*, to draw the distribution of the normalized surface displacement of each point. A simple way is as follows: The normalized value of the surface displacements of the selected mode with certain *fd* is expressed using the *3D* diagram curves, where the *x*-coordinate corresponds to *fd*, the *y*-coordinate corresponds to C_p, and the *z*-coordinate corresponds to the normalized surface displacement component. Based on the solutions of the dispersion curve and the wave structures, the results of the symmetric modes are given in Fig. 4.6, where Fig. 4.6a represents the normalized distribution of the in-plane displacement value u_N and Fig. 4.6b represents the normalized distribution of the out-of-plane displacement value v_N. Figure 4.7 represents the result of the antisymmetric modes.

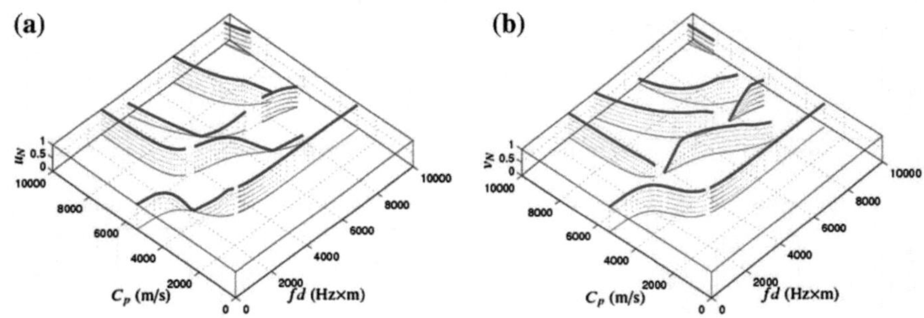

Fig. 4.6 The distributions of the normalized surface displacement of the symmetric modes. **a** Displacement *u*. **b** Displacement *v*

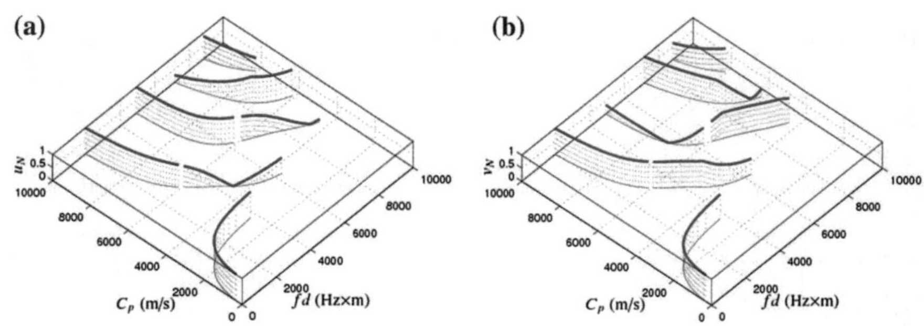

Fig. 4.7 The distributions of the normalized surface displacement of the antisymmetric modes

4.2 The Characteristics of Dispersion and Wave Structures of SH Guided Waves in the Plate

SH guided waves in the plate only have the displacements that are parallel to the surface of the planar waveguide; therefore, many aspects of SH guided wave propagation are quite valuable for the applications. As discussed in the last section, *Lamb waves* in the plate can be taken as the sufficient superpositions of the longitudinal wave and *SV* wave, while SH guided waves in the plate can also be considered as the superposition of up- and down-reflecting bulk shear waves in the plate.

4.2.1 Dispersion of SH Guided Waves in the Plate

As shown in Fig. 4.8, the SH guided waves in the plate only have the displacement w that is perpendicular to the cross section of the plate. SH guided waves in the plate can be considered as an antiplane problem.

The vibration of SH guided waves in the plate is relatively simple, so it is not necessary to use decomposition of displacement potential functions to solve the dispersion equations. There are analytical solutions of the dispersion equation for SH guided waves; thus, it is not necessary to obtain the numerical solutions. Furthermore, we can also obtain the analytical expression of the group velocity. The dispersion curves of SH guided waves in a steel plate with 1 mm thickness are shown in Fig. 4.9. Material parameters of the specimen are provided in Table 4.1. It can be seen from Fig. 4.9 that there is no dispersion phenomenon for the SH_0 mode, which means that both the phase velocity and the group velocity have no relation to the frequency. The SH_0 guided wave mode is more frequently selected for the special feature to detect defects.

4.2.2 Wave Structure of SH Guided Waves in the Plate

Similar to *Lamb waves*, we can also provide the wave structure of displacement and stress for the SH guided wave, as shown in Fig. 4.10. In Fig. 4.10, w^{S} represents the

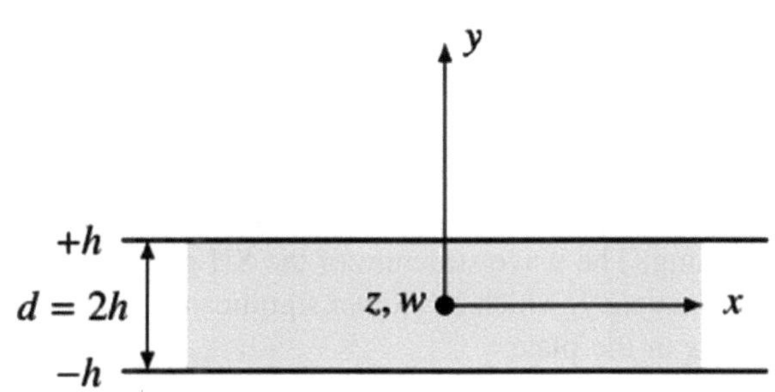

Fig. 4.8 The geometry and displacement component of SH guided waves in the free plate

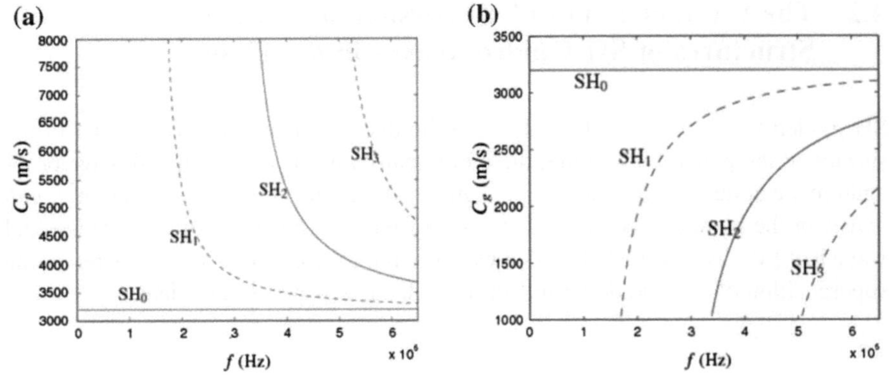

Fig. 4.9 The dispersion curves of SH guided waves in the plate. **a** Phase velocity dispersion curves. **b** Group velocity dispersion curves

Table 4.1 The parameters used in the calculation of dispersion curves

Parameter	Parameter value
Density (kg/m^3)	7800
The velocity of transverse wave C_T (m/s)	3200
The velocity of longitudinal wave C_L (m/s)	5940

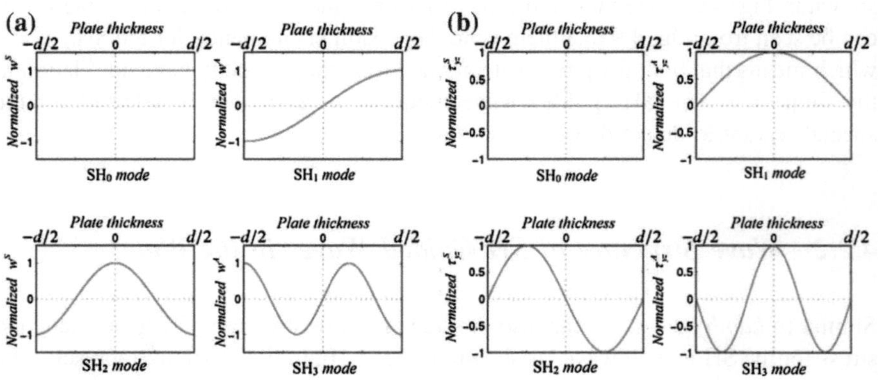

Fig. 4.10 The wave structure of SH guided waves in the plate. **a** The displacement wave structure. **b** The stress wave structure

z axis displacement of the symmetric mode and w^A represents the z axis displacement of the antisymmetric mode. The superscript of the stress has the same meaning. The wave structure of the SH guided wave in the plate has no relation to the frequency, which is another significantly different feature compared to the *Lamb waves* in the plate.

4.3 Dispersion and Wave Structure of Circumferential Lamb Waves in Pipe [1]

4.3.1 Dispersion Equations and Their Solution of Circumferential Lamb Waves in Pipe

Similar to the *Lamb waves* in the plate, the ultrasonic guided wave can propagate along the circumferential direction in the pipe, and this is called the circumferential *Lamb waves* in the pipe. The definition of the coordinates in the pipeline's circumferential direction is shown in Fig. 4.11, as well as the direction of the particle displacement components of the circumferential *Lamb waves* propagating in the pipe.

Comparing the circumferential guided wave in the pipe shown in Fig. 4.11 with the guided wave in the plate shown in Fig. 4.2, the r-direction of the pipeline is equivalent to the y-direction of the plate, and the θ-direction of the pipeline is equivalent to the x-direction of the plate. In the case of the *Lamb waves* in the elastic plate, the particle displacement component u in the x-direction and v in the y-direction are not zero. Correspondingly, in the case of the circumferential *Lamb waves* in the pipe, u_θ and u_r are not zero, while the displacement in the z-direction is zero. The physical quantities u_θ and u_r are both functions of the coordinates r and θ, as shown in the following equation.

$$u_\theta = u_\theta(r, \theta), \quad u_r = u_r(r, \theta), \quad u_z = 0$$

Accordingly, the potential functions in the cylindrical coordinate can be expanded as follows:

$$u_r = \frac{\partial \varphi}{\partial r} + \frac{1}{r}\frac{\partial \psi_Z}{\partial \theta}, \quad u_\theta = \frac{1}{r}\frac{\partial \varphi}{\partial \theta} - \frac{\partial \psi_Z}{\partial r}, \quad u_z = 0$$

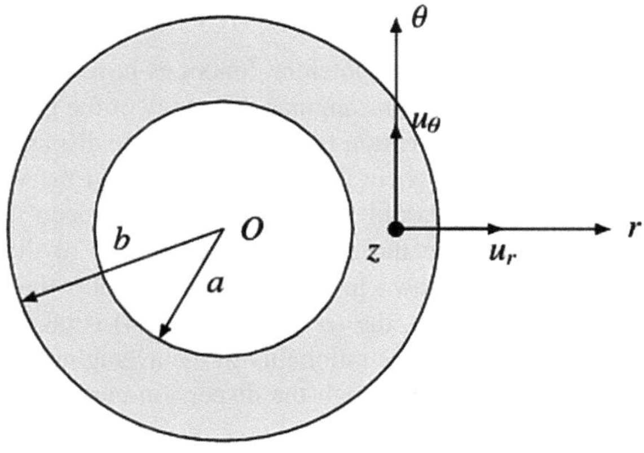

Fig. 4.11 The coordinates and particle displacement component of the circumferential *Lamb* waves in the pipe

Ignoring the subscript z in ψ_Z, we can get (4.1).

$$u_r = \frac{\partial \varphi}{\partial r} + \frac{1}{r}\frac{\partial \psi}{\partial \theta}, \quad u_\theta = \frac{1}{r}\frac{\partial \varphi}{\partial \theta} - \frac{\partial \psi}{\partial r} \tag{4.1}$$

Similar with that in the rectangular coordinate system, the wave equations are listed in (4.2).

$$\nabla^2 \varphi = \frac{1}{C_L^2}\frac{\partial^2 \varphi}{\partial t^2}, \quad \nabla^2 \psi_Z = \frac{1}{C_T^2}\frac{\partial^2 \psi_Z}{\partial t^2} \tag{4.2}$$

In the cylindrical coordinate, we have the following equation:

$$\nabla^2 \varphi = \frac{1}{r}\frac{\partial}{\partial r}\left(r\frac{\partial \varphi}{\partial r}\right) + \frac{1}{r^2}\frac{\partial^2 \varphi}{\partial \theta^2} + \frac{\partial^2 \varphi}{\partial z^2}$$

For the time-harmonic waves, we have the following conclusion:

$$\frac{\partial^2 \varphi}{\partial t^2} = -\omega^2 \varphi, \quad \frac{\partial^2 \psi_Z}{\partial t^2} = -\omega^2 \psi_Z$$

Ignoring the subscript z in ψ_z, the wave equations (4.2) can be written as in (4.3).

$$\begin{cases} \frac{1}{r}\frac{\partial \varphi}{\partial r} + \frac{\partial^2 \varphi}{\partial r^2} + \frac{1}{r^2}\frac{\partial^2 \varphi}{\partial \theta^2} = \frac{1}{C_L^2}(-\omega^2 \varphi) \\ \frac{1}{r}\frac{\partial \psi}{\partial r} + \frac{\partial^2 \psi}{\partial r^2} + \frac{1}{r^2}\frac{\partial^2 \psi}{\partial \theta^2} = \frac{1}{C_T^2}(-\omega^2 \psi) \end{cases} \tag{4.3}$$

The solutions of the potential functions are assumed as shown in (4.4).

$$\begin{cases} \varphi = \Phi(r)e^{i\left(k\frac{a+b}{2}\theta - \omega t\right)} \\ \psi = \Psi(r)e^{i\left(k\frac{a+b}{2}\theta - \omega t\right)} \end{cases} \tag{4.4}$$

It is found that the potential functions have a distribution in the cross section of the wall thickness. The angular frequency of the guided wave propagation along the circumferential direction is ω. Since particle displacements of the guided waves are distributed everywhere in the cross section of the wall thickness, the center of the wall thickness should be taken as the reference to describe its propagation arc length to average the effect. The arc length of the circumferential guided wave propagation is shown in Fig. 4.12. In the following calculations of the circumferential *Lamb waves*, the $(a+b)\theta/2$ in (4.4) is taken as the arc length in the propagation direction. The rationality of the averaging effect will be further illustrated in the following part through the dispersion curves.

Substituting (4.4) into wave equations (4.3), we have the following equation:

$$\begin{cases} \Phi''(r) + \frac{1}{r}\Phi'(\mathrm{r}) + \left[\left(\frac{\omega}{C_\mathrm{L}}\right)^2 - \left(\frac{k\frac{a+b}{2}}{r}\right)^2\right]\Phi(r) = 0 \\ \Psi''(r) + \frac{1}{r}\Psi'(\mathrm{r}) + \left[\left(\frac{\omega}{C_\mathrm{T}}\right)^2 - \left(\frac{k\frac{a+b}{2}}{r}\right)^2\right]\Psi(r) = 0 \end{cases}$$

The solutions of the equations contain *Bessel functions*.

$$\begin{cases} \Phi(r) = A_1 J_M\left(\frac{\omega r}{C_\mathrm{L}}\right) + A_2 Y_M\left(\frac{\omega r}{C_\mathrm{L}}\right) \\ \Psi(r) = B_1 J_M\left(\frac{\omega r}{C_\mathrm{T}}\right) + B_2 Y_M\left(\frac{\omega r}{C_\mathrm{T}}\right) \end{cases}$$

In the above equations, J is the first-order *Bessel function, Y* is the second-order *Bessel function*, and $M = k(a+b)/2$ is the order of the *Bessel function.*

According to the recursion equation of the *Bessel function*, we obtain the following equation:

$$\frac{\mathrm{d}\left(J_M\left(\frac{\omega r}{C_\mathrm{L}}\right)\right)}{\mathrm{d}r} = \frac{\omega}{C_\mathrm{L}} J'_M\left(\frac{\omega r}{C_\mathrm{L}}\right) = \frac{\omega}{2C_\mathrm{L}}\left(J_{M-1}\left(\frac{\omega r}{C_\mathrm{L}}\right) - J_{M+1}\left(\frac{\omega r}{C_\mathrm{L}}\right)\right)$$

$$\frac{\mathrm{d}^2\left(J_M\left(\frac{\omega r}{C_\mathrm{L}}\right)\right)}{\mathrm{d}r^2} = \frac{\omega^2}{4C_\mathrm{L}^2}\left(J_{M-2}\left(\frac{\omega r}{C_\mathrm{L}}\right) - 2J_M\left(\frac{\omega r}{C_\mathrm{L}}\right) + J_{M+2}\left(\frac{\omega r}{C_\mathrm{L}}\right)\right)$$

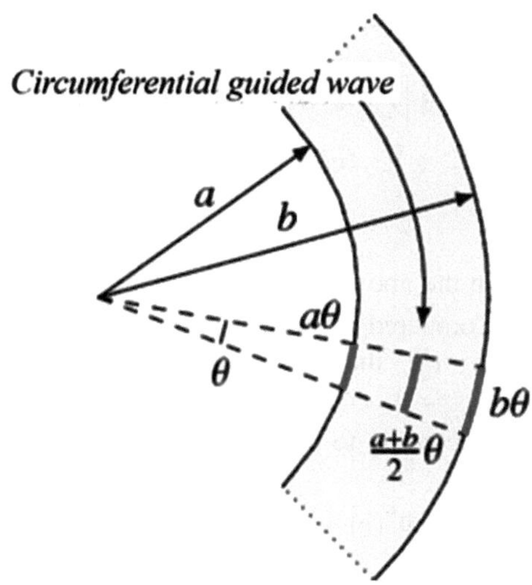

Fig. 4.12 The arc length of the circumferential guided wave propagation

So, we get (4.5).

$$\begin{cases} \Phi(r) = J_M A_1 + Y_M A_2 \\ \Phi(r) = \frac{\omega}{2C_L}(J_{M-1} - J_{M+1})A_1 + \frac{\omega}{2C_L}(Y_{M-1} - Y_{M+1})A_2 \\ \Phi''(r) = \left(\frac{\omega}{2C_L}\right)^2 (J_{M-2} - 2J_M + J_{M+2})A_1 + \left(\frac{\omega}{2C_L}\right)^2 (Y_{M-2} - 2Y_M + Y_{M+2})A_2 \\ \Psi(r) = J_M B_1 + Y_M B_2 \\ \Psi'(r) = \frac{\omega}{2C_T}(J_{M-1} - J_{M+1})B_1 + \frac{\omega}{2C_T}(Y_{M-1} - Y_{M+1})B_2 \\ \Psi''(r) = \left(\frac{\omega}{2C_T}\right)^2 (J_{M-2} - 2J_M + J_{M+2})B_1 + \left(\frac{\omega}{2C_T}\right)^2 (Y_{M-2} - 2Y_M + Y_{M+2})B_2 \end{cases} \tag{4.5}$$

In (4.5), $\omega r/C_L$ is ignored in the *Bessel function* of the first three equations, while $\omega r/C_T$ is ignored in the *Bessel function* for the last three equations.

According to the assumption of plane strain, we have the following equation:

$$\begin{cases} \sigma_r = \lambda\left(\frac{\partial u_r}{\partial r} + \frac{u_r}{r} + \frac{1}{r}\frac{\partial u_\theta}{\partial \theta}\right) + 2\mu\frac{\partial u_r}{\partial r} \\ \tau_{r\theta} = \mu\left(\frac{\partial u_\theta}{\partial r} - \frac{u_\theta}{r} + \frac{1}{r}\frac{\partial u_r}{\partial \theta}\right) \end{cases}$$

Substituting (4.4) into (4.1), we have (4.6).

$$\begin{cases} u_r = \left[\frac{\partial \Phi(r)}{\partial r} + \frac{ik\frac{a+b}{2}}{r}\Psi(r)\right] e^{i\left(k\frac{a+b}{2}\theta - \omega t\right)} \\ u_\theta = \left[\frac{ik\frac{a+b}{2}}{r}\Phi(r) - \frac{\partial \Psi(r)}{\partial r}\right] e^{i\left(k\frac{a+b}{2}\theta - \omega t\right)} \end{cases} \tag{4.6}$$

So, the expression of σ_r and $\tau_{r\theta}$ is given as follows:

$$\begin{cases} \sigma_r = \frac{\mu}{r^2}\left[\chi^2 r^2 \Phi''(r) + (\chi^2 - 2)\left(r\Phi'(r) - k^2\left(\frac{a+b}{2}\right)^2 \Phi(r)\right) + 2ik\frac{a+b}{2}(r\Psi'(r) - \Psi(r))\right] \\ \tau_{r\theta} = \frac{\mu}{r^2}\left[-r^2\Psi''(r) + r\Psi'(r) - k^2\left(\frac{a+b}{2}\right)^2 \Psi(r) + 2ik\frac{a+b}{2}(r\Phi'(r) - \Phi(r))\right] \end{cases} \tag{4.7}$$

In the above equation, $\chi = \frac{C_L}{C_T}$.

Compared with the *Lamb waves* in the plate, the traction-free boundary conditions of the circumferential guided wave in the pipe require that $\sigma_r|_{r=a,b} = 0,\ \tau_{r\theta}|_{r=a,b} = 0$.

According to the boundary conditions, we can get the following equation.

$$\begin{cases} \chi^2 r^2 \Phi''(r) + (\chi^2 - 2)\left(r\Phi'(r) - k^2\left(\frac{a+b}{2}\right)^2 \Phi(r)\right) + 2ik\frac{a+b}{2}(r\Psi'(r))\Big|_{r=a,b} = 0 \\ -r^2\Psi''(r) + r\Psi'(r) - k^2\left(\frac{a+b}{2}\right)^2 \Psi(r) + 2ik\frac{a+b}{2}(r\Phi'(r) - \Phi(r))\Big|_{r=a,b} = 0 \end{cases}$$

The above equation set is actually composed of four equations, and the value of *r*, *a*, and *b*, corresponds to two equations, respectively. A_1, A_2, B_1, and B_2 in each equation are the unknowns. The determinant composed by the coefficients A_1, A_2, B_1, and B_2 is set as *zero*, and the frequency dispersion equation can be solved. The dispersion equation should be solved by a numerical method.

If the ratio of the inner and outer diameters ($\eta = a/b$), which is used as the representation of the pipe curvature, takes different values, the results of the dispersion curves will also be different. In this book, investigations and discussions are made at the condition of the value of η at 80 and 95 %, respectively. For different values of η, the thickness of the pipe is kept as 1 mm, as shown:

1. When η = 80 %, a = 4 mm and b = 5 mm;
2. When η = 95 %, a = 19 mm and b = 20 mm.

The velocities of the bulk waves used in the calculations C_L and C_T are also provided in Table 4.1. The dispersion curves obtained based on two different values of η are shown in Figs. 4.13 and 4.14, respectively.

The frequency thickness products in the dispersion curves are the same as that of the *Lamb waves* in the plate, as shown in Fig. 4.3. It can be seen from the derivation of the dispersion equations and calculation of the dispersion curves that there is no symmetric and antisymmetric modes like that of the *Lamb waves* in plate. From the dispersion curves, the modes $CLamb_0$, $CLamb_1$, and $CLamb_2$ of the circumferential *Lamb wave* in the pipe are similar to the modes A_0, S_0, and A_1 of the *Lamb waves* in the plate. Another important finding is as follows: The dispersion curves of the circumferential *Lamb waves* in the pipe gradually tend to those of the *Lamb waves* in the plate as the value of η increases. Relevant data show that the wall thickness of a steel pipeline with an external diameter of 1016 mm is 14.6–26.2 mm; these are used in the *China* national gas transmission project. The corresponding value of the ratio of the outer and inner diameters is 94.84–97.13 %. In this situation ($\eta \approx 100$ %), the dispersion curves of the circumferential guided waves in the pipe

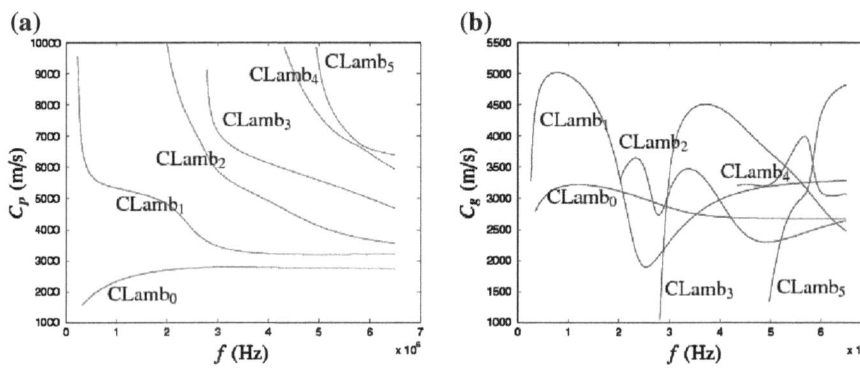

Fig. 4.13 Dispersion curves of the circumferential *Lamb waves* in the pipe (η = 80 %). **a** Phase velocity dispersion curves. **b** Group velocity dispersion curves

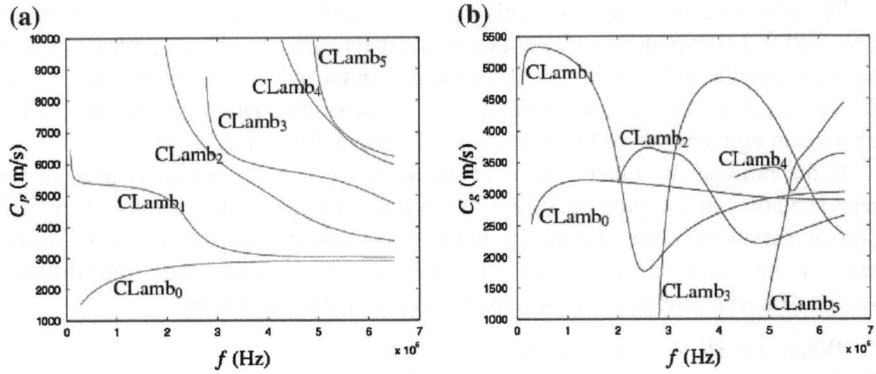

Fig. 4.14 Dispersion curves of the circumferential *Lamb waves* in the pipe (η = 95 %). **a** Phase velocity dispersion curves. **b** Group velocity dispersion curves

is very close to that of the *Lamb wave* in the plate. Thus, it is practicable to apply the methods used in the plate in the pipelines, which will make the theoretical analysis and actual physical experiments much more convenient.

In the derivation of the dispersion equation of the circumferential *Lamb waves* in the pipe, the three parameters $a\theta$, $(a+b)\theta/2$, and $b\theta$ can be used to determine the arc length, where $a\theta$ and $b\theta$ represent the inner and outer arc lengths of the pipe, respectively. Both of them are the extreme case of the arc length calculation; both these two extreme cases cannot deny each other, since if one of the two cases is reasonable, the other should also be feasible. The method to solve this conflict uses the average value $(a+b)/2$ to calculate the arc length along the propagation direction. For the case (η = 80 %), the comparison of three phase velocity dispersion curves calculated with three different arc lengths is shown in Fig. 4.15. It is found that there are some differences among the three sets of dispersion curves, and the dispersion curves calculated with the $(a+b)/2$ arc length is located in the middle of the curves calculated with the arc length of the two extreme cases. Overall, it is obviously more reasonable to use the average value for the calculation.

From another view point, the reasonability of using the average value of the inner and outer diameter of the pipe for the arc length calculation can also be verified. Figure 4.16 shows that the dispersion curve of the *CLamb*$_0$ mode of the circumferential *Lamb waves* in the pipe is approaching the A_0 *Lamb wave* mode in the dispersion curve of the *Lamb wave* in the plate. It can be seen from Fig. 4.16, when the value η remains constant, the *CLamb*$_0$ mode frequency dispersion curve calculated with the average radius is in the middle of the other two dispersions obtained with the two extreme cases. There is a good agreement between this phenomenon and that of the former case; in addition, when the average radius is selected, the *CLamb*$_0$ mode in the dispersion curve shows a better approach to the

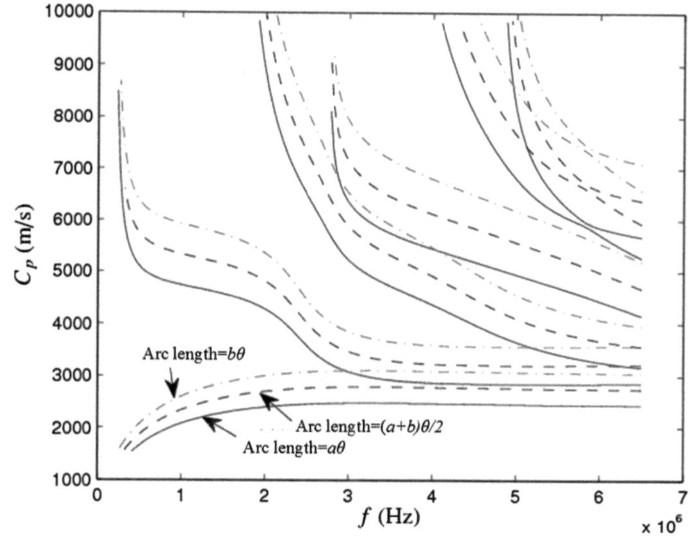

Fig. 4.15 Comparisons of dispersion curves when η = 80 %

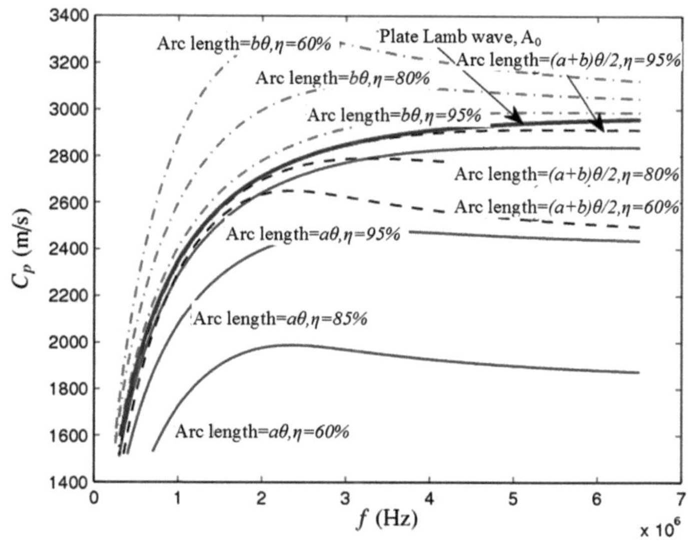

Fig. 4.16 Dispersion curve of circumferential $CLamb_0$ guided wave modes approaching A_0 *Lamb wave* mode in the plate with the increase in η

A_0 mode in the dispersion curve with the increase in η compared to the other two extreme cases, which also illustrates the reasonability of selecting the average value of the inner and outer diameters to do the calculation.

4.3.2 Wave Structure of Circumferential Lamb Waves in the Pipe

According to (4.5) and (4.6), the expressions of displacement can be calculated using the following equation:

$$\frac{u_r}{B_2}=\left[\frac{\omega}{2C_L}\left(J_{M-1}\left(\frac{\omega r}{C_L}\right)-J_{M+1}\right)\right]\frac{A_1}{B_2}+\left[\frac{\omega}{2C_L}(Y_{M-1}-Y_{M+1})\right]\frac{A_2}{B_2}+\frac{ik\frac{a+b}{2}}{r}\left[J_M\left(\frac{\omega r}{C_T}\right)\frac{B_1}{B_2}+Y_M\right]$$

$$\frac{u_\theta}{B_2}=\frac{ik\frac{a+b}{2}}{r}\left[J_M\left(\frac{\omega r}{C_L}\right)\frac{A_1}{B_2}+Y_M\frac{A_2}{B_2}\right]-\left[\frac{\omega}{2C_T}\left(J_{M-1}\left(\frac{\omega r}{C_T}\right)-J_{M+1}\right)\right]\frac{B_1}{B_2}-\left[\frac{\omega}{2C_T}(Y_{M-1}-Y_{M+1})\right]$$

According to (4.5) and (4.7), the expressions of the stress can be calculated using the following equation:

$$\frac{\sigma_r}{\mu}=\frac{1}{r^2}\left[\begin{array}{l}\chi^2r^2\left[\left(\frac{\omega}{2C_L}\right)^2(J_{M-2}-2J_M+J_{M+2})A_1+\left(\frac{\omega}{2C_L}\right)^2(Y_{M-2}-2Y_M+Y_{M+2})A_2\right]\\+(\chi^2-2)r\left[\frac{\omega}{2C_L}(J_{M-1}-J_{M+1})A_1+\frac{\omega}{2C_L}(Y_{M-1}-Y_{M+1})A_2\right]\\-(\chi^2-2)k^2\left(\frac{a+b}{2}\right)^2(J_MA_1+Y_MA_2)\\+2ik\frac{a+b}{2}\left(r\left[\frac{\omega}{2C_T}(J_{M-1}-J_{M+1})B_1+\frac{\omega}{2C_T}(Y_{M-1}-Y_{M+1})B_2\right]-[J_MB_1+Y_MB_2]\right)\end{array}\right]$$

$$\frac{\tau_{r\theta}}{\mu}=\frac{1}{r^2}\left[\begin{array}{l}-r^2\left[\left(\frac{\omega}{2C_T}\right)^2(J_{M-2}-2J_M+J_{M+2})B_1+\left(\frac{\omega}{2C_T}\right)^2(Y_{M-2}-2Y_M+Y_{M+2})B_2\right]\\+r\left[\frac{\omega}{2C_T}(J_{M-1}-J_{M+1})B_1+\frac{\omega}{2C_T}(Y_{M-1}-Y_{M+1})B_2\right]\\-k^2\left(\frac{a+b}{2}\right)^2(J_MB_1+Y_MB_2)\\+2ik\frac{a+b}{2}\left(r\left[\frac{\omega}{2C_L}(J_{M-1}-J_{M+1})A_1+\frac{\omega}{2C_L}(Y_{M-1}-Y_{M+1})A_2\right]-[J_MA_1+Y_MA_2]\right)\end{array}\right]$$

In the expressions of stress, the parameter of the *Bessel function* with the terms A_1 and A_2 is wr/C_L, while the *Bessel function* with the terms B_1 and B_2 is wr/C_T. The above displacement expressions can be normalized by the maximum value of two directions to obtain the wave structures of the displacement. The expressions of stress are divided by B_2 to get $\sigma_r/\mu B_2$ and $\sigma_{r\theta}/\mu B_2$ to obtain the stress wave structures. The wave structures of stress can be obtained by normalizing the maximum value in two directions.

Similar to the dispersion curves of circumferential *Lamb waves* in the pipe, the wave structure will also be different if the η (the ratio of the inner and outer diameters of the pipe) used to illustrate the curvature of the pipe is different. Two cases with 80 and 95 % of the η value, as well as 1 mm thickness of the pipe wall, are investigated and discussed in this book. The frequencies selected are 600, 1800, 3000, and 4200 kHz, respectively.

The wave structures expressed by the displacement and stress for the two cases with different values of η are shown in Figs. 4.17, 4.18, 4.19, and 4.20. Compared with the displacement wave structure of the A_0 *Lamb wave* mode in the plate shown in Fig. 4.5a, the wave structure of the circumferential $CLamb_0$ *Lamb* wave mode in the pipe shown in 4.17a and 4.19b clearly shows that the displacement wave structure of *Lamb waves* in the plate is symmetric within the range of the plate thickness, and the symmetry is independent of the product of frequency and thickness. But the displacement wave structures of the circumferential *Lamb wave* in the pipe are not strictly symmetric, and the symmetry will get worse with the increase in frequency. In addition, the displacement wave structure of the circumferential *Lamb wave* in the pipe is approaching that of the *Lamb wave* in the

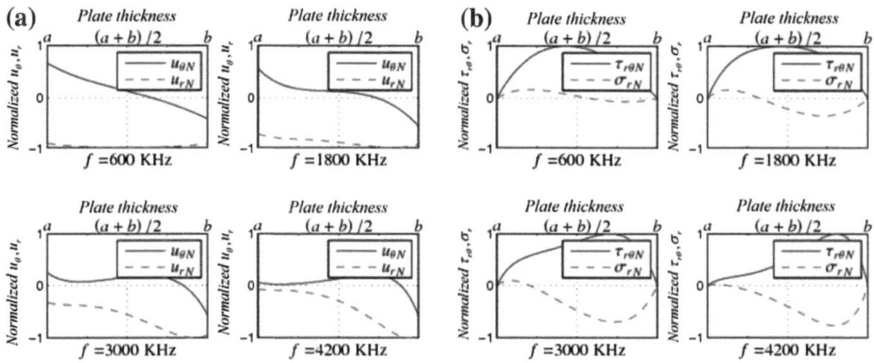

Fig. 4.17 The wave structure of the $CLamb_0$ mode in the pipe when $\eta = 80\ \%$. **a** The displacement wave structure. **b** The stress wave structure

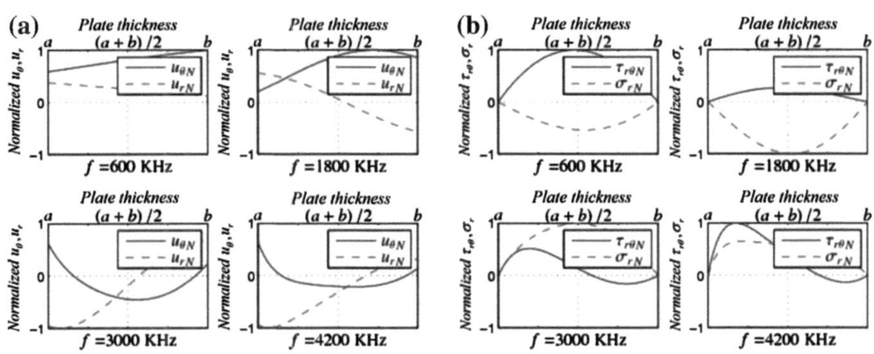

Fig. 4.18 The wave structure of the $CLamb_1$ mode in the pipe when $\eta = 80\ \%$. **a** The displacement wave structure. **b** The stress wave structure

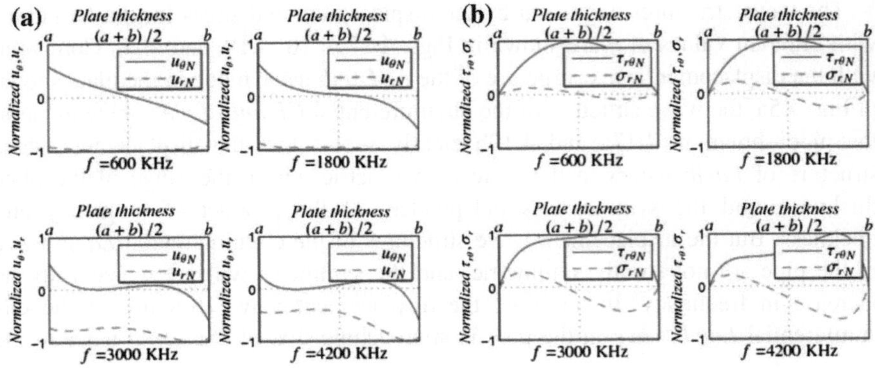

Fig. 4.19 The wave structure of the $CLamb_0$ mode in the pipe when $\eta = 95$ %. **a** The displacement wave structure. **b** The stress wave structure

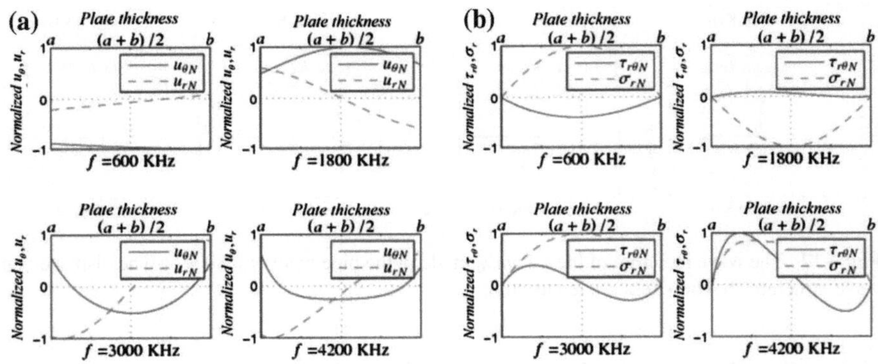

Fig. 4.20 The wave structure of the $CLamb_1$ mode in the pipe when $\eta = 95$ %. **a** The displacement wave structure. **b** The stress wave structure

plate when the value η is close to 100 %. The stress wave structure has the same trend. Similar results can also be obtained by the comparison between the S_0 *Lamb wave* mode in the plate and the $CLamb_1$ circumferential *Lamb wave* mode in the pipe. Besides, it is found that the distribution of the stress of the circumferential *Lamb wave* is *zero* at both the inner and outer surfaces of the wall, which satisfies the traction-free boundary conditions.

Based on *Lamb waves* in the plate, the normalized displacement components distributed in the inner surface of the pipe are shown in Fig. 4.21, at the condition of $\eta = 95$ %. These *3D* curve diagrams play a significant role in the mode selection and frequency tuning of testing.

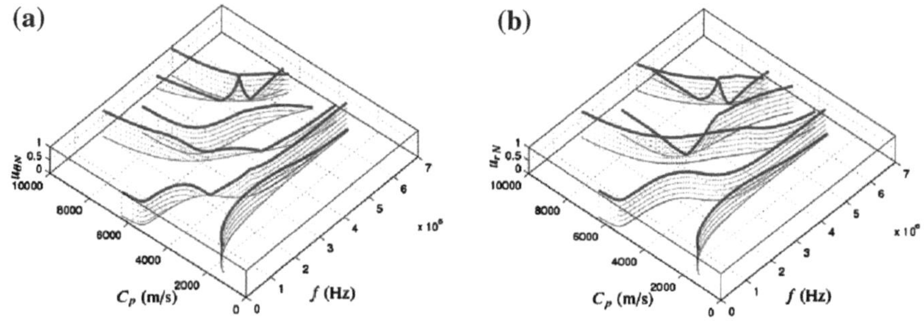

Fig. 4.21 Normalized displacement distributions of the modes in the dispersion curves of circumferential *Lamb waves* in the pipe when η = 95 %. **a** Displacement u_θ. **b** Displacement u_r

4.4 Dispersion and Wave Structure of Circumferential SH Guided Waves in the Pipe [2]

4.4.1 *The Dispersive Equations and Solutions of the Circumferential SH Guided Waves in the Pipe*

The ultrasonic guided waves can propagate along the circumferential direction, which is similar to the SH guided wave in the plate and is called the circumferential SH guided wave in the pipe. Similar to that of circumferential *Lamb waves* in pipe, the propagation arc length of the derived expression is modified to get more accurate results.

The circumferential coordinate system in the pipe and the displacement direction of the particle vibration of the circumferential SH guided wave are shown in Fig. 4.22.

In the case of the circumferential SH guided wave, particle displacement would be in the z-direction with no displacement in the r- or θ-directions,

$$u_r = u_\theta = 0, \quad u_z \neq 0$$

As u_z is the function of r, θ, and t, the equation of particle motion is assumed as shown in (4.8).

$$u_z(r,\theta,t) = f(r)e^{i\left(k\frac{a+b}{2}\theta-\omega t\right)} \tag{4.8}$$

This means that there is a certain distribution of u_z in the direction of r and it propagates along the direction of θ; the angular frequency of the particle vibration in the process of propagation is ω. For the calculation of the arc length along the propagation direction, $b\theta$ is used in some references. However, this value of the arc length corresponds to the outer diameter of the pipe, which is an extreme case; the

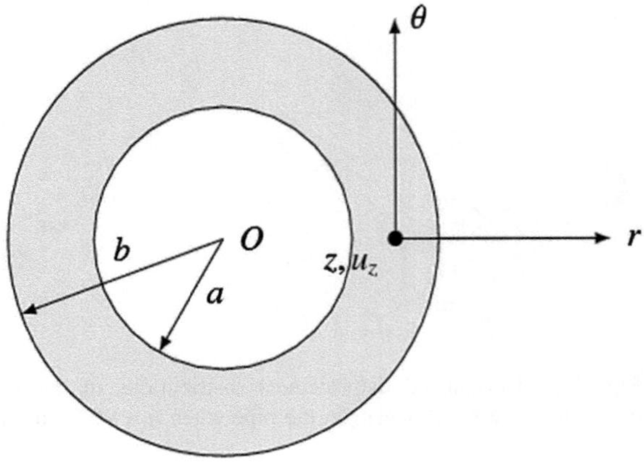

Fig. 4.22 The circumferential coordinate system and displacement direction of particle vibration of the circumferential SH guided wave in a pipe structure

propagation arc length is corrected as $(a+b)\theta/2$, which is more reasonable on average. This point will be illustrated later by the dispersion curves.

Since $u_r = u_\theta = 0$, and u_z is not the function of z, the divergence can be expressed as shown in (4.9).

$$\nabla \cdot u = \frac{1}{r}u_r + \frac{\partial u_r}{\partial r} + \frac{1}{r}\frac{\partial u_\theta}{\partial \theta} + \frac{\partial u_z}{\partial z} = \frac{\partial u_z}{\partial z} = 0 \tag{4.9}$$

The curl of displacement can be expressed as follows:

$$\begin{aligned}\nabla \times u &= \frac{1}{r}\begin{vmatrix} e_r & re_\theta & e_z \\ \frac{\partial}{\partial r} & \frac{\partial}{\partial \theta} & \frac{\partial}{\partial z} \\ u_r & ru_\theta & u_z \end{vmatrix} \\ &= e_r\frac{1}{r}\left[\frac{\partial u_Z}{\partial \theta} - \frac{\partial(ru_\theta)}{\partial z}\right] + e_\theta\left(\frac{\partial u_r}{\partial z} - \frac{\partial u_Z}{\partial r}\right) + e_z\frac{1}{r}\left[\frac{\partial(ru_\theta)}{\partial r} - \frac{\partial u_r}{\partial \theta}\right] \\ &= e_r\frac{1}{r}\left(\frac{\partial u_Z}{\partial \theta}\right) + e_\theta\left(-\frac{\partial u_Z}{\partial r}\right)\end{aligned}$$

So, we have the following equation:

$$\nabla \times \nabla \times u = e_z\left(-\frac{1}{r}\frac{\partial u_z}{\partial r} - \frac{\partial^2 u_z}{\partial r^2} - \frac{1}{r^2}\frac{\partial^2 u_z}{\partial \theta^2}\right)$$

Now, the *Naiver* governing equation can be transformed into the following equation:

$$\mu\left(\frac{1}{r}\frac{\partial u_z}{\partial r} + \frac{\partial^2 u_z}{\partial r^2} + \frac{1}{r^2}\frac{\partial^2 u_z}{\partial \theta^2}\right) = \rho\frac{\partial^2 u_z}{\partial t^2}$$

Substituting (4.8) into the above equation, we get the following equation:

$$f''(r) + \frac{1}{r}f'(r) + \left[\frac{\rho\omega^2}{\mu} - \frac{k^2\left(\frac{a+b}{2}\right)^2}{r^2}\right]f(r) = 0$$

It can also expressed as follows:

$$f''(r) + \frac{1}{r}f'(r) + \left[\left(\frac{\omega}{C_{\mathrm{T}}}\right)^2 - \left(\frac{k\frac{a+b}{2}}{r}\right)\right]f(r) = 0$$

The above function is a *Bessel equation,* and its solution is a *Bessel function*, as shown in the following form, where $M = k(a+b)/2$ is the order of the *Bessel function*.

$$f(r) = AJ_M\left(\frac{wr}{C_{\mathrm{T}}}\right) + BY_M\left(\frac{wr}{C_{\mathrm{T}}}\right)$$

Compared with the SH guided wave in the infinite plate, the boundary condition is given as follows:

$$\tau_{rz}|_{r=a,b} = 0$$

So, we get the following equation:

$$\tau_{rz} = \mu\left(\frac{\partial u_r}{\partial z} + \frac{\partial u_z}{r}\right)$$

where $u_r = 0$ is known, and (4.8) is given, so we can obtain the following equation:

$$\tau_{rz} = \mu f'(r)\mathrm{e}^{i\left(k\frac{a+b}{2}\theta - \omega t\right)}$$

So, the traction-free boundary condition is equivalent to $f'(r)|_{r=a,b} = 0$, as given as follows:

$$f'(r) = A\frac{\mathrm{d}\left(J_M\left(\frac{\omega r}{C_{\mathrm{T}}}\right)\right)}{\mathrm{d}r} + B\frac{\mathrm{d}\left(Y_M\left(\frac{\omega r}{C_{\mathrm{T}}}\right)\right)}{\mathrm{d}r}$$

According to the recurrence relation of the *Bessel function*, we have the following equation:

$$\frac{\mathrm{d}\left(J_M\left(\frac{\omega r}{C_{\mathrm{T}}}\right)\right)}{\mathrm{d}r} = \frac{\omega}{C_{\mathrm{T}}}\left[\frac{M}{\frac{\omega r}{C_{\mathrm{T}}}}J_M\left(\frac{\omega r}{C_{\mathrm{T}}}\right) - J_{M+1}\left(\frac{\omega r}{C_{\mathrm{T}}}\right)\right] = \frac{M}{r}J_M\left(\frac{\omega r}{C_{\mathrm{T}}}\right) - \frac{\omega}{C_{\mathrm{T}}}J_{M+1}\left(\frac{\omega r}{C_{\mathrm{T}}}\right)$$

So, we get the following:

$$f'(r) = \left[\frac{M}{r}J_M\left(\frac{\omega r}{C_T}\right) - \frac{\omega}{C_T}J_{M+1}\left(\frac{\omega r}{C_T}\right)\right]A + \left[\frac{M}{r}Y_M\left(\frac{\omega r}{C_T}\right) - \frac{\omega}{C_T}Y_{M+1}\left(\frac{\omega r}{C_T}\right)\right]B$$

The expression of stress is as follows:

$$\mu\left\{\left[\frac{M}{r}J_M\left(\frac{\omega r}{C_T}\right) - \frac{\omega}{C_T}J_{M+1}\left(\frac{\omega r}{C_T}\right)\right]A + \left[\frac{M}{r}Y_M\left(\frac{\omega r}{C_T}\right) - \frac{\omega}{C_T}Y_{M+1}\left(\frac{\omega r}{C_T}\right)\right]B\right\}e^{i\left(k\frac{a+b}{2}\theta-\omega t\right)}$$

According to the boundary condition, we can get a homogeneous equation set, as shown in (4.10).

$$\begin{cases}\left[\frac{M}{a}J_M\left(\frac{\omega a}{C_T}\right) - \frac{\omega}{C_T}J_{M+1}\left(\frac{\omega a}{C_T}\right)\right]A + \left[\frac{M}{a}Y_M\left(\frac{\omega a}{C_T}\right) - \frac{\omega}{C_T}Y_{M+1}\left(\frac{\omega a}{C_T}\right)\right]B = 0 \\ \left[\frac{M}{a}J_M\left(\frac{\omega b}{C_T}\right) - \frac{\omega}{C_T}J_{M+1}\left(\frac{\omega b}{C_T}\right)\right]A + \left[\frac{M}{b}Y_M\left(\frac{\omega b}{C_T}\right) - \frac{\omega}{C_T}Y_{M+1}\left(\frac{\omega b}{C_T}\right)\right]B = 0\end{cases} \tag{4.10}$$

The determinant of their coefficient matrix equals zero and expressed as follows:

$$\begin{vmatrix}\left[\frac{M}{a}J_M\left(\frac{\omega a}{C_T}\right) - \frac{\omega}{C_T}J_{M+1}\left(\frac{\omega a}{C_T}\right)\right] & \left[\frac{M}{a}Y_M\left(\frac{\omega a}{C_T}\right) - \frac{\omega}{C_T}Y_{M+1}\left(\frac{\omega a}{C_T}\right)\right] \\ \left[\frac{M}{b}J_M\left(\frac{\omega b}{C_T}\right) - \frac{\omega}{C_T}J_{M+1}\left(\frac{\omega b}{C_T}\right)\right] & \left[\frac{M}{b}Y_M\left(\frac{\omega b}{C_T}\right) - \frac{\omega}{C_T}Y_{M+1}\left(\frac{\omega b}{C_T}\right)\right]\end{vmatrix} = 0$$

The above equation is the dispersion equation of the circumferential SH guided wave in the pipe.

The numerical method is used to solve the dispersion equation by solving the corresponding phase velocity c_p for a certain frequency f. The group velocity dispersion curve can also be calculated by a method of numerical calculation. Similar to the circumferential *Lamb waves* in the pipe, the dispersion curves of the circumferential SH guided waves in the pipe are also related to the value of η. The dispersion curves when $\eta = 80$ and 95 % are shown in Figs. 4.23 and 4.24 with a 1 mm thickness of the pipe wall. Based on the derivation of dispersion equations and the calculation of the dispersion curves, it is found that, unlike the SH guided wave in the plate, there is no symmetric and antisymmetric modes for the circumferential SH guided wave in the pipe. Considering the shape of the dispersion curves, the modes CSH_0, CSH_1, and CSH_2 of the circumferential SH guided waves in the pipe correspond to the modes SH_0, SH_1, and SH_2 of the SH guided waves in the plate. Another important rule is that the dispersion curves of the circumferential SH guided waves in the pipe approache the SH guided wave in the plate as η increases.

In the derivation of the dispersion equation of the circumferential SH guided waves in the pipe, the three parameters $a\theta$, $(a+b)\theta/2$, and $b\theta$ can be used to determine the arc length, where $a\theta$ and $b\theta$ represent the inner and outer arc lengths of the pipe, respectively. For the case (η = 80 %), a comparison of the three stes of phase velocity dispersion curves calculated with the three different arc lengths is

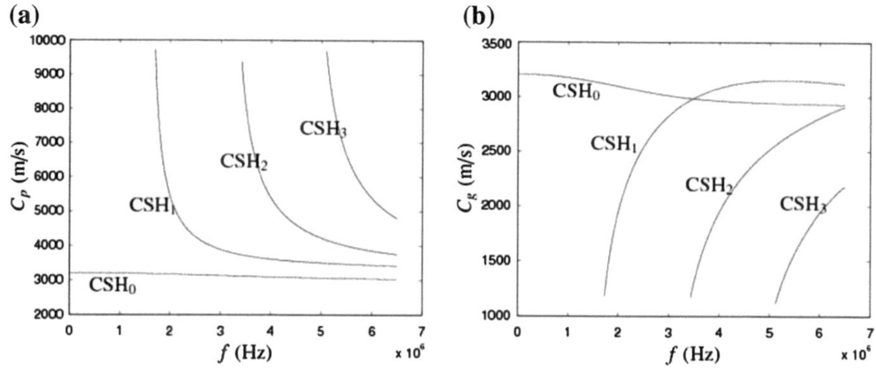

Fig. 4.23 The dispersion curves of circumferential SH guided waves in the pipe when $\eta = 80\,\%$. **a** The phase velocity dispersion curves. **b** The group velocity dispersion curves

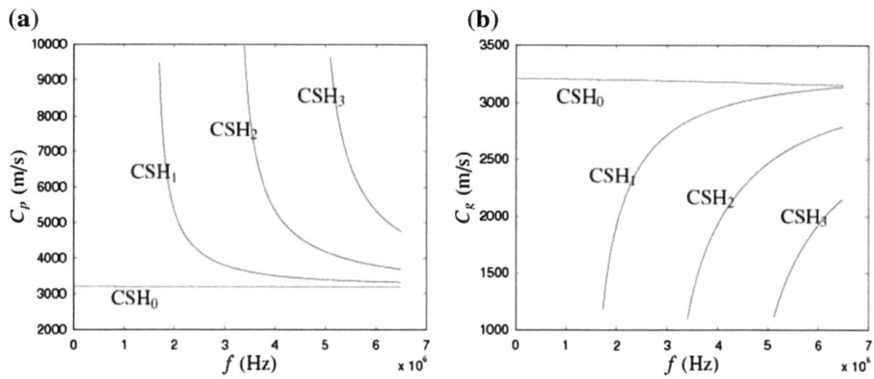

Fig. 4.24 The dispersion curves of circumferential SH guided waves in the pipe when $\eta = 95\,\%$. **a** The phase velocity dispersion curves. **b** The group velocity dispersion curves

shown in Fig. 4.25. It is discovered that similar results can be found to those of the circumferential *Lamb waves* in the pipe. It is obviously more reasonable to use the average value $(a + b)/2$ for the calculation.

From another view point, the reasonability of using the average value of the inner and outer diameters of the pipe for the arc length calculation can also be verified. Figure 4.26 shows that the dispersion curve of the CSH_0 mode of the circumferential SH guided wave in the pipe approaches the dispersion curve of the SH_0 wave mode of the SH guided wave in the plate. It can be seen from Fig. 4.26, when the value η is constant the CSH_0 mode frequency dispersion curve, calculated with the average radius, is in the middle of the other two curves obtained from the two extreme cases. There is a good agreement between this phenomenon and that of former case; in addition, when the average radius is selected, the CSH_0 mode in the dispersion curve shows a better approach to the SH_0 mode in the dispersion curve

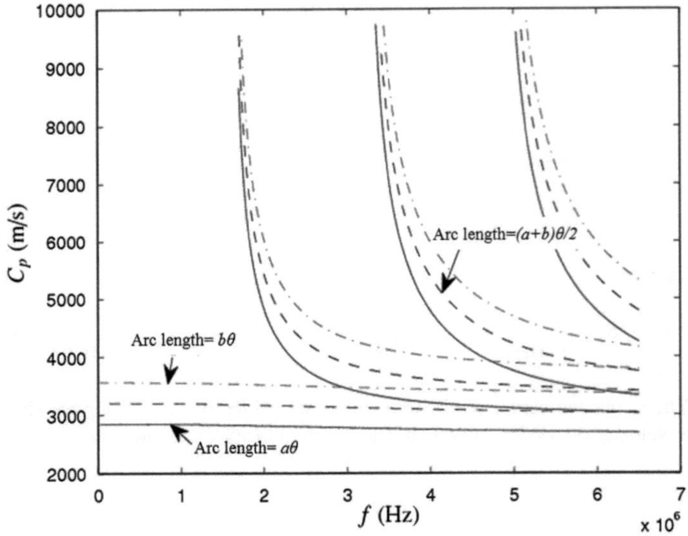

Fig. 4.25 The dispersion curves of circumferential SH guided waves in the pipe when $\eta = 80\ \%$

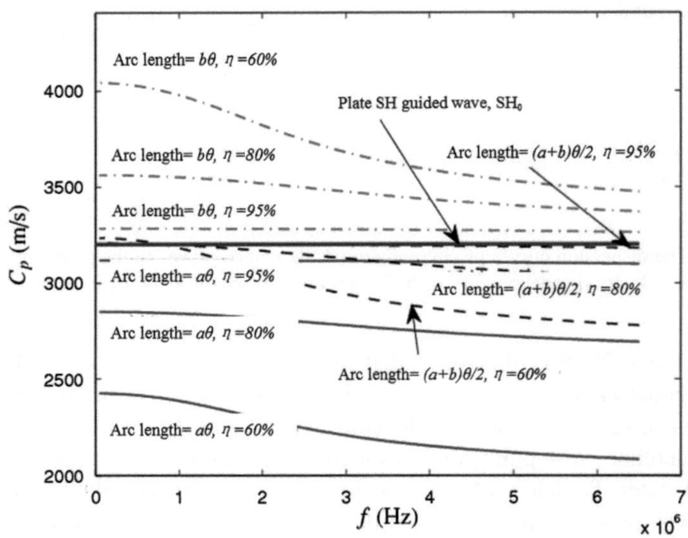

Fig. 4.26 The dispersion curves of the circumferential CSH_0 guided wave mode in the pipe with three different arc length values and increases of η

as η increases, compared with the other two extreme cases. This also illustrates the reasonability of selecting the average value of the inner and outer diameters to do the calculation.

4.4.2 Wave Structure of Circumferential SH Guided Waves in the Pipe

The expressions of displacement for the circumferential SH guided waves are given in (4.8), and we have the following equation, where $M = k(a + b)/2$ is the order of the *Bessel function*:

$$f(r) = AJ_M\left(\frac{\omega r}{C_\mathrm{T}}\right) + BY_M\left(\frac{\omega r}{C_\mathrm{T}}\right)$$

From the system of homogeneous equations (4.10), we can get the following:

$$\frac{A}{B} = -\frac{\frac{M}{a}Y_M\left(\frac{\omega a}{C_\mathrm{T}}\right) - \frac{\omega}{C_\mathrm{T}}Y_{M+1}\left(\frac{\omega a}{C_\mathrm{T}}\right)}{\frac{M}{a}J_M\left(\frac{\omega a}{C_\mathrm{T}}\right) - \frac{\omega}{C_\mathrm{T}}J_{M+1}\left(\frac{\omega a}{C_\mathrm{T}}\right)}$$

Therefore, the displacement distribution can be given as follows:

$$\frac{f(r)}{B} = -\frac{\frac{M}{a}Y_M\left(\frac{\omega a}{C_\mathrm{T}}\right) - \frac{\omega}{C_\mathrm{T}}Y_{M+1}\left(\frac{\omega a}{C_\mathrm{T}}\right)}{\frac{M}{a}J_M\left(\frac{\omega a}{C_\mathrm{T}}\right) - \frac{\omega}{C_\mathrm{T}}J_{M+1}\left(\frac{\omega a}{C_\mathrm{T}}\right)}J_M\left(\frac{\omega r}{C_\mathrm{T}}\right) + Y_M\left(\frac{\omega r}{C_\mathrm{T}}\right)$$

The stress distribution is given as follows:

$$\begin{aligned}\frac{\tau_{rz}}{\mu B} = & \left[\frac{M}{r}Y_M\left(\frac{\omega r}{C_\mathrm{T}}\right) - \frac{\omega}{C_\mathrm{T}}Y_{M+1}\left(\frac{\omega r}{C_\mathrm{T}}\right)\right] \\ & - \frac{\frac{M}{a}Y_M\left(\frac{\omega a}{C_\mathrm{T}}\right) - \frac{\omega}{C_\mathrm{T}}Y_{M+1}\left(\frac{\omega a}{C_\mathrm{T}}\right)}{\frac{M}{a}J_M\left(\frac{\omega a}{C_\mathrm{T}}\right) - \frac{\omega}{C_\mathrm{T}}J_{M+1}\left(\frac{\omega a}{C_\mathrm{T}}\right)}\left[\frac{M}{r}J_M\left(\frac{\omega r}{C_\mathrm{T}}\right) - \frac{\omega}{C_\mathrm{T}}J_{M+1}\left(\frac{\omega r}{C_\mathrm{T}}\right)\right]\end{aligned}$$

The expressions of stress are divided by B_2 to get $\sigma_r/\mu B_2$ and $\sigma_{r\theta}/\mu B_2$ to obtain the displacement wave structures. The wave structures of stress can be obtained by normalizing the maximum value in two directions.

The maximum of the absolute values of $f(r)/B$ and $\tau_{rz}/\mu B$ is normalized to obtain the displacement and stress wave structures of the circumferential SH guided waves in the pipe. Similar to the circumferential *Lamb waves*, the wave structures under different diameter ratio η will be different. The following discussions of two cases are based on $\eta = 85\ \%$ and $\eta = 95\ \%$. The values of the frequencies under test are listed in Table 4.2.

Figures 4.27a and 4.27b show the curves of the CSH_0 mode displacement and stress wave structures of the circumferential SH guided waves in the pipe, respectively. Compared to the SH guided waves in the plate shown in Fig. 4.10a, b, the displacement and stress distributions of the mode in the pipe are similar to those

Table 4.2 The tested frequencies of the pipeline circumferential SH guided wave structure (η = 80, 95 %)

Number	CSH_0 frequency (KHz)	CSH_1 frequency (KHz)
1	600	2100
2	1800	3000
3	3000	3900
4	4200	4800

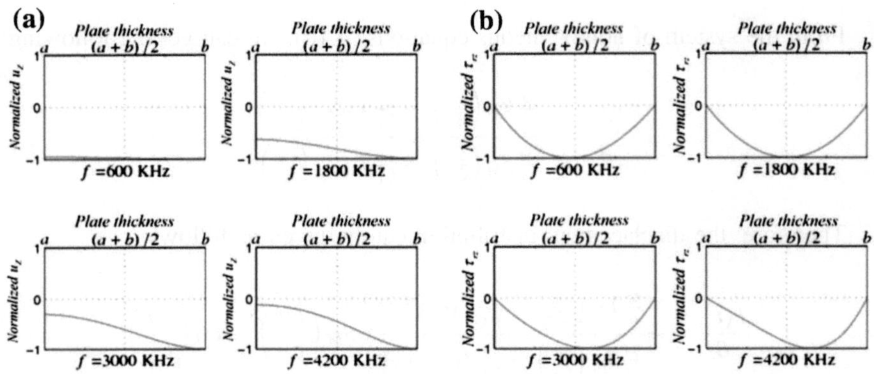

Fig. 4.27 The wave structures of the CSH_0 mode in the pipe when η = 80 %. **a** The displacement wave structure. **b** The stress wave structure

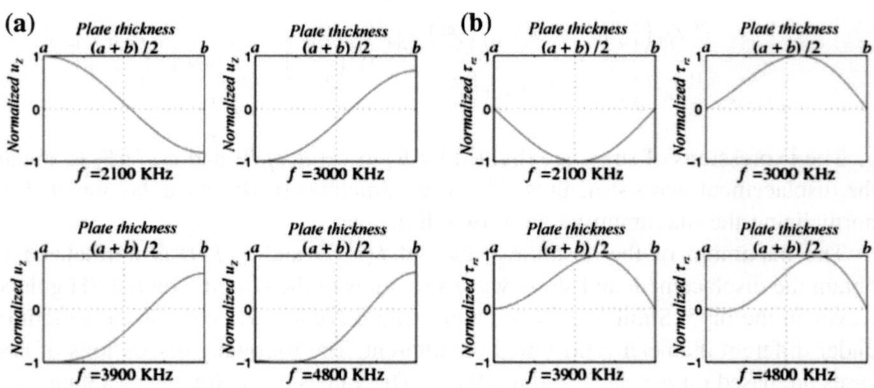

Fig. 4.28 The wave structures of the CSH_1 mode in the pipe when η = 80 %. **a** The displacement wave structure. **b** The stress wave structure

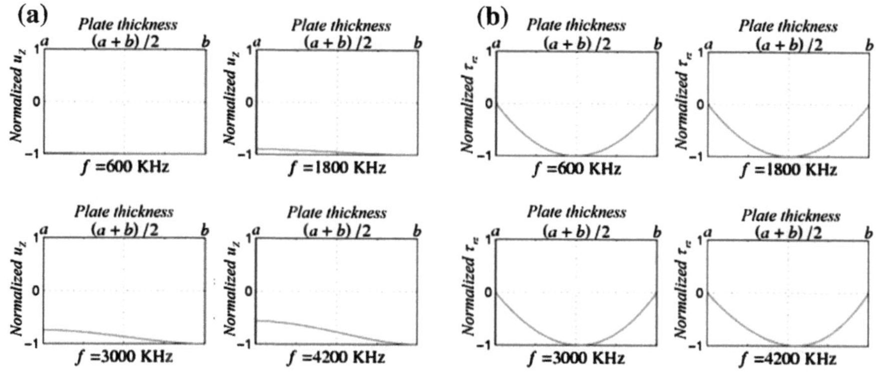

Fig. 4.29 The wave structures of the CSH_0 mode in the pipe when $\eta = 95$ %. **a** The displacement wave structure. **b** The stress wave structure

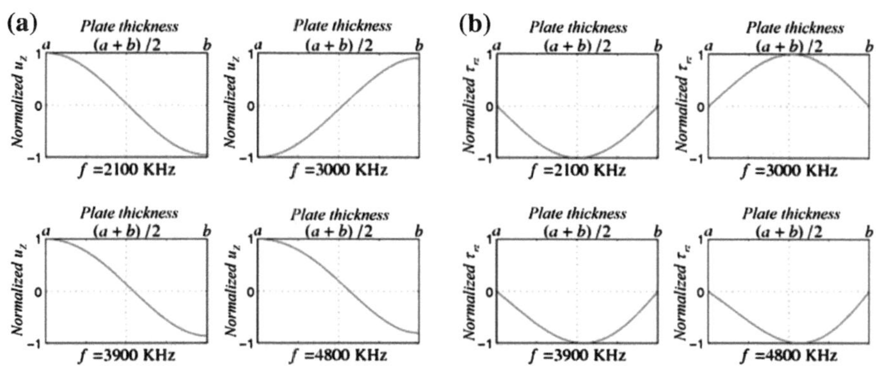

Fig. 4.30 The wave structures of the CSH_1 mode in the pipe when $\eta = 95$ %. **a** The displacement wave structure. **b** The stress wave structure

in the plate when the frequency is relatively low, and the difference increases with the increase in frequency between that in the plate and the pipe. As shown in Fig. 4.28, there is a similar conclusion for the high-order CSH_1 mode. For $\eta = 95$ %, the wave structures of the CSH_0 and CSH_1 modes are shown in Figs. 4.29 and 4.30. It is found that the wave structure of the circumferential SH guided wave in the pipe is more similar with that of the plate in this situation.

In the curves of stress wave structures of the circumferential SH guided wave, the stresses on the inner and outer wall surface are zero, which is consistent with the traction-free boundary condition in solving the dispersion equation.

4.5 Comparison of the Propagation Characteristics Between Guided Waves in the Plate and Circumferential Guided Waves in the Pipe

It can be concluded, based on the above theoretical derivation and numerical calculation, that *Lamb waves* in the plate have some characteristics in common with circumferential *Lamb waves* in the pipe, and the SH guided waves in the plate have some of the characteristics of the circumferential SH guided waves in the pipe. The similarities are listed as follows:

1. Their directions of displacement and stress are similar.

It can be seen from the derivation of the dispersion equations that both the *Lamb waves* in the plate and circumferential *Lamb waves* in the pipe have two displacement components in the cross section; both the SH guided waves in the plate and circumferential SH guided waves in the pipe have only one displacement component, which is perpendicular to the cross section. Furthermore, the stress directions of the guided waves in the plate and circumferential guided waves in the pipe are also similar.

2. The dispersion curves are similar.

It can be seen from the dispersion curves that the shapes of the dispersion curves of circumferential guided waves in the pipe approach those of guided waves in the plate as the parameter η (the ratio of inner and outer diameters in pipe) get close to 100 %.

3. The wave structures are similar.

The characteristics of wave structures of circumferential guided waves in the pipe approach those of guided waves in the plate as the parameter η (the ratio of inner and outer diameters in pipe) get close to 100 %.

A guided wave mode with a certain frequency (or frequency thickness product *fd*) is determined as a working point for the guided wave test, the working point corresponding to a certain point on the curves of the dispersion curve.

The purposes of the dispersion curve are as follows: the phase velocity dispersion curve is used to design the parameters of transducer; the group velocity dispersion curves are used to verify the working conditions, and the suitable working point is selected based on the shape of the dispersion curves. Wave structure also plays a significant role in the guided wave test. For example, when the piezoelectric transducer is used to generate the *Lamb waves* in the plate, the out-of-plane displacement on the surface of the plate should be as large as possible so that the energy can be coupled smoothly into the tested specimen; but for the immersion testing, the out-of-plane displacement on the surface of the plate should be as small as possible to minimize the leakage of energy.

References

1. Shen, W., Songling, H., Wei, Z.: Plate and pipe circumferential Lamb waves dispersion and wave structure characteristics. J Tsinghua Univ **49**(7), 925–928 (2009). (In Chinese)
2. Shen, W., Songling, H., Wei, Z.: Study on pipe circumferential SH guided wave characteristics. China Mech Eng **20**(2), 146–149 (2009). (In Chinese)

Chapter 5
Simulation of Interactions Between Guided Waves and the Defects by Boundary Element Method

Dispersion curves and wave structures of guided wave propagation are just the fundamental knowledge for detecting defects by guided waves, since the dispersion curves and wave structures only provide the information of guided wave propagation in an ideal situation without any disturbance from defects and guided wave reflectors. For defect detection by guided waves generated by the electromagnetic acoustic transducers, the studies of the interactions between ultrasonic guided waves and defects, in other words, the investigations of mode conversion and scattering when the guided waves meet the defects, are also important for guided wave nondestructive testing. There are two methods of studying the interactions between guided waves and the defects: theoretical and experimental approaches. Since there are many unexpected factors in experimental tests that could affect the stability and repeatability of results, it is difficult to get the results with good reproducibility, consistency, and accuracy just based on the experiments. However, pure analytical methods are severely limited and difficult to use to analyze the complex problems. So, in general, numerical simulation methods are considered to be the best way to study the interactions between guided waves and defects.

In this chapter, the boundary element method is used to perform the simulations of the interactions between *Lamb waves* and the defects, which are all crack-like defects, such as rectangular and triangular grooves on the surface of the plate as well as rectangular grooves inside the plate. The boundary element model used here is specific to the circumstance of the elastic plate. The boundary element model follows the hybrid boundary element methods which Younho Cho used in earlier works, and modifications and improvements are made based on it to solve the current problems. Some critical issues of guided wave testing are investigated based on the simulated results.

S. Huang et al., *Electromagnetic Ultrasonic Guided Waves*, Springer Series in Measurement Science and Technology, DOI 10.1007/978-981-10-0564-0_5

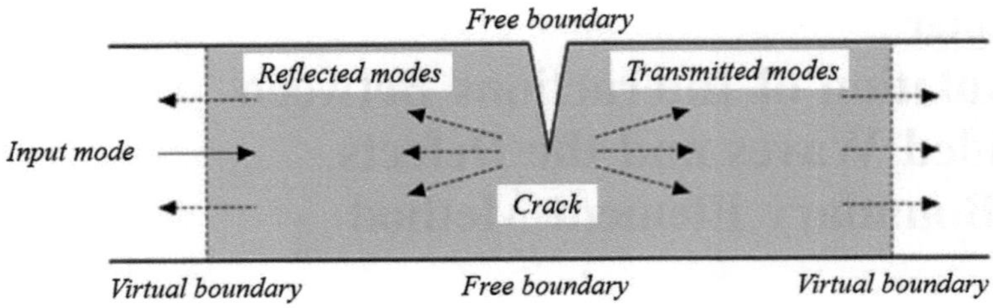

Fig. 5.1 BEM model for the external defect

5.1 Hybrid BEM Model of the External Defects in a Plate

The hybrid BEM model used to study the interactions of *Lamb waves* with the external defects in an elastic plate is shown in Fig. 5.1. As for the assumptions in the derivation of the *Lamb waves* in the plate, the model of the hybrid BEM corresponds to the problem of plane strain. A cross section of an infinite plate is given in Fig. 5.1. In the direction perpendicular to the cross section, the plate and triangle defects all stretch infinitely and the shape remains unchanged in this direction, which means that the triangle defect on the cross section is actually the infinitely extending triangle groove defect in the direction that is perpendicular to the cross section. The displacement components of the input and the reflected and transmitted *Lamb waves* are all in the cross section, and those distributions can be extended infinitely in the direction that is perpendicular to the cross section.

As shown in Fig. 5.1, a closed region for the BEM formulation consisting of the top and bottom traction-free boundaries, two virtual boundaries, and the defect surface is formed. The left and right virtual boundaries are adopted to form a closed region for the BEM simulation. A chosen pure mode *Lamb wave* is input into the left boundary and then scattering occurs at the crack defect. The target of the simulation is to find out how much of the input energy has been distributed to the reflected and transmitted modes, respectively, during this process of mode conversion.

5.2 Elastodynamic Integration Equation and Its Fundamental Solution

Lamb wave propagation is related to wave motion, so the elastodynamic BEM method should be used to solve the problem. Neglecting the effect of body force, the elastodynamic integration equation is provided from the weighted residual method, as shown in (5.1).

$$\begin{bmatrix} u(P) \\ v(P) \end{bmatrix} + \int_{\Gamma} \begin{bmatrix} T_{11}(q,P) & T_{12}(q,P) \\ T_{21}(q,P) & T_{22}(q,P) \end{bmatrix} \begin{bmatrix} u(q) \\ v(q) \end{bmatrix} \mathrm{d}\Gamma_q$$
$$= \int_{\Gamma} \begin{bmatrix} U_{11}(q,P) & U_{12}(q,P) \\ U_{21}(q,P) & U_{22}(q,P) \end{bmatrix} \begin{bmatrix} t_x(q) \\ t_y(q) \end{bmatrix} \mathrm{d}\Gamma_q \tag{5.1}$$

In the above equation, T_{jk} and U_{jk} (j, k = 1, 2) are the elastodynamic traction force and displacement fundamental solutions, representing the generated traction force and displacement at the observation point q in k direction when a unit force is applied at the source point p in j direction; u and v are the displacement components in the x and y directions and t_x and t_y are the traction force components in the x and y directions.

Two different forms of elastodynamic fundamental solutions can be found. The first form is expressed using a kind of Hankel function, and the displacement solution is given in (5.2).

$$U_{jk} = \frac{i}{4\mu}\left(U_1\delta_{jk} - U_2 r_{,j} r_{,k}\right), \quad j,k = 1,2 \tag{5.2}$$

The traction force fundamental solution is given in (5.3).

$$\begin{aligned} T_{jk} = \frac{i}{4}\Bigg\{ & \left[\left(\delta_{jk} r_{,n} + n_j r_{,k}\right) + \frac{\lambda}{\mu} n_k r_{,j}\right]\frac{\mathrm{d}U_1}{\mathrm{d}r} \\ & - \left(2 r_{,j} r_{,k} r_{,n} + \frac{\lambda}{\mu} n_k r_{,j}\right)\frac{\mathrm{d}U_2}{\mathrm{d}r} \\ & - \left[\left(\delta_{jk} r_{,n} + n_j r_k\right) + 2\left(n_k r_{,j} - 2 r_{,j} r_{,k} r_{,n}\right) + \frac{\lambda}{\mu} n_k r_{,j}\right]\frac{U_2}{r}\Bigg\} \end{aligned} \tag{5.3}$$

In the above equation, we have

$$U_1 = H_0^{(1)}(k_\mathrm{T} r) - \frac{1}{k_\mathrm{T} r} H_1^{(1)}(k_\mathrm{T} r) + \left(\frac{k_\mathrm{L}}{k_\mathrm{T}}\right)^2 \frac{1}{k_\mathrm{L} r} H_1^{(1)}(k_\mathrm{L} r)$$

$$U_2 = -H_2^{(1)}(k_\mathrm{T} r) + \left(\frac{k_\mathrm{L}}{k_\mathrm{T}}\right)^2 H_2^{(1)}(k_\mathrm{L} r)$$

$$k_\mathrm{T} = \frac{\omega}{C_\mathrm{T}}, \quad k_\mathrm{L} = \frac{\omega}{C_\mathrm{L}}$$

r is the distance from the source point to the field point; the "," in the subscripts refers to the derivative, for example, $r_{,j} = \partial r/\partial x_j$, $j = 1, 2$; it is also possible to use x and y to represent the two directions in the two-dimensional coordinates; that is, x is the same as x_1; y is the same at x_2; λ and μ are the *Lame constants*; C_T is the velocity of the shear wave; C_L is the velocity of the longitudinal wave; and $H_n^{(1)}$ is the n rank *Hankel function*.

The second forms uses the second kind of modified *Bessel function* K as the fundamental solution of the elastodynamic frequency domain. The displacement fundamental solution is given in (5.4).

$$U_{jk} = \frac{1}{2\pi\mu}\left(\psi\delta_{jk} - \chi r_{,j} r_{,k}\right) \tag{5.4}$$

The traction force fundamental solution is given in (5.5).

$$T_{jk} = \frac{1}{2\pi}\left[\left(\psi_{,r} - \frac{\chi}{r}\right)\left(\delta_{jk} r_{,n} + r_{,k} n_j\right) - \frac{2\chi}{r}\left(n_k r_{,j} - 2 r_{,j} r_{,k} r_{,n}\right) - 2\chi_{,r} r_{,j} r_{,k} r_{,n} + \left(\psi_{,r} - \chi_{,r} - \frac{\chi}{r}\right)\left(\frac{C_L^2}{C_T^2} - 2\right)\left(r_{,j} n_k\right)\right] \tag{5.5}$$

In the above equation,

$$\psi = K_0\left(\frac{i\omega r}{C_T}\right) + \frac{C_T}{i\omega r}\left[K_1\left(\frac{i\omega r}{C_T}\right) - \frac{C_T}{C_L} K_1\left(\frac{i\omega r}{C_L}\right)\right]$$

$$\chi = K_2\left(\frac{i\omega r}{C_T}\right) - \left(\frac{C_T}{C_L}\right)^2 K_2\left(\frac{i\omega r}{C_L}\right)$$

where K_n is the n rank second kind of modified *Bessel function K*.

In the hybrid BEM method introduced in this chapter, the two forms of elastodynamic fundamental solutions lead to similar results. The form containing the second kind modified *Bessel function* K_n would be used in the simulation examples in this study.

5.3 Boundary Integration Equation and Its Discretized Numerical Solution

The integration equation is used to express the relationship between the internal points and the boundary points in the plate. If the internal points are moved to the boundary, the integration equation must be adjusted to the boundary integration equation as given in (5.6).

$$\begin{bmatrix} \frac{1}{2} & 0 \\ 0 & \frac{1}{2} \end{bmatrix}\begin{bmatrix} u(p) \\ v(p) \end{bmatrix} + \int_\Gamma \begin{bmatrix} T_{11}(p,q) & T_{12}(p,q) \\ T_{21}(p,q) & T_{22}(p,q) \end{bmatrix}\begin{bmatrix} u(q) \\ v(q) \end{bmatrix} \mathrm{d}\Gamma_q = \int_\Gamma \begin{bmatrix} U_{11}(p,q) & U_{12}(p,q) \\ U_{21}(p,q) & U_{22}(p,q) \end{bmatrix}\begin{bmatrix} t_x(q) \\ t_y(q) \end{bmatrix} \mathrm{d}\Gamma_q \tag{5.6}$$

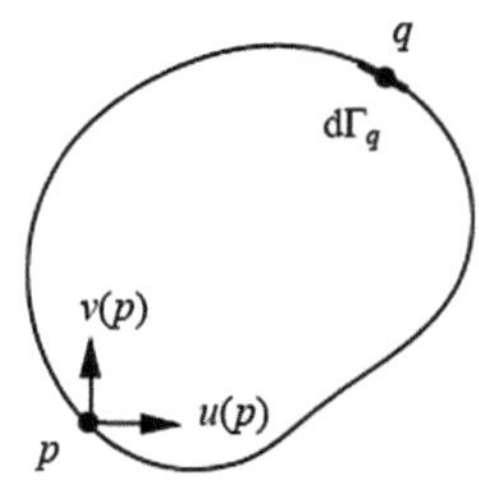

Fig. 5.2 The source point and observation point on the boundary

In the above equation, p is the source point on the boundary and q is the observation point on the boundary, as shown in Fig. 5.2. The coefficient matrix $\begin{bmatrix} \frac{1}{2} & 0 \\ 0 & \frac{1}{2} \end{bmatrix}$ corresponds to the smooth boundary element.

After obtaining the boundary integration equation, the next step is the discretization of this equation, forming the linear equation system, which includes the unknown values of unit displacement and traction force.

In the next step, it is assumed that the boundary is divided into N elements. The integration on the whole boundary can be divided into N integrations of these elements, and we get (5.7), where Γ_j represents the jth boundary element.

$$\frac{1}{2}\begin{bmatrix} u(p) \\ v(p) \end{bmatrix} + \sum_{j=1}^{N} \int\limits_{\Gamma_j} \begin{bmatrix} T_{11} & T_{12} \\ T_{21} & T_{22} \end{bmatrix} \begin{bmatrix} u(q) \\ v(q) \end{bmatrix} d\Gamma_j = \sum_{j=1}^{N} \int\limits_{\Gamma_j} \begin{bmatrix} U_{11} & U_{12} \\ U_{21} & U_{22} \end{bmatrix} \begin{bmatrix} t_x(q) \\ t_y(q) \end{bmatrix} d\Gamma_j \tag{5.7}$$

Several approaches can be used for the boundary discretization, each method corresponding to a unique geometry of the boundary elements and the interpolating functions with certain orders. For the 2D cases, the constant element with one nodal point included in each element is the simplest situation. This type of element transforms the boundary discretization of the solution domain into the line segments that are connected together, and the displacement and traction force on the line elements are assumed to be constant, which means that the physical quantities within the elements are constants. The constant elements are the rough description of the physical model. The linear elements that have two nodal points are also line segments, but the physical quantity within the elements has a linear variation; the geometry of the quadratic elements that have three nodal points are a quadratic curve segment, and the physical quantities of the elements will also change according to the quadratic polynomial. The three elements mentioned above are shown in Fig. 5.3.

Constant elements are used in the BEM program of this book for the reason of the relative simple establishment of the equations based on the constant elements; and the boundaries are line segments when only triangle defects and rectangular defects are considered. From the view point of the geometry, the constant elements

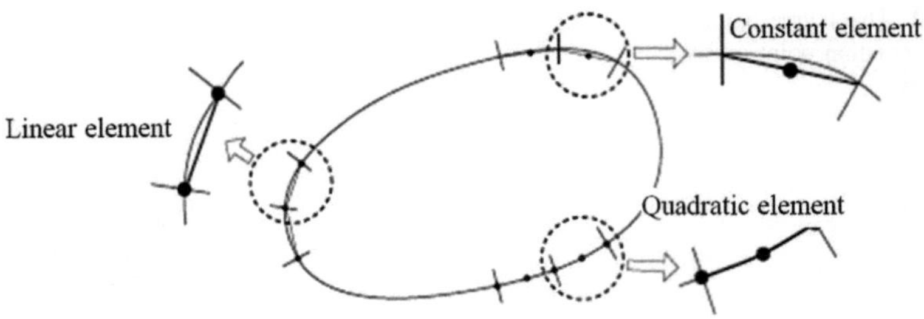

Fig. 5.3 The constant element, linear element, and quadratic element

can describe accurately the problem; the lower accuracy caused by using the constant elements can be recovered by increasing the number of boundary elements appropriately. Previous studies have shown that the constant elements are good enough to characterize and solve the problems of general wave motion.

As mentioned above, for the constant elements, the displacement and traction force within each boundary element are assumed to be unchanged, so when we solve integrations of each boundary element according to (5.7), the constant displacement and traction force in the element can be extracted from the integration, as given in (5.8).

$$\frac{1}{2}u^i + \sum_{j=1}^{N} \hat{H}_{ij}u^j = \sum_{j=1}^{N} G_{ij}t^j \tag{5.8}$$

In the above equation, the items are as follows:

$$u^i = \begin{bmatrix} u \\ v \end{bmatrix}^i, \quad t^j = \begin{bmatrix} t_x \\ t_y \end{bmatrix}^j,$$
$$G_{ij} = \begin{bmatrix} \int\limits_{\Gamma_j} U_{11}\mathrm{d}\Gamma_j & \int\limits_{\Gamma_j} U_{12}\mathrm{d}\Gamma_j \\ \int\limits_{\Gamma_j} U_{21}\mathrm{d}\Gamma_j & \int\limits_{\Gamma_j} U_{22}\mathrm{d}\Gamma_j \end{bmatrix}, \quad H_{ij} = \begin{bmatrix} \int\limits_{\Gamma_j} T_{11}\mathrm{d}\Gamma_j & \int\limits_{\Gamma_j} T_{12}\mathrm{d}\Gamma_j \\ \int\limits_{\Gamma_j} T_{21}\mathrm{d}\Gamma_j & \int\limits_{\Gamma_j} T_{22}\mathrm{d}\Gamma_j \end{bmatrix} \tag{5.9}$$

And in the above equations, i is the element index. In fact, (5.8) contains two equations in the x and y directions. Now we traverse i in the range $1 \leq i \leq N$, then $2N$ equations can be obtained and arranged as $\mathbf{HU} = \mathbf{GT}$, where $\mathbf{U}$ is the column vector composed by the displacement on every element node and $\mathbf{T}$ is the column vector composed by the traction force on every element node, as shown in (5.10).

$$\begin{cases} U = [u_1\ v_1\ u_2\ v_2 \ldots u_N\ v_N]^{\mathrm{T}} \\ T = [t_{x1}\ t_{y1}\ t_{x2}\ t_{y2} \ldots t_{xN}\ t_{yN}]^{\mathrm{T}} \end{cases} \tag{5.10}$$

Also, we have (5.11), where I is the identity matrix.

$$\mathbf{H} = \hat{\mathbf{H}} + \frac{1}{2}\mathbf{I} \tag{5.11}$$

After obtaining (5.9), the elements of matrix $\hat{\mathbf{H}}$ and $\mathbf{G}$ could be calculated. For the constant elements, the calculation is actually to solve the integration of the fundamental solution.

5.3.1 *The Solution of the Elements in Matrix G*

Matrix $\mathbf{G}$ consists of the 2×2 submatrix $\mathbf{G}_{ij}$, the subscripts i and j correspond to the excitation point and field point, respectively. Singular integrals exist in the diagonal submatrix $\mathbf{G}_{kk}$, ($k = 1, 2, \ldots, N$), and they need special treatment. Two methods are described here to solve the singular integrals in $\mathbf{G}_{kk}$.

(1) If $i \neq j$: the non-diagonal blocks of $\mathbf{G}$ should be calculated. In this case, the nodal point is not located on element j and the calculation is a standard Gauss integration. Taking the 4 points *Gaussian* quadrature as an example, as shown in Fig. 5.4, the used integral point local coordinates and coefficients are listed as follows:

$$\begin{cases} \xi_1 = -\xi_2 = 0.86113631, & \xi_3 = -\xi_4 = 0.33998104 \\ w_1 = w_2 = 0.34785485, & w_3 = w_4 = 0.65214515 \end{cases}$$

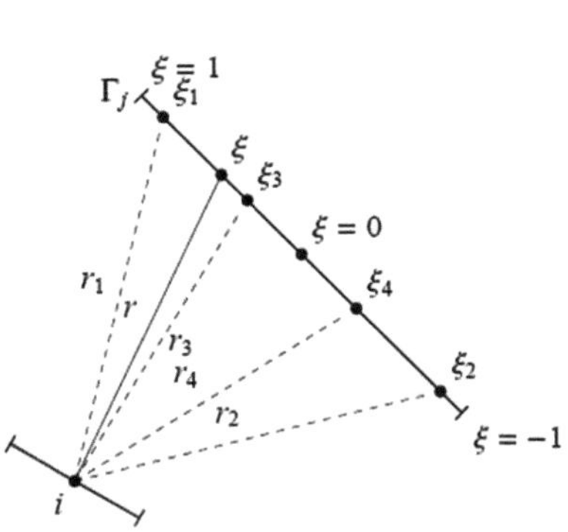

Fig. 5.4 The singular integration when $i \neq j$

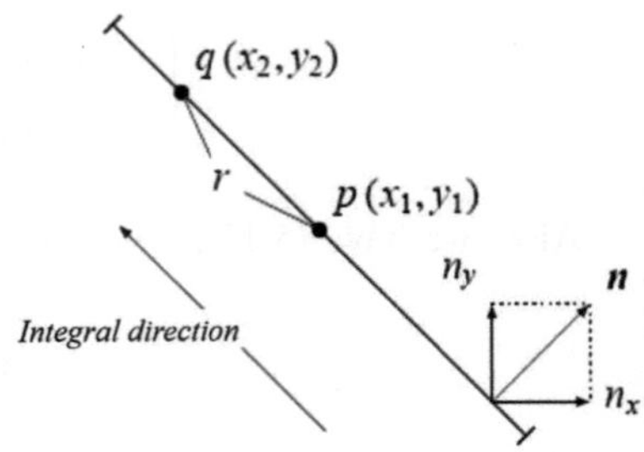

Fig. 5.5 The circumstance of the singular integral when $i = j$

(2) If $i = j$: as shown in Fig. 5.5, the nodal point i is located on the element j. Considering the fundamental solutions of the dynamic displacement (5.2) or (5.4), in the process of integration, if the distance r between the integral point and nodal point (the center point of the element) is approaching zero, the integrand function will tend to infinity; therefore, the integration has singularity. Some special treatment should be made to solve the singular integrals, especially for the complex situation when the *Bessel function* is involved in the integrand function.

Two different methods are used to solve the singular integrals in the BEM program, one of which is based on the expansion of the fundamental solution of dynamic displacement and the other is based on the principle of singularity elimination by differencing between the singular dynamic fundamental solution and the singular static fundamental solution.

The expression of the dynamic displacement fundamental solution which involves the *Hankel function* is given in (5.12).

$$U_{jk} = \frac{i}{4\mu}\left(\hat{U}_1\delta_{jk} - \hat{U}_2 r_{,j} r_{,k}\right), \quad \hat{U}_1 = U_1^0 + U_1^D, \; \hat{U}_2 = U_2^0 + U_2^D \tag{5.12}$$

In the above equation, we have the following relationship:

$$U_0^1 = \frac{i(\lambda+3\mu)}{\pi(\lambda+2\mu)}\ln r, \quad U_2^0 = \frac{i(\lambda+\mu)}{\pi(\lambda+2\mu)}$$

$$\begin{aligned} U_1^D = {} & \frac{1}{2(\lambda+2\mu)}\left\{\lambda+3\mu+\frac{i}{\pi}\left[\lambda+\mu+2(\lambda+3\mu)\gamma+2(\lambda+2\mu)\ln\frac{k_\mathrm{T}}{2}+2\mu\ln\frac{k_\mathrm{L}}{2}\right]\right\} \\ & +\sum_{n=1}^{\infty}\frac{(-1)^n}{(n!)^2}\left(\frac{k_\mathrm{T} r}{2}\right)^{2n}\left\{\frac{1}{2(n+1)}\left[2n+1+\left(\frac{k_\mathrm{L}}{k_\mathrm{T}}\right)^{2n+2}\right]A+B \right. \\ & \left. -\frac{1}{2(n+1)}D+\frac{1}{2(n+1)}D+\frac{1}{2(n+1)}\left[1-\left(\frac{k_\mathrm{L}}{k_\mathrm{T}}\right)^{2n+2}\right]E\right\} \end{aligned}$$

$$U_2^D = \sum_{n=1}^{\infty} \frac{(-1)^n}{(n!)^2} \left(\frac{k_{\mathrm{T}} r}{2} \right)^{2n} \left\{ \frac{n}{n+1} \left[1 - \left(\frac{k_{\mathrm{L}}}{k_{\mathrm{T}}} \right)^{2n+2} \right] A + B \right.$$
$$\left. - \left(\frac{k_{\mathrm{L}}}{k_{\mathrm{T}}} \right)^{2n+2} C - \frac{n}{n+1} D + \frac{1}{n+1} \left[1 - \left(\frac{k_{\mathrm{L}}}{k_{\mathrm{T}}} \right)^{2n+2} \right] E \right\}$$

$$A = 1 + \frac{2i}{\pi}(\ln r + \gamma), \quad B = \frac{2i}{\pi}\left[\ln\frac{k_{\mathrm{T}}}{2} - \varphi(n) \right],$$
$$C = \frac{2i}{\pi}\left[\ln\frac{k_{\mathrm{L}}}{2} - \varphi(n) \right], \quad D = \frac{2i}{\pi}\left[\ln\frac{k_{\mathrm{T}}}{2} - \left(\frac{k_{\mathrm{L}}}{k_{\mathrm{T}}} \right)^{2n+2} \ln\frac{k_{\mathrm{L}}}{2} \right]$$
$$E = \frac{i}{\pi}[\varphi(n) + \varphi(n+1)], \quad \varphi(n) = \sum_{m=1}^{n} \frac{1}{m},$$
$$\gamma = 0.57721$$

Then the original displacement fundamental solution can be decomposed into two parts, as shown in (5.13).

$$\begin{aligned} U_{jk} &= U_{jk}^0 + U_{jk}^D \\ U_{jk}^0 &= \frac{i}{4\mu}\left(U_1^0 \delta_{jk} - U_2^0 r_{,j} r_{,k} \right) \\ U_{jk}^D &= \frac{i}{4\mu}\left(U_1^D \delta_{jk} - U_2^D r_{,j} r_{,k} \right) \end{aligned} \tag{5.13}$$

It can be verified that U_{jk}^0, determined by U_1^0 and U_2^0 as one part of the elastodynamic displacement fundamental solution, is the same as the following elastostatic displacement fundamental solution.

As a reference, the fundamental solution of the elastostatic displacement and traction force is given as shown in (5.14) and (5.15).

$$U_{jk}^0 = \frac{1}{8\pi\mu(1-v)} \left[(3-4v) \ln\frac{1}{r} \delta_{jk} + r_{,j} r_{,k} \right] \tag{5.14}$$

$$T_{jk}^0 = -\frac{1}{4\pi(1-v)r} \left[\frac{\partial r}{\partial n} \left[(1-2v)\delta_{jk} + 2 r_{,j} r_{,k} \right] + (1-2v)\left(n_j r_{,k} - n_k r_{,j} \right) \right] \tag{5.15}$$

In the above equation, v is the *Poisson ratio* of the elastic materials. The relations between the *Lame constants*, *Poisson ratio* v, and elastic modulus E are given in (5.16).

$$\mu = \frac{E}{2(1+v)}, \quad \lambda = \frac{vE}{(1+v)(1-2v)} \tag{5.16}$$

Based on (5.16), the consistency of U^0_{jk} in (5.14) and (5.13) can be proved.

Because of ln$r$, the part U^0_{jk} that is the same as the elastostatic fundamental solution in the elastodynamic fundamental solution shows a singularity in the solving process of integration. The integral of the elastostatic displacement fundamental solution has an analytical solution.

From (5.13) and the expressions of U^D_1 and U^D_2, U^D_{jk} can be expressed as the summation of an infinite number of terms. The integral of any of these items is non-singular and can be solved analytically. For small r values, we can fetch the first finite number of items, solve their integrals analytically, and then add the results together as an approximate solution of the integral of U^D_{jk}. Sufficient accuracy can be achieved in this way.

Except for the expanded expression of the elastodynamic displacement fundamental solution as an approximate calculation method, another approach is indirect, and also based on the fact that the singular elastodynamic displacement fundamental solution U_{jk} is composed of the singular elastostatic fundamental solution U^0_{jk} and the non-singular elastodynamic part U^D_{jk}. From (5.13), we obtain the following equations:

$$\int U_{jk}\mathrm{d}r = \int U^0_{jk}\mathrm{d}r + \int U^D_{jk}\mathrm{d}r$$

$$\int U_{jk}\mathrm{d}r = \int U^0_{jk}\mathrm{d}r + \int \left(U_{jk} - U^0_{jk}\right)\mathrm{d}r$$

The first integral on the right of the equality sign $\int U^0_{jk}\mathrm{d}_r$ can be solved analytically; in the second integral $\int \left(U_{jk} - U^0_{jk}\right)\mathrm{d}r$, the item $(U_{jk} - U^0_{jk})$ is already non-singular and its detailed expression is known, so this integral can be solved using *Gaussian* integration.

This shows that the results obtained by the two singular integrals are consistent.

5.3.2 *The Solution of the Elements in Matrix* $\hat{\mathbf{H}}$

Two different conditions should be considered to solve the matrix $\hat{\mathbf{H}}$.

(1) If $i \neq j$

Similar to the solution of the non-diagonal sub-blocks of matrix **G**, there is no singular integral solution involved and the *Gaussian* quadratic equation can be used to do the calculation.

(2) If $i = j$

The integral of the traction force fundamental solution must be *zero*, which can be confirmed without any calculation. A more detailed analysis can be referenced by Fig. 5.5.

Another method is similar to the indirect approach mentioned above, as listed in (5.17).

$$\int T_{jk}\mathrm{d}r = \int T_{jk}^{0}\mathrm{d}r + \int \left(T_{jk} - T_{jk}^{0}\right)\mathrm{d}r \tag{5.17}$$

In the above equation, the first integral term on the right side of the equation is the integral of the elastostatic traction force and its value is 0. The integrand function in the subsequent integral term is the difference between the two singular functions. The margin after differencing does not have singularity and can be solved by a *Gaussian equation*. It is shown that the results obtained from those two methods are consistent.

After obtaining $\hat{\mathbf{H}}$, the *matrix* $\mathbf{H}$ can be solved by (5.11).

5.4 Construction of the Boundary Condition Based on Mode Expansion

The foregoing steps are simply to get the equation system $\boldsymbol{HU} = \boldsymbol{GT}$ which is used to express the relationship between displacement and traction force on all the boundary elements. But $\boldsymbol{U}$ and $\boldsymbol{T}$ are unknown quantities; so, we have to combine the boundary conditions to solve the equation system. The boundary conditions of traction-free on the top and bottom, as well as the external surface, and the left and right virtual boundaries, should be solved separately.

As mentioned before, some analytical solving procedures are included in the hybrid BEM method; that is, the boundary conditions on the virtual boundaries are established according to the classical *Lamb wave* equation. The establishment of the boundary conditions of those parts is based on the concept of mode expansion.

A pure symmetrical or non-symmetrical wave mode at a certain frequency and order is input in the left virtual boundary as the incident mode. The modes of reflection and transmission after the interaction between incident mode and defects might contain any mode which exists at this frequency. The existing modes can be checked by the dispersion curves and the given frequency. The total displacement or traction force on the left boundary should be the summation of the displacement or traction force of incident mode and all reflected modes. The displacement or traction force on the right boundary should be the summation of all displacement or traction force of the transmitted mode. The required boundary conditions for the computation can be obtained from the constrained relationships as mentioned above.

Unknown coefficients are involved in the expressions of displacement and stress given by the fundamental equation of the *Lamb waves*. In order to solve the equations by the BEM method, the expressions of displacement and stress should be first normalized into the form of mode functions. The concept of this normalization is to multiply the expressions of displacement and traction force of each mode with their coefficients so that all the modes have the same mean energy, and then to solve the ratio of each mode in the total displacement or traction force on the condition of the same mean energy.

$$p = \sqrt{\left(\frac{\omega}{C_{\mathrm{L}}}\right)^2 - k^2}, \quad q = \sqrt{\left(\frac{\omega}{C_{\mathrm{T}}}\right)^2 - k^2} \tag{5.18}$$

In the equation, ω is the angular frequency; $k = \omega/C_p$ is the wave number; C_{p} is the phase velocity; C_{L} is the longitudinal wave velocity of bulk waves; and C_{T} is the shear wave velocity of the bulk waves.

The symmetric modes of the *Lamb waves* propagating along the positive direction of x are contained in the following equations:

$$\begin{cases} u = [ikA_2\cos(py) + qB_1\cos(qy)]\mathrm{e}^{i(kx-\omega t)} \\ v = [-pA_2\sin(py) - ikB_1\sin(qy)]\mathrm{e}^{i(kx-\omega t)} \\ \tau_{yx} = \mu[-2ikpA_2\sin(py) + (k^2 - q^2)B_1\sin(qy)]\mathrm{e}^{i(kx-\omega t)} \\ \sigma_y = \{-\lambda(k^2+p^2)A_2\cos(py) - 2\mu[p^2A_2\cos(py) + ikqB_1\cos(qy)]\}\mathrm{e}^{i(kx-\omega)} \end{cases}$$

In addition, the normal stress σ_x can be calculated by the displacement u and v of the symmetric mode, as given in the following equation:

$$\begin{aligned} \sigma_x &= \lambda\left(\frac{\partial u}{\partial x} + \frac{\partial v}{\partial y}\right) + 2\mu\frac{\partial u}{\partial x} \\ &= \{[-\lambda(k^2+p^2) - 2\mu k^2]\cos(py)A_2 + 2\mu ikq\cos(qy)B_1\}\mathrm{e}^{i(kx-\omega t)} \end{aligned}$$

If the expressions of displacement and stress are divided by A_2, we can get (5.19).

$$\begin{cases} \hat{u} = \frac{u}{A_2} = \left[ik\cos(py) + q\frac{B_1}{A_2}\cos(qy)\right]\mathrm{e}^{i(kx-\omega t)} \\ \hat{v} = \frac{v}{A_2} = \left[-p\sin(py) - ik\frac{B_1}{A_2}\sin(qy)\right]\mathrm{e}^{i(kx-\omega t)} \\ \hat{\tau}_{xy} = \frac{\tau_{xy}}{A_2} = \mu\left[-2ikp\sin(py) + (k^2 - q^2)\frac{B_1}{A_2}\sin(qy)\right]\mathrm{e}^{i(kx-\omega t)} \\ \hat{\sigma}_x = \frac{\sigma_x}{A_2} = \left\{[-\lambda(k^2+p^2) - 2\mu k^2]\cos(py) + 2\mu ikq\cos(qy)\frac{B_1}{A_2}\right\}\mathrm{e}^{i(kx-\omega t)} \end{cases} \tag{5.19}$$

Similar to the procedure to solve the dispersion equation, the coefficient B_1/A_2 in (5.19) can be obtained by satisfying the boundary condition (traction-free boundary conditions) on the top and bottom surfaces of the plate, as listed in the following equation:

$$\frac{B_1}{A_2} = \frac{2ikp\sin(ph)}{(k^2 - q^2)\sin(qh)}$$

The above equations are available to the symmetric modes of the *Lamb wave* propagating along the positive direction of x, if the propagation direction is along the negative direction of x, then all the k should be replaced by $-k$, as follows:

$$\begin{cases} \hat{u} = \left[-ik\cos(py) + q\frac{B_1}{A_2}\cos(qy)\right]e^{i(-kx-\omega t)} \\ \hat{v} = \left[-p\sin(py) + ik\frac{B_1}{A_2}\sin(qy)\right]e^{i(-kx-\omega t)} \\ \hat{\tau}_{xy} = \mu\left[2ikp\sin(py) + (k^2 - q^2)\frac{B_1}{A_2}\sin(qy)\right]e^{i(-kx-wt)} \\ \hat{\sigma}_x = \left\{[-\lambda(k^2+p^2) - 2\mu k^2]\cos(py) - 2\mu ikq\cos(qy)\frac{B_1}{A_2}\right\}e^{i(-kx-\omega t)} \end{cases} \tag{5.20}$$

For the above equation,

$$\frac{B_1}{A_2} = \frac{-2ikp\sin(ph)}{(k^2 - q^2)\sin(qh)}$$

For the antisymmetric modes propagating along the positive direction of x,

$$\begin{cases} u = [ikA_1\sin(py) - qB_2\sin(qy)]e^{i(kx-\omega t)} \\ v = [pA_1\cos(py) - ikB_2\cos(qy)]e^{i(kx-\omega t)} \\ \tau_{yx} = \mu[2ikpA_1\cos(py) + (k^2 - q^2)B_2\cos(qy)]e^{i(kx-\omega t)} \\ \sigma_y = \{-\lambda(k^2+p^2)A_1\sin(py) - 2\mu[p^2A_1\sin(py) - ikqB_2\sin(qy)]\}e^{i(kx-\omega t)} \end{cases}$$

The normal stress σ_x can be calculated by the displacement u and v, as expressed in the following equation:

$$\begin{aligned} \sigma_x &= \lambda\left(\frac{\partial u}{\partial x} + \frac{\partial v}{\partial y}\right) + 2\mu\frac{\partial u}{\partial x} \\ &= \{[-\lambda(k^2+p^2) - 2\mu k^2]\sin(py)A_1 - 2\mu ikq\sin(qy)B_2\}e^{i(kx-\omega t)} \end{aligned}$$

If the expressions of displacement and stress are divided by A_1, we can get (5.21).

$$\begin{cases} \hat{u} = \frac{u}{A_1} = \left[ik\sin(py) - q\frac{B_2}{A_1}\sin(qy)\right]\mathrm{e}^{i(kx-\omega t)} \\ \hat{v} = \frac{v}{A_1} = \left[p\cos(py) - ik\frac{B_2}{A_1}\cos(qy)\right]\mathrm{e}^{i(kx-\omega t)} \\ \hat{\tau}_{xy} = \frac{\tau_{xy}}{A_1} = \mu\left[2ikp\cos(py) + (k^2-q^2)\frac{B_2}{A_1}\cos(qy)\right]\mathrm{e}^{i(kx-\omega t)} \\ \hat{\sigma}_x = \frac{\sigma_x}{A_1} = \left\{[-\lambda(k^2+p^2) - 2\mu k^2]\sin(py) - 2\mu ikq\sin(qy)\frac{B_2}{A_1}\right\}\mathrm{e}^{i(kx-\omega t)} \end{cases} \tag{5.21}$$

In the above equation,

$$\frac{B_2}{A_1} = \frac{-2ikp\cos(ph)}{(k^2-q^2)\cos(qh)}$$

For the antisymmetric modes of the *Lamb waves* propagating along the negative direction x, all k should be replaced by $-k$, as expressed in (5.22).

$$\begin{cases} \hat{u} = \left[-ik\sin(py) - q\frac{B_2}{A_1}\sin(qy)\right]\mathrm{e}^{i(-kx-\omega t)} \\ \hat{v} = \left[p\cos(py) + ik\frac{B_2}{A_1}\cos(qy)\right]\mathrm{e}^{i(-kx-\omega t)} \\ \hat{\tau}_{xy} = \mu\left[-2ikp\cos(py) + (k^2-q^2)\frac{B_2}{A_1}\cos(qy)\right]\mathrm{e}^{i(-kx-\omega t)} \\ \hat{\sigma}_x = \left\{[-\lambda(k^2+p^2) - 2\mu k^2]\sin(py) + 2\mu ikq\sin(qy)\frac{B_2}{A_1}\right\}\mathrm{e}^{i(-kx-\omega t)} \end{cases} \tag{5.22}$$

In the above equation,

$$\frac{B_2}{A_1} = \frac{2ikp\cos(ph)}{(k^2-q^2)\cos(qh)}$$

And the expression of the mean energy in the plane is given in (5.23).

$$p_x = -\frac{1}{2}\mathrm{Re}\left\{-i\omega\int_{-h}^{h}\left[(\hat{\sigma}_x)^*\hat{u} + (\hat{\tau}_{xy})^*\hat{v}\right]\mathrm{d}y\right\} \tag{5.23}$$

In the above equation, h is the half thickness of the plate, "Re" represents taking the real part, and "*" represents taking the conjugate stress. The analytical solution for the integral term of the power expression can be obtained directly, but the calculation of the conjugate stress should be analyzed first. The details are listed as follows:

Equation (5.18) shows the definitions of p and q. If $C_P < C_T$, both p and q are pure imaginary numbers; if $C_T < C_P < C_L$, p is a pure imaginary number, then q is a real number. Besides, the sine function and cosine function have the following properties: The real number can be obtained from the sine function sin(x) if x is a real number, and the pure imaginary number can be obtained if x is a pure imaginary number; but, the computed result of cosine function cos(x) is a real number if

x is a real number or an imaginary number. According to these features, the following results can be obtained by analyzing (5.19), (5.20), (5.21), and (5.22).

(1) For the symmetric mode illustrated in (5.19) and (5.20), $\hat{\sigma}_x$ is always a real number and $\hat{\tau}_{xy}$ is a pure imaginary number.
(2) For the antisymmetry mode illustrated in (5.21) and (5.22), the virtuality of $\hat{\sigma}_x$ is the same as P and the virtuality of $\hat{\tau}_{xy}$ is the opposite of P.

Based on the analytical results in the above, when solving the conjugate stress, if the stress is a real number, the conjugation will be itself; if the stress is an imaginary number, then the conjugation is its opposite number. Additionally, when solving the mean power, the exponential term in (5.19) should be neglected.

The mean power of mode n is written as P_x^n and the power of incident mode is P_x^I. The coefficient is defined in the following equation:

$$N^n = \sqrt{\frac{P_x^I}{P_x^n}}$$

The final normalized mode function is listed in (5.24).

$$\bar{u}_x = \bar{u} = \hat{u} \times N, \quad \bar{u}_y = \bar{v} = \hat{v} \times N, \quad \bar{\tau}_{xy} = \hat{\tau}_{xy} \times N, \quad \bar{\sigma}_x = \hat{\sigma}_x \times N \qquad (5.24)$$

In the calculation of (5.24), the $\mathrm{e}^{i(\pm kx)}$ in $\mathrm{e}^{i(\pm kx - \omega t)}$ should be reserved. The mode function obtained from the process mentioned above has the same power as the incident mode.

When mode P is selected as the incident wave, the displacement field is expressed as $\mathbf{u}^I$, the amplitude is expressed as α^p, the normalized mode function of the displacement is $\bar{\mathbf{u}}^p$, and the wave number of the incident mode p is k^p, as expressed in the following equation:

$$\mathbf{u}^I = \alpha^P \bar{\mathbf{u}}^p = \alpha^p \begin{pmatrix} \bar{u}_x^p \\ \bar{u}_y^p \end{pmatrix} \mathrm{e}^{ik^p x}$$

For the traction force, the expression is similar to the displacement and expressed as follows:

$$\mathbf{t}^I = \alpha^p \bar{\mathbf{t}}^p = \alpha^p \begin{pmatrix} \bar{t}_c^p \\ \bar{t}_y^p \end{pmatrix} \mathrm{e}^{ik^p x}$$

When the angular frequency ω is selected, the p mode phase velocity C_p^p can be obtained from the dispersion curves of the phase velocity, and $k^p = \omega / C_p^p$ is also known already.

$\begin{pmatrix} \bar{u}_x^p \\ \bar{u}_y^p \end{pmatrix}$ and $\begin{pmatrix} \bar{t}_x^p \\ \bar{t}_y^p \end{pmatrix}$ are the displacement and traction force mode functions, which can be solved. Thus, the amplitude α^p is the only unknown term.

The reflected displacement and traction force on the left virtual boundary can be expressed in the following equation:

$$\begin{cases} \mathbf{u}^{\mathrm{BS}} = \sum_{j=1}^{J} \beta^j \bar{\mathbf{u}}^j = \sum_{j=1}^{J} \beta^j \begin{pmatrix} \bar{u}_x^j \\ \bar{u}_y^j \end{pmatrix} \mathrm{e}^{-ik^j x} \\ \mathbf{t}^{\mathrm{BS}} = \sum_{j=1}^{J} \beta^j \bar{\mathbf{t}}^j = \sum_{j=1}^{J} \beta^j \begin{pmatrix} \bar{t}_x^j \\ \bar{t}_y^j \end{pmatrix} \mathrm{e}^{-ik^j x} \end{cases}$$

Assuming that there are J modes obtained from the reflection of the left boundary, β^j is the amplitude coefficient of the jth reflected mode and k^j is the wave number of the jth reflected mode.

According to the discussion mentioned above, k^j, $\begin{pmatrix} \bar{u}_x^j \\ \bar{u}_y^j \end{pmatrix}$, and $\begin{pmatrix} \bar{t}_x^j \\ \bar{t}_y^j \end{pmatrix}$ are all calculable, and the unknowns in the expressions are the J coefficients β^j. The value of J is determined by the mode number of the *Lamb waves* at the selected frequency. The mode number of the *Lamb waves* can be calculated from the dispersion curves.

The transmitted displacement and traction force on the right boundary can be expressed as the following:

$$\begin{cases} \mathbf{u}^{\mathrm{FS}} = \sum_{l=1}^{L} \beta^l \bar{\mathbf{u}}^l = \sum_{l=1}^{L} \beta^l \begin{pmatrix} \bar{u}_x^l \\ \bar{u}_y^l \end{pmatrix} \mathrm{e}^{ik^l x} \\ \mathbf{t}^{\mathrm{FS}} = \sum_{l=1}^{L} \beta^l \bar{\mathbf{t}}^l = \sum_{l=1}^{L} \beta^l \begin{pmatrix} \bar{t}_x^l \\ \bar{t}_y^l \end{pmatrix} \mathrm{e}^{ik^l x} \end{cases}$$

Assuming that there are L modes obtained from the transmission on the right boundary, β^l is the amplitude coefficient of the lth transmitted mode and the value of L is also provided from the phase velocity dispersion curves. Actually, $L = J$.

The total displacement on the left boundary is the summation of the input displacement and reflected displacement, $\mathbf{u} = \mathbf{u}^{\mathrm{I}} + \mathbf{u}^{\mathrm{BS}}$.

That is, $\begin{pmatrix} u_x \\ u_y \end{pmatrix} = \alpha^p \begin{pmatrix} \bar{u}_x^p \\ \bar{u}_y^p \end{pmatrix} \mathrm{e}^{ik^p x} + \sum_{j=1}^{J} \beta^j \begin{pmatrix} \bar{u}_x^j \\ \bar{u}_y^j \end{pmatrix} \mathrm{e}^{-ik^j x}$

Assuming that there are k elements on the boundary, applying the above equation to every element on the left boundary, we can get the following equation:

$$[u]_{2m\times 1} = [\bar{u}]_{2m\times L}^{\mathrm{FS}} \times \left\{ \beta^l \mathrm{e}^{ik^l x} \right\}_{L\times 1}$$

So, we get the equation as follows:

$$\begin{pmatrix} u_x(1) \\ u_y(1) \\ u_x(2) \\ u_y(2) \\ \vdots \\ u_x(k) \\ u_y(k) \end{pmatrix} = \begin{pmatrix} \bar{u}_x^1(1) & \bar{u}_x^2(1) & \cdots & \bar{u}_x^J(1) \\ \bar{u}_y^1(1) & \bar{u}_y^2(1) & \cdots & \bar{u}_y^J(1) \\ \bar{u}_x^1(2) & \bar{u}_2^2(2) & \cdots & \bar{u}_x^J(2) \\ \bar{u}_y^1(2) & \bar{u}_y^2(2) & \cdots & \bar{u}_y^J(2) \\ \vdots & \vdots & \ddots & \vdots \\ \bar{u}_x^1(k) & \bar{u}_x^2(k) & \cdots & \bar{u}_x^J(k) \\ \bar{u}_y^1(k) & \bar{u}_y^2(k) & \cdots & \bar{u}_y^J(k) \end{pmatrix}_{2k\times J}^{I} \begin{pmatrix} \alpha^p \delta_{p1} e^{ik^1 x} \\ \alpha^p \delta_{p2} e^{ik^2 x} \\ \vdots \\ \alpha^p \delta_{pJ} e^{ik^J x} \end{pmatrix}_{J\times 1}$$

$$+ \begin{pmatrix} \bar{u}_x^1(1) & \bar{u}_x^2(1) & \cdots & \bar{u}_x^J(1) \\ \bar{u}_y^1(1) & \bar{u}_y^2(1) & \cdots & \bar{u}_y^J(1) \\ \bar{u}_x^1(2) & \bar{u}_2^2(2) & \cdots & \bar{u}_x^J(2) \\ \bar{u}_y^1(2) & \bar{u}_y^2(2) & \cdots & \bar{u}_y^J(2) \\ \vdots & \vdots & \ddots & \vdots \\ \bar{u}_x^1(k) & \bar{u}_x^2(k) & \cdots & \bar{u}_x^J(k) \\ \bar{u}_y^1(k) & \bar{u}_y^2(k) & \cdots & \bar{u}_y^J(k) \end{pmatrix}_{2k\times J}^{\mathrm{BS}} \begin{pmatrix} \beta^1 e^{-ik^1 x} \\ \beta^2 e^{-ik^2 x} \\ \vdots \\ \beta^J e^{-ik^J X} \end{pmatrix}_{J\times 1}$$

The above equation can be reduced to the form expressed in (5.25).

$$[u]_{2k\times 1} = [\bar{u}]_{2k\times J}^{I} \times \left\{ \alpha^p \delta_{pj} e^{ik^j x} \right\}_{J\times 1} + [\bar{u}]_{2k\times J}^{\mathrm{BS}} \times \left\{ \beta^j e^{-ik^j x} \right\}_{J\times 1} \tag{5.25}$$

In the equation given above, the $[\bar{u}]_{2k\times J}^{I}$ obtained from the expression of forward-propagating along the x-axis is different from the $[\bar{u}]_{2k\times J}^{\mathrm{BS}}$ obtained from that used to describe the counter-propagation.

For traction force, a similar *equation* can be expressed as shown in (5.26).

$$[t]_{2k\times 1} = [\bar{t}]_{2k\times J}^{I} \times \left\{ \alpha^p \delta_{pj} e^{ik^j x} \right\}_{J\times 1} + [\bar{t}]_{2k\times J}^{\mathrm{BS}} \times \left\{ \beta^j e^{-ik^j x} \right\}_{J\times 1} \tag{5.26}$$

The mode function of traction force can be determined based on the following relationship between the mode function of stress and the component of the boundary normal unit vector, as shown in (5.27).

$$\begin{cases} t_x = \sigma_x n_x + \tau_{xy} n_y \\ t_y = \tau_{xy} n_x + \sigma_y n_y \end{cases} \tag{5.27}$$

In the above equation, n_x and n_y are two components of the boundary normal unit vector. For the left boundary, $n_x = -1$ and $n_y = 0$, so we have the following equation:

$$\begin{cases} t_x = -\sigma_x \\ t_y = -\tau_{xy} \end{cases}$$

In (5.25) and (5.26), when the angular frequency is given, the k^j of each mode can be calculated, $[\bar{u}]^I_{2k\times J}$, $[\bar{u}]^{\mathrm{BS}}_{2k\times J}$, $[\bar{t}]^I_{2k\times J}$ and $[\bar{t}]^{\mathrm{BS}}_{2k\times J}$ are already known; α^p is also known, but β^j is unknown.

The generalized inverse matrix is introduced and defined as the following equation:

$$[\bar{u}^{-1}] = \left([\bar{u}^*]^{\mathrm{T}}[\bar{u}]\right)^{-1}[\bar{u}^*]^{\mathrm{T}}$$

In the above equation, "*" represents taking the conjugation and "*T*" represents the matrix transposition, and then according to (5.25), we can get (5.28).

$$\left\{\beta^j \mathrm{e}^{-ik^j x}\right\}_{J\times 1} = \left[\bar{u}^{-1}\right]^{\mathrm{BS}}_{J\times 2k}[u]_{2k\times 1} - \left[\bar{u}^{-1}\right]^{\mathrm{BS}}_{J\times 2k}[\bar{u}]^I_{2k\times J}\times\left\{\alpha^p\delta_{pj}\mathrm{e}^{ik^j x}\right\}_{J\times 1} \tag{5.28}$$

Substituting the above equation into (5.26), we obtain (5.29).

$$\begin{aligned}[t]_{2k\times 1} = {} & [\bar{t}]^I_{2k\times J}\times\left\{\alpha^p\delta_{pj}\mathrm{e}^{ik^j x}\right\}_{J\times 1} + [\bar{t}]^{\mathrm{BS}}_{2k\times J} \\ & \times\left\{\left[\bar{u}^{-1}\right]^{\mathrm{BS}}_{J\times 2k}[u]_{2k\times 1} - \left[\bar{u}^{-1}\right]^{\mathrm{BS}}_{J\times 2k}[\bar{u}]^I_{2k\times J}\times\left\{\alpha^p\delta_{pj}\mathrm{e}^{ik^j x}\right\}_{J\times 1}\right\}\end{aligned} \tag{5.29}$$

The above equation can be simplified into the form given in (5.30), where matrix **C** and vector **d** are all calculable, and then we can see the relationship between the traction force and displacement of each element on the left boundary.

$$[t]_{2k\times 1} = \mathbf{C}_{2k\times 2k}[u]_{2k\times 1} + \mathbf{d}_{2k\times 1} \tag{5.30}$$

The total displacement on the right boundary is the transmitted displacement, that is, $\mathbf{u} = \mathbf{u}^{\mathrm{FS}}$, then we have the following equation:

$$\begin{pmatrix} u_x \\ u_y \end{pmatrix} = \sum_{l=1}^{L}\beta^l\begin{pmatrix} \bar{u}^l_x \\ \bar{u}^l_y \end{pmatrix}\mathrm{e}^{ik^l x}$$

Assuming there are m elements on the right boundary, applying the above equation on every element, we can obtain the following equations:

$$\begin{pmatrix} u_x(1) \\ u_y(1) \\ u_x(2) \\ u_y(2) \\ \vdots \\ u_x(m) \\ u_y(m) \end{pmatrix} = \begin{pmatrix} \beta^1 e^{ik^1x}\bar{u}_x^1(1) + \beta^2 e^{ik^2x}\bar{u}_x^2(1) + \cdots \beta^L e^{ik^Lx}\bar{u}_x^L(1) \\ \beta^1 e^{ik^1x}\bar{u}_y^1(1) + \beta^2 e^{ik^2x}\bar{u}_y^2(1) + \cdots \beta^L e^{ik^Lx}\bar{u}_y^L(1) \\ \beta^1 e^{ik^1x}\bar{u}_x^1(2) + \beta^2 e^{ik^2x}\bar{u}_x^2(2) + \cdots \beta^L e^{ik^Lx}\bar{u}_y^L(2) \\ \vdots \\ \beta^1 e^{ik^1x}\bar{u}_x^1(m) + \beta^2 e^{ik^2x}\bar{u}_x^2(m) + \cdots \beta^L e^{ik^Lx}\bar{u}_x^L(m) \\ \beta^1 e^{ik^1x}\bar{u}_y^1(m) + \beta^2 e^{ik^2x}\bar{u}_y^2(m) + \cdots \beta^L e^{ik^Lx}\bar{u}_y^L(m) \end{pmatrix}$$

So, we have the following equations:

$$\begin{pmatrix} u_x(1) \\ u_y(1) \\ u_x(2) \\ u_y(2) \\ \vdots \\ u_x(m) \\ u_y(m) \end{pmatrix}_{2m\times 1} = \begin{pmatrix} \bar{u}_x^1(1) & \bar{u}_x^2(1) & \cdots & \bar{u}_x^L(1) \\ \bar{u}_y^1(1) & \bar{u}_y^2(1) & \cdots & \bar{u}_y^L(1) \\ \bar{u}_x^1(2) & \bar{u}_x^2(2) & \cdots & \bar{u}_x^L(2) \\ \bar{u}_y^1(2) & \bar{u}_y^2(2) & \cdots & \bar{u}_y^L(2) \\ \vdots & \vdots & \ddots & \vdots \\ \bar{u}_x^1(m) & \bar{u}_x^2(m) & \cdots & \bar{u}_x^L(m) \\ \bar{u}_y^1(m) & \bar{u}_y^2(m) & \cdots & \bar{u}_y^L(m) \end{pmatrix}_{2m\times L}^{\mathrm{FS}} \begin{pmatrix} \beta^1 e^{ik^1x} \\ \beta^2 e^{ik^2x} \\ \vdots \\ \beta^L e^{ik^Lx} \end{pmatrix}_{L\times 1}$$

The above equation can be simplified into the following form:

$$[u]_{2m\times 1} = [\bar{u}]_{2m\times L}^{\mathrm{FS}} \times \left\{\beta^l e^{ik^lx}\right\}_{L\times 1}$$

For the traction force, we have the expressions as follows:

$$[t]_{2m\times 1} = [\bar{t}]_{2m\times L}^{\mathrm{FS}} \times \left\{\beta^l e^{ik^lx}\right\}_{L\times 1}$$

The mode function of the traction force is still determined by (5.27). For the right boundary, $n_x = 1$ and $n_y = 0$, so we have the following equation:

$$\begin{cases} t_x = \sigma_x \\ t_y = \tau_{xy} \end{cases}$$

Similarly, (5.31) can be obtained using the concept of the generalized inverse matrix.

$$\left\{\beta^l e^{ik^lx}\right\}_{L\times 1} = \left[\bar{u}^{-1}\right]_{L\times 2m}^{\mathrm{FS}} [u]_{2m\times 1} \tag{5.31}$$

So, we have the following equation:

$$[t]_{2m\times 1} = [\bar{t}]^{\mathrm{FS}}_{2m\times L} \times \left[\bar{u}^{-1}\right]^{\mathrm{FS}}_{L\times 2m} [u]_{2m\times 1}$$

The above equation can be simplified into the form shown in (5.32).

$$[t]_{2m\times 1} = \mathbf{D}_{2m\times 2m} [u]_{2m\times 1} \tag{5.32}$$

In (5.32), the matrix **D** is calculable, which means that the relationship between the traction force and displacement of every element on the right boundary can be provided in (5.32).

The relationships between the traction force and displacement on the left and right virtual boundary are given in (5.30) and (5.32), respectively. Those relationships are considered to be the boundary conditions on the left and right boundaries. These boundary conditions and the traction-free boundary conditions of the top and bottom surface of the plate can be grouped as the whole boundary conditions of the model as well as the matrix form (5.9). So, finally, the linear equation set to be solved is eventually obtained, as shown in (5.33).

$$\mathbf{AU} = \mathbf{R} \tag{5.33}$$

After solving $\boldsymbol{U}$ in the above equation, extracting the $[u]_{2k\times 1}$ with respect to the left boundary and the $[u]_{2m\times 1}$ with respect to the right boundary, the reflected and transmitted coefficients can be calculated based on (5.28) and (5.31), as shown in (5.34), where $|R|^j$ is the reflected coefficient of mode j and $|T|^l$ is the transmitted coefficient of mode l.

$$|R|^j = \left|\frac{\beta^j}{\alpha^p}\right|, \quad |T|^l = \left|\frac{\beta^l}{\alpha^p}\right| \tag{5.34}$$

5.5 Structure of the BEM Program

Based on the above theoretical derivations, the elastodynamic frequency domain BEM program is written to simulate the interactions between the *Lamb waves* and the cracks. The BEM program is realized by hybrid programming based on Visual C++ and MATLAB R2008a. Most of the calculations are realized by coding in Visual C++, while MATLAB R2008a is used to draw the curves and figures and prepare the boundary conditions. The strategy of dynamic distribution of memory is used in the program, which means that any of the large memory unit that is not used will be released at any time in the calculation process so that the program can run smoothly, even when there are many elements.

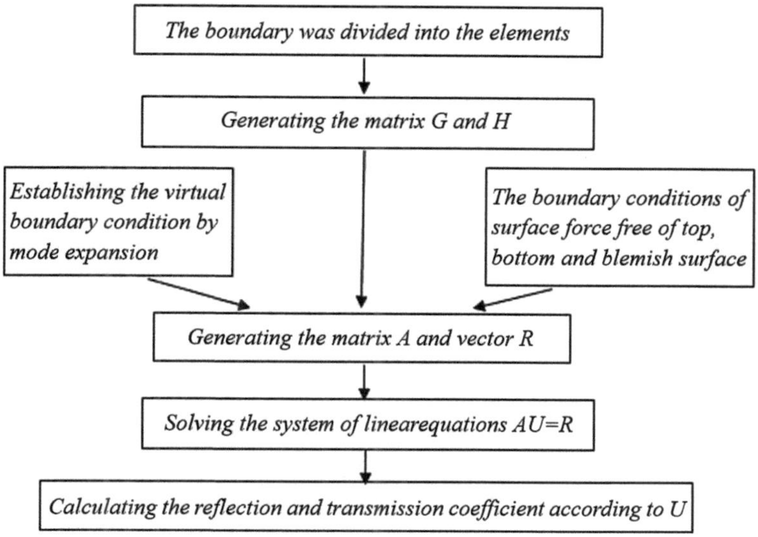

Fig. 5.6 The structure of the BEM program

The program flowchart of the BEM that is used to calculate the interactions between the ultrasonic guided waves and the defects is shown in Fig. 5.6. The configuration and input parameters of the BEM program are read from the files. The number of the elements on each boundary line segment is determined automatically based on the largest element size assigned by users, to ensure the element size is as large as possible within the specified range; in addition, the size of every element on each boundary line segment is the same. After the division of element units, matrices $\boldsymbol{G}$ and $\boldsymbol{H}$ can be generated. The linear equations set (5.33) can be obtained by the traction-free boundary conditions on the top, bottom, and blemished surface, as well as the boundary conditions on the left and right virtual boundaries obtained from the mode expansion of the *Lamb wave*. The $\boldsymbol{U}$ in (5.33) is the vector composed from the displacements on every constant element and needs to be solved for the definition of (5.10). After obtaining the displacements, the reflected and transmitted coefficients of each mode at certain frequency thickness products are calculated.

The above program is just for the first-order BEM solution. The function of sweeping is also provided in the BEM program. A certain amount of initial values, step lengths, and steps can be assigned by the user, and the simulated results under these conditions can be calculated automatically by the program. The items can be studied using the sweeping function provided by the program including the length of model, the size of the boundary element, the frequency thickness product, the depth and width of the defect. In the process of the sweeping of one item, all other items are assumed to be constant.

The formation of the boundary element mesh is determined by the structure of the boundary element geometry model and the selected boundary elements. So, for the sweeping of model length, the size of boundary elements and the depth and width of the defects, it is necessary to redistribute the boundary element meshes for the calculation by BEM after every update of the items; for the sweeping of the frequency thickness product, it is necessary to mesh at the beginning of the programming.

The validity of the simulated results can be checked by the principle of energy conservation. Energy redistribution will happen in the process of interactions between the *Lamb waves* and the defects. The energy conservation in this process is that the sum of the energies contained in the forward and back-scattered fields must equal the energy of the incident field. Once the wave structure of each mode is normalized by the relative power ratio between the incident mode and a scattered mode, the normalized transported energy fluxes can be simply expressed as the squares of the normalized displacement, as shown in (5.35).

$$E=\sum_{j}\left(|R|^{j}\right)^{2}+\sum_{l}\left(|T|^{l}\right)^{2}=1 \tag{5.35}$$

This shows that the sum of the squares of all reflected and transmitted coefficients should be *one*. The value of E in (5.35) is only for the result of the first-order BEM simulation. In the case of parameter scanning, plotting a curve by connecting every value of E to every value calculated with the scanning parameters, the curve should be similar to the straight line $E = 1$. Thus, in order to verify the correctness of the simulated calculation, the corresponding curves of energy conversion of the most offered examples are given in the latter part of this book.

Beside the energy conversion curves of some examples in the latter part of this book, which are used as the inspection standard to solve inaccuracies, another method used to verify the correctness of the calculation is to study the situation where the guided wave propagation in the plate is without defect. Since there is no mode conversion of guided wave propagation in this situation, the mode of the transmitted wave is the same as the incident mode. Meanwhile, there is no reflected mode; thus, the simulated calculation value of the boundary elements distributed along the thickness direction of the plate should be similar to the displacement structure of the wave. Figure 5.7 shows the comparison between the wave structures of the theoretical and simulated values of the displacement distribution when $fd = 1000$ Hz × m for each incidence of the S_0 and A_0 mode *Lamb waves* in the defect-free plate. The subscript "Theo" of the displacement in this figure represents the theoretical value and the subscript "Simu" represents the simulated value of the displacement by BEM. It is found from Fig. 5.7 that the wave structure of theoretical displacements is very close to that of simulated displacements.

Unless otherwise stated, for all the BEM calculations in the below study, a steel plate with a 1 mm thickness and the other parameters shown in Table 5.1 such as density and wave velocities are used.

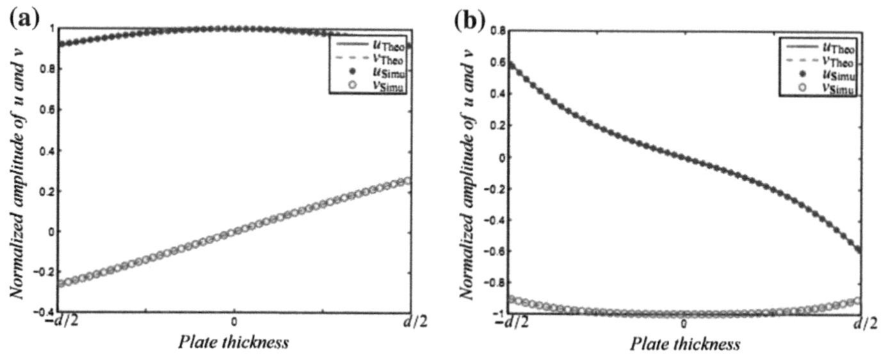

Fig. 5.7 The comparison between the calculated and theoretical value at $fd = 1000$ Hz × m, for the wave structures of the S_0 and A_0 mode in the defect-free plate. **a** S_0 mode. **b** A_0 mode

Table 5.1 The quantitative calculated examples of the rectangular groove cracks

Width (mm)	Depth (%)	Reflection R	Transmission T	Width (mm)	Depth (%)
0.26	40	0.3975	0.7848	0.2614	39.9995
0.45	23	0.2124	0.9318	0.4761	22.7661
0.15	55	0.5506	0.6154	0.1501	55.0013
0.56	22	0.2100	0.9365	0.5849	21.8495
0.67	46	0.4945	0.7231	0.6744	46.0250

5.6 Factors of Computational Accuracy

In this section, the calculation is based on the constant element and the model of the external defect on the plate is used to investigate the two factors that affect the accuracy of the BEM simulation: the length of the model and the size of chosen boundary elements. The reason for studying the factors that affect the calculation accuracy is that appropriate parameters in the BEM simulation can be selected based on the investigations.

5.6.1 Sweeping of the Model Length

In the derivation of the *Lamb wave* dispersion equation and the calculation of boundary conditions on the virtual boundary of the BEM model, the wave number k is always taken as a real number. But, in fact, k can be a complex number, $k = k_r + ik_{im}$, where k_r and k_{im} are the real part and imaginary part of the wave number, respectively.

If $k_{im} > 0$, the *Lamb wave* will show an exponential decay with increases in the propagation distance in the form of a dissipative wave. If $k_{im} < 0$, the *Lamb wave* will show an exponential increase with increases in the propagation distance, but this situation actually does not exist. If $k_{im} = 0$, which means k is a real number, the *Lamb wave* will propagate without any decay of amplitude, which is the case stated in previous discussions. Thus, if the wave number k is taken as a real number, the exponential decay of the dissipative wave will be neglected. It is obvious that this situation is only reasonable when the distance between the dissipative wave source and the inspection point is far enough that the dissipative wave will dissipate entirely. Thus, for the BEM model of the interactions between *Lamb wave* and defect mentioned in this book, the distance between the defect and the left or right boundaries should be far enough; that is, the accuracy result can be obtained only in the situation where the model length is large enough. The model length can be increased gradually to check when the reflected and transmitted coefficients of the waves tend toward stability, and to evaluate if the distance between the scattering centers and the boundaries is large enough.

The sweeping variations of the model length are shown in Fig. 5.8. The defects under each length are all triangular defects with a thickness of 50 % of the plate thickness and an opening width of 0.1 mm, and the defects are always located at the horizontal midpoint of the plate model.

As an example to investigate the effect of the model length on the simulation, Fig. 5.9 shows reflection and transmission coefficient variations with the increasing model length. The S_0 *Lamb wave* mode at $fd = 500$ Hz × m is taken as the incident wave. According to the dispersion curves, only S_0 and A_0 modes may exist under this fd. It is shown in Fig. 5.9 that the coefficients become stable and constant if the length of the plate model is bigger than 6 mm.

In Fig. 5.10, the S_0 *Lamb wave* mode is taken as the incident wave, $fd = 1000$ Hz × m, and only S_0 and A_0 modes exist. It can be seen that the coefficients become stable and constant if the length of the plate model is bigger that 8 mm.

Consider the situation where the A_0 mode *Lamb wave* is taken as the incident wave. In Fig. 5.11, the A_0 *Lamb wave* mode at $fd = 1000$ Hz × m is taken as the incident wave. If the length of the plate model is bigger than 8 mm, this will enable us to achieve stable numerical results for the reflected and transmitted coefficients. The two coefficients, written together as $|R(S_0)|/|T(S_0)|$, indicate that the two coefficient curves almost overlap with each other.

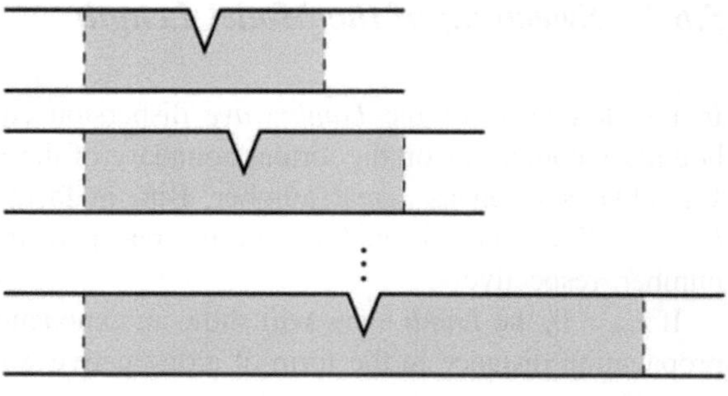

Fig. 5.8 The sweep of the boundary element model length

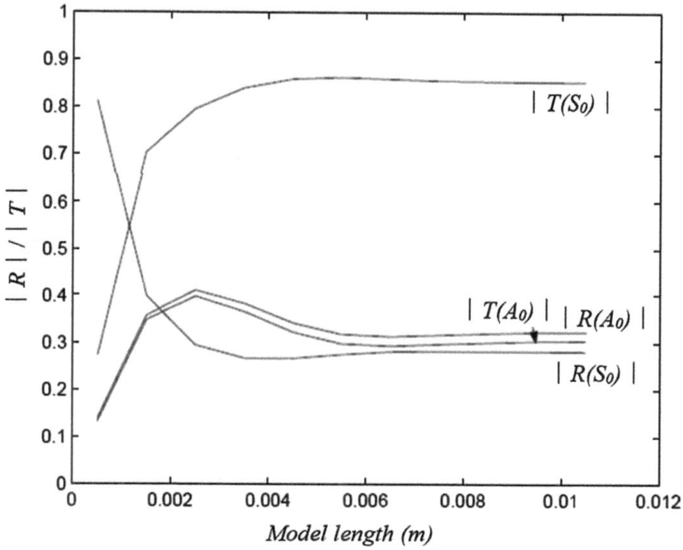

Fig. 5.9 BEM convergence test for the S_0 mode (fd = 500 Hz × m) incidence to the triangular defect with a width of 0.1 mm

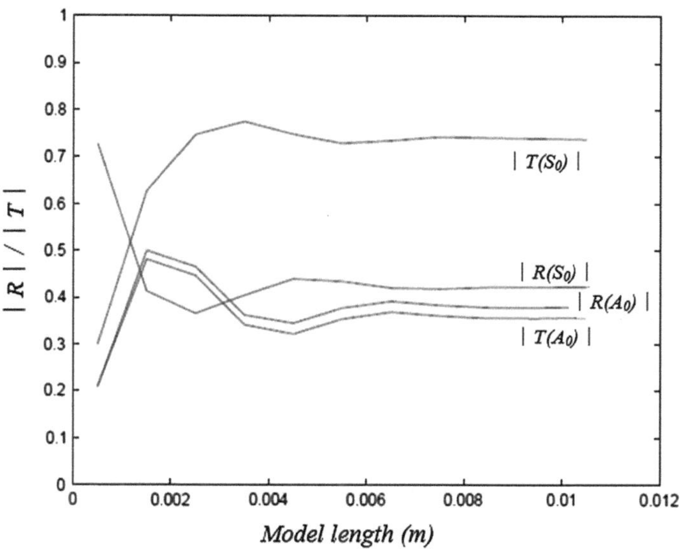

Fig. 5.10 BEM convergence test for the S_0 mode (fd = 1000 Hz × m) incidence to the triangular defect with a width of 0.1 mm

In Fig. 5.12, the A_0 *Lamb wave* mode is taken as the incident wave. It can be seen that the coefficients become stable and constant if the length of the plate model is larger than 8 mm.

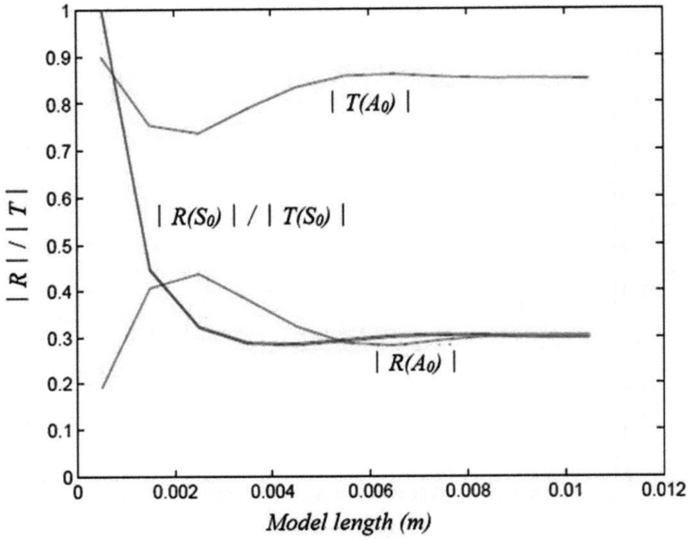

Fig. 5.11 BEM convergence test for the A_0 mode (fd = 500 Hz × m) incidence to the triangular defect with a width of 0.1 mm

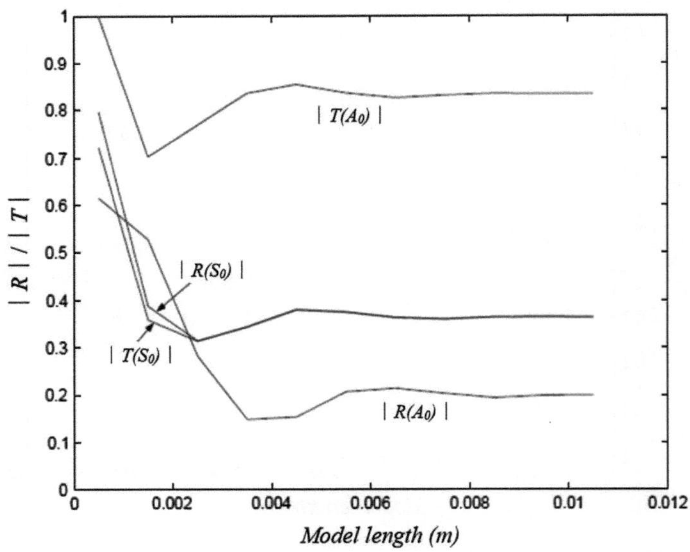

Fig. 5.12 BEM convergence test for the A_0 mode (fd = 1000 Hz × m) incidence to the triangular defect with a width of 0.1 mm

5.6.2 Sweeping of the Boundary Elements Size

The size of the boundary elements has an important influence on the accuracy of the BEM technique. If the size of the boundary elements is too big, variations in the acoustic field cannot be described accurately; when the boundary conditions on the left and right virtual boundaries are constructed by mode expansion, the wave structure cannot be described accurately if the elements' size is too big, which will cause inaccuracies in the boundary conditions. In addition, the problem of wave propagation has some requirements of the elements' size; that is, a certain number of elements should be included within a wave length.

In order to exclude the effect of the description's accuracy of the wave structure on the left and right virtual boundaries, the elements' size on the left and right boundaries is fixed as a smaller value, only the size of the elements in the other parts is flexible. In the calculated example, the length of all models are chosen as 8 mm, and the simulated defect is a triangular slot with an opening width of 0.1 mm and a depth of 50 % thickness of the plate.

Figures 5.13 and 5.14 show the coefficients and energy curves for the S_0 *Lamb wave* mode when fd = 500 Hz × m and fd = 1000 Hz × m are taken as the incidence, respectively; Figs. 5.15 and 5.16 show the coefficients and energy curves for the A_0 *Lamb wave* mode when fd = 500 Hz × m and fd = 1000 Hz × m are taken as the incidence, respectively. It is shown in Figs. 5.13, 5.14, 5.15, 5.16 that the changes in each coefficient are very small with increasing element size. Thus, a relatively accurately simulated result can be obtained using the elements with a size of 0.04 mm.

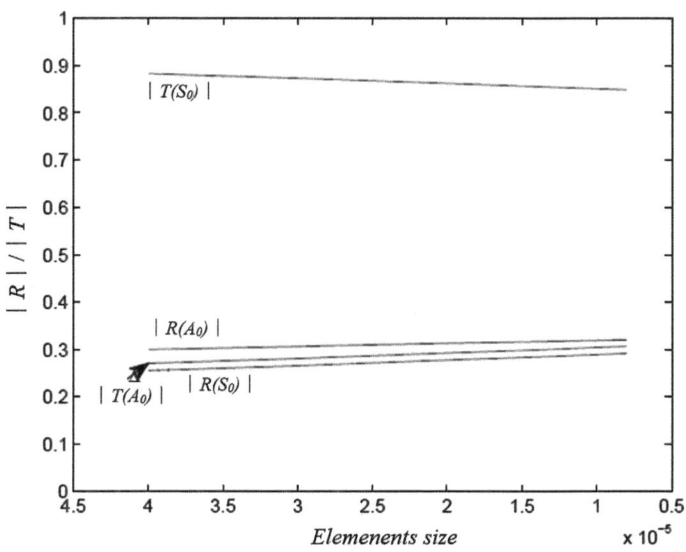

Fig. 5.13 The triangular defect with a 0.1 mm width when the S_0 mode incidence is at fd = 500 Hz × m

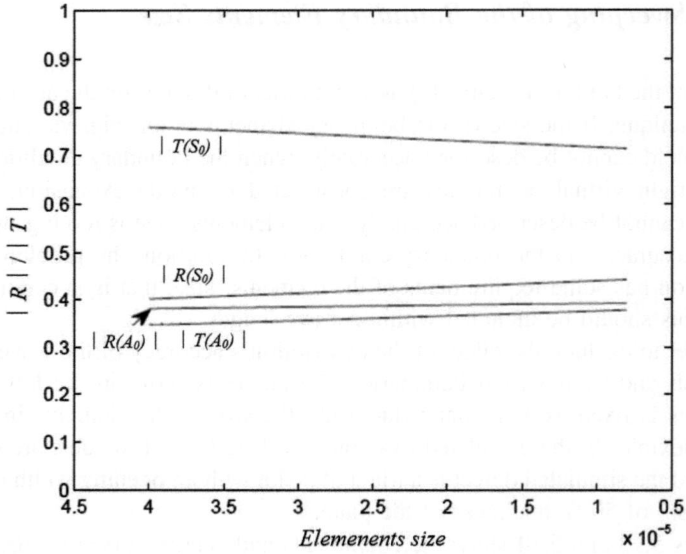

Fig. 5.14 The triangular defect with a 0.1 mm width when the S_0 mode incidence is at fd = 1000 Hz × m

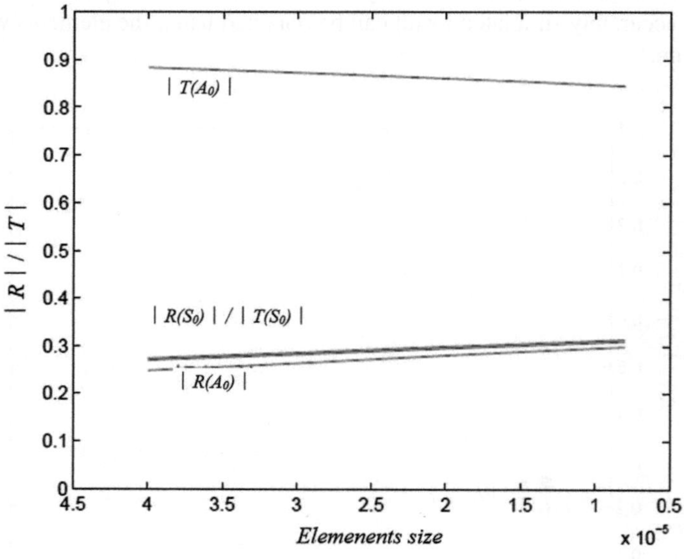

Fig. 5.15 The triangular defect with a 0.1 mm width when the A_0 mode incidence is at fd = 500 Hz × m

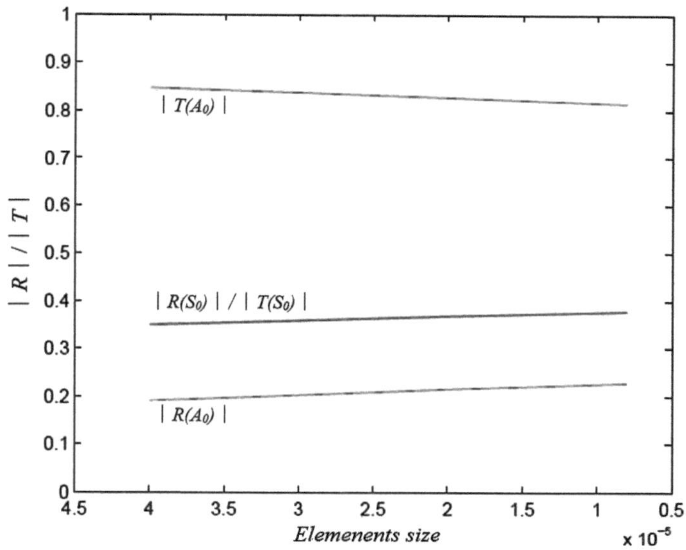

Fig. 5.16 The triangular defect with a 0.1 mm width when the A_0 mode incidence is at $fd = 1000$ Hz × m

5.7 Calculation of the Reflections at the End of the Plate

The guided wave propagation along the circumferential direction of the defect-free pipe will not encounter the scatterers, while the reflection and mode conversion of the guided waves in a finite defect-free plate will appear at the end of the plate, so the end of the plate will play the same role as that of the defect.

In the view of the simulation realization, the reflection at the end of the defect-free plate is the simplification of the two virtual boundaries of the ends. For the established boundary condition based on the mode expansion, the only issue to be considered is the reflection on the left virtual boundary; thus, the result of the reflection at the end of the plate can be taken as the preliminary validation of the correctness of the program. The end of the plate has the same traction-free boundary condition as the top and bottom surface of the plate. The model of reflection at the end of plate is given in Fig. 5.17.

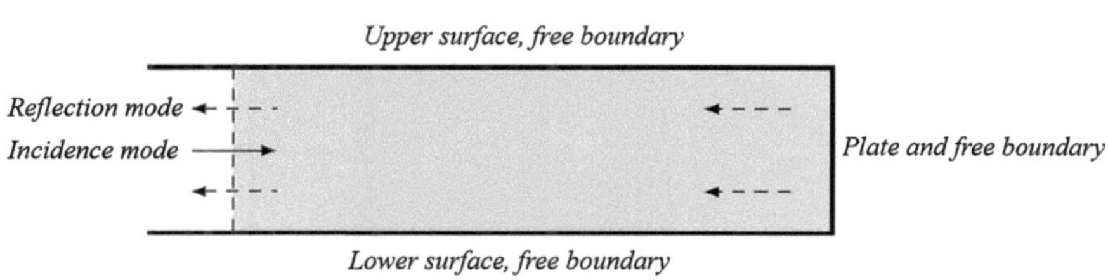

Fig. 5.17 The model of reflection at the end of the defect-free plate

For the model of reflection at the end of the plate, the sweeping of the frequency thickness product fd is the only thing to be considered, and in the sweeping process of fd, we just change the frequency while the plate thickness is fixed at 1 mm. The model length is 8 mm and the maximum size of the elements is 0.02 mm, both of them are kept unchanged. The horizontal section at the vertical center of the plate is selected, and the geometric model of the reflection at the end of the defect-free plate is symmetrical about the cross section. This situation determines that there is no mode conversion of the guided wave during its reflection; that is, if the mode of the incident *Lamb wave* is symmetric, the reflected wave mode is also symmetric. If the mode of the incident *Lamb wave* is antisymmetric, the reflected wave mode is also antisymmetric. The conclusion can be verified by calculations: The reflected coefficient of the guided wave mode that is opposite to the incident mode is approaching *zero*; thus, the reflected coefficient curve of the guided wave mode that is opposite to the incident mode is ignored in the curve chart of the reflected coefficient-fd.

Figure 5.18 shows the variations in the reflected coefficient at the end of the plate, with the frequency thickness product fd, for the S_0 *Lamb wave* mode incidence. Figure 5.18 indicates that the change in the reflected coefficient $|R(S_1)|$ is very small with increases in the frequency thickness product, while $|R(S_2)|$ increases first and then decreases after a certain value, but $|R(S_0)|$ decreases first and then increases after a certain point. $|R(S_1)|$ and $|R(S_2)|$ will intersect at $fd = 4900$ Hz × m. The variations in $|R(S_2)|$ and $|R(S_0)|$ are quite complementary, so the energy conservation is validated indirectly.

If the S_1 *Lamb wave* mode is incident in the plate, the variations in its reflected coefficient at the end of the plate with the frequency thickness product are given in Fig. 5.19. It is shown in Fig. 5.19 that all the reflected coefficients are very small; that is, the majority of the reflected energies are in the form of the S_1 mode.

When the A_0 mode is incident in the plate, the variations in its reflected coefficient at the end of the plate with the frequency thickness product is given in

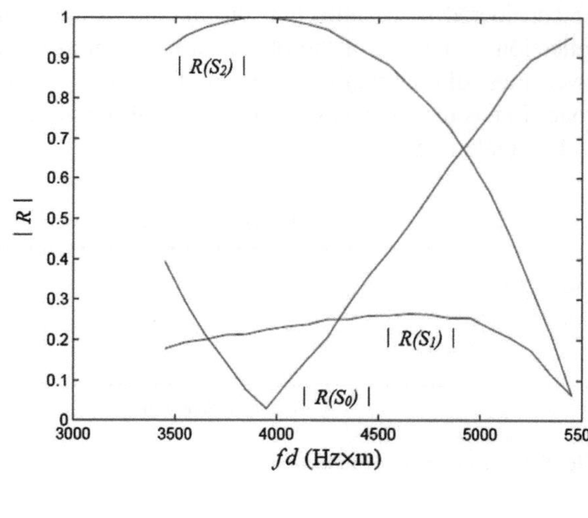

Fig. 5.18 Reflection of the S_0 *Lamb wave* mode at the end of the plate

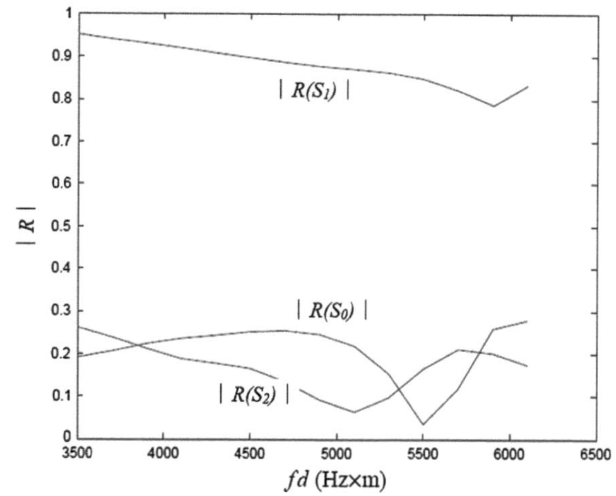

Fig. 5.19 Reflection of the S_1 *Lamb wave* mode at the end of the plate

Fig. 5.20. It is shown in Fig. 5.20 that $|R(A_0)|$ approximately equals 1 at the low frequency range, which means that there is only the A_0 mode reflected in this situation; after overing the cut-off frequency of the A_1 mode, $|R(A_0)|$ will have an abrupt change and $|R(A_1)|$ begins to appear, and then $|R(A_0)|$ will increase from the point that is close to 0 and $|R(A_1)|$ will decrease from the point that is close to 1. These two coefficients will intersect with each other at the point $fd = 3000$ Hz × m.

If the A_1 *Lamb wave* mode is incident in the plate, the variations in its reflected coefficient at the end of plate with the frequency thickness product are given in Fig. 5.21. It can be seen that $|R(A_0)|$ will decrease from the point that is close to 1 and $|R(A_1)|$ will increase from the point that is close to 0. These two coefficients will intersect with each other at the point $fd = 3100$ Hz × m.

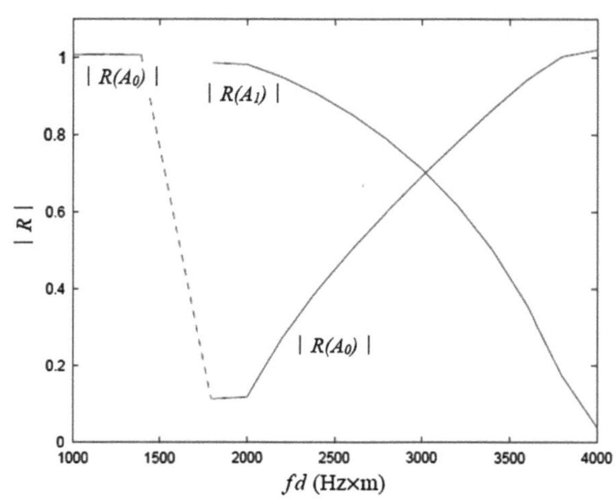

Fig. 5.20 Reflection of the A_0 *Lamb wave* mode at the end of the plate

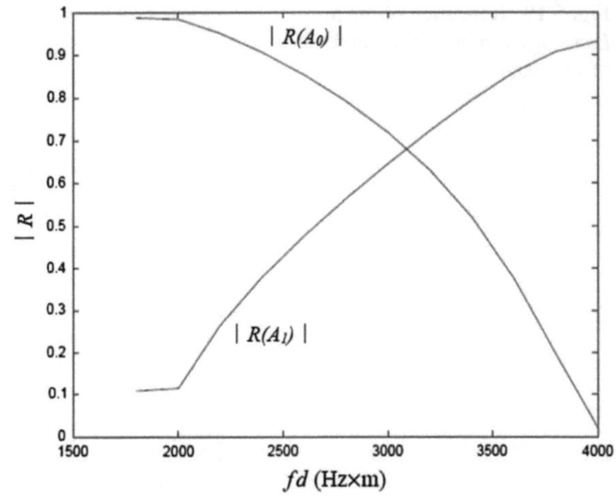

Fig. 5.21 Reflection of the A_1 *Lamb wave* mode at the end of the plate

5.8 Simulation of the External Defect in the Plate [1]

There are two kinds of defect models that are used to simulate the crack on the external surface of the steel plate in the calculation: a triangular groove and a rectangular groove.

5.8.1 Sweeping of the Crack Depth on the External Surface of the Plate

Firstly, the variations in each reflected and transmitted coefficient with the change in defect depth are investigated. The sweeping results of the defect depth on the external surface are significantly important for the quantization of the defect. The crack models of the triangular and rectangular grooves are shown in Fig. 5.22.

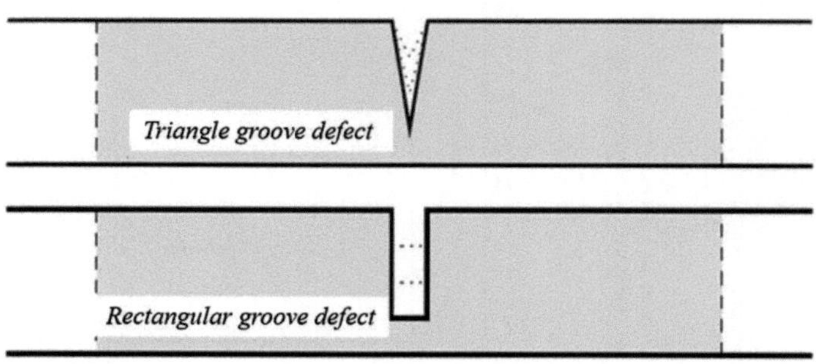

Fig. 5.22 The sweeping model of the external defect

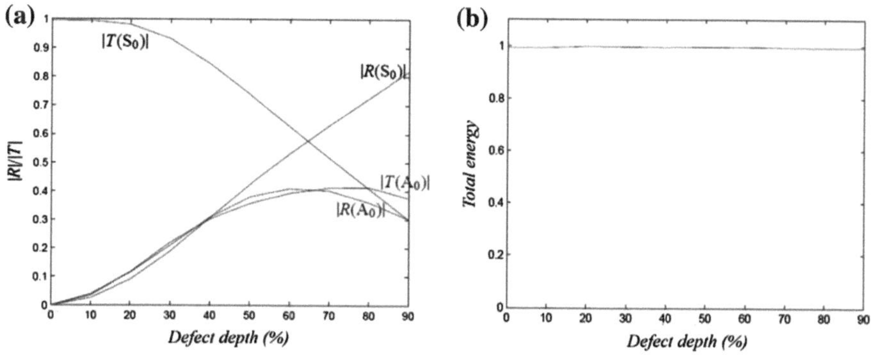

Fig. 5.23 The $fd = 1000$ Hz × m S_0 mode incidence for the external defect depth. **a** Reflection and transmission coefficient. **b** Total energy

The corresponding parameters used in these two sweeping models are provided as follows: The length of the model is 10 mm, the maximum length of the elements is 0.02 mm, and the opening width of the defect is 0.1 mm.

For the triangular defect, if the S_0 *Lamb* wave mode at fd = 1000 Hz × m is taken as the incident wave, the coefficients of reflection and transmission are given in Fig. 5.23. Specifically, Fig. 5.23a, b shows the curves describing the changes in the reflection and transmission coefficients and the total energy defined in (5.35), with the increase in the defect depth percentage from 0 %. The initial defect depth is 0 %, which is equivalent to the defect-free situation and corresponds to the ideal circumstance of the free-propagation. Also, there is no mode conversion, only the coefficient $|T(S_0)| = 1$. With the increase in defect depth, $|T(S_0)|$ will decrease and $|R(S_0)|$ will increase, while at the same time, the A_0 mode will be generated. $|R(A_0)|$ will increase first and then tend to smooth out, with a small decrease at the end. The energy is always near 1 in the depth sweeping process, so the accuracy of the solutions on the boundary is verified.

As shown in Fig. 5.24, the A_0 *Lamb* wave mode at fd = 1000 Hz × m is taken as the incident wave. It is shown in the figure that with the increase in the depth, $|T(A_0)|$ will decrease and $|R(A_0)|$ will increase. However, the change of $|R(A_0)|$ is not monotonic: $|T(S_0)|$ and $|R(S_0)|$ will increase first and then there will be a slight decrease.

The situation of the rectangular groove is similar to that of the triangular defect. The calculated results are given in Figs. 5.25 and 5.26, which show that fd = 1000 Hz × m and S_0 and A_0 *Lamb wave* modes are taken as the incident wave in the specimen with a rectangular groove defect.

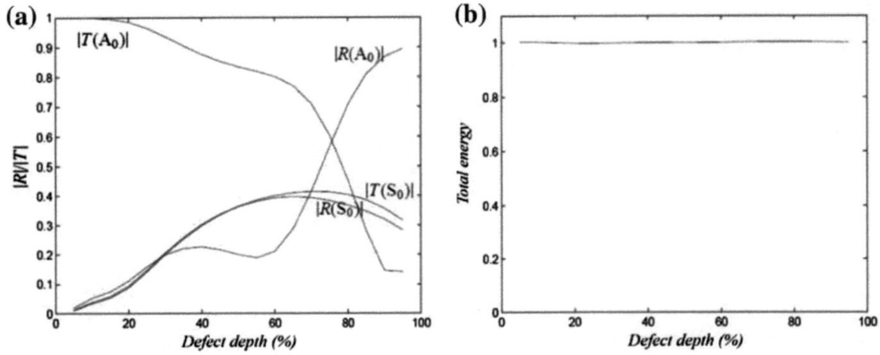

Fig. 5.24 The $fd = 1000$ Hz × m A_0 mode incidence for the external defect depth. **a** Reflection and transmission coefficient. **b** Total energy

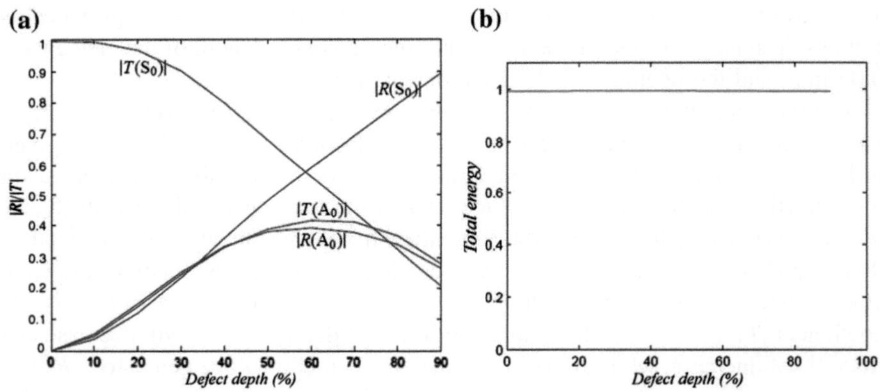

Fig. 5.25 The $fd = 1000$ Hz × m S_0 mode incidence for the external defect depth. **a** Reflection and transmission coefficient. **b** Total energy

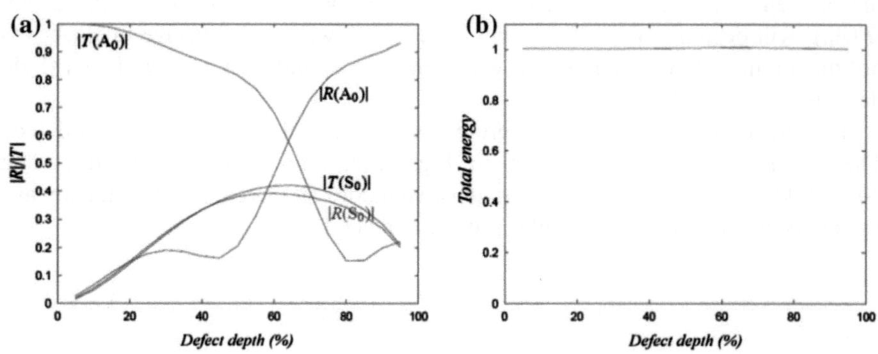

Fig. 5.26 The $fd = 1000$ Hz × m A_0 mode incidence for the external defect depth. **a** Reflection and transmission coefficient. **b** Total energy

5.8.2 Sweeping of the Crack Width on the External Surface of the Plate

As with the sweeping of the defect depth, the sweeping results of the crack width are equally important to the quantization of the defect size. When the sweeping of the crack width is made on the external surface of the specimen, the types of the defects are also triangular and rectangular grooves. The model of the width sweeping is given in Fig. 5.27. The model with a length of 10 mm and element with an upper limited size of 0.02 mm are used in each example of the sweeping calculations, and the depth of the defects are all 50 % of the plate thickness.

The results of the S_0 and A_0 *Lamb wave* mode incidence in the specimen with the triangular defect are shown in Figs. 5.28 and 5.29, respectively. It can be seen that variations in all the coefficients are very small within the examined range. Under the same value of fd, the width sweeping result of the rectangular groove is given in

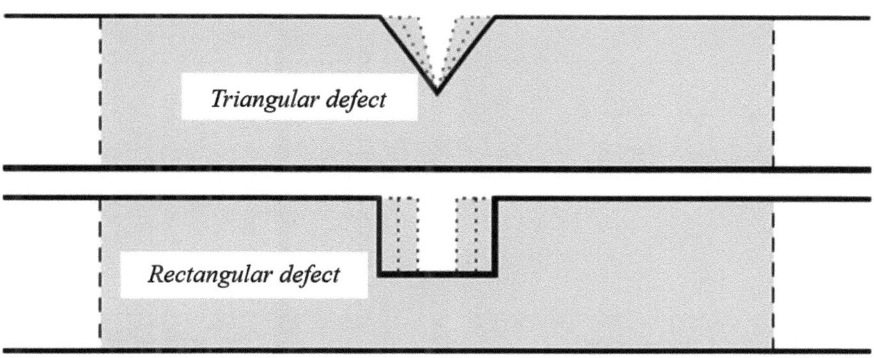

Fig. 5.27 The sweeping model of the external defect

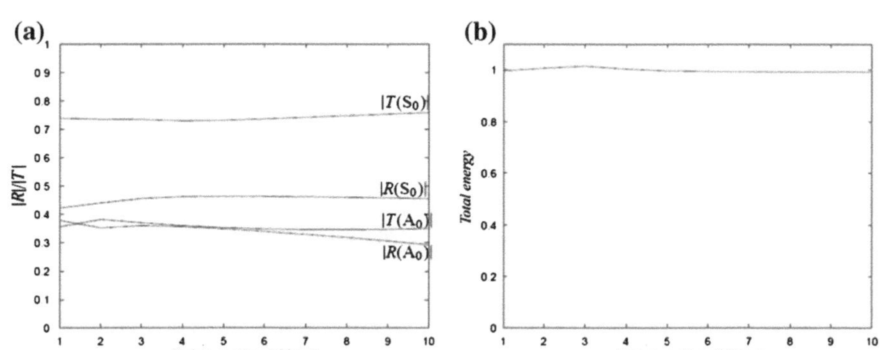

Fig. 5.28 The fd = 1000 Hz × mm S_0 mode incidence for the external triangular defect. **a** Reflection and transmission coefficient. **b** Total energy

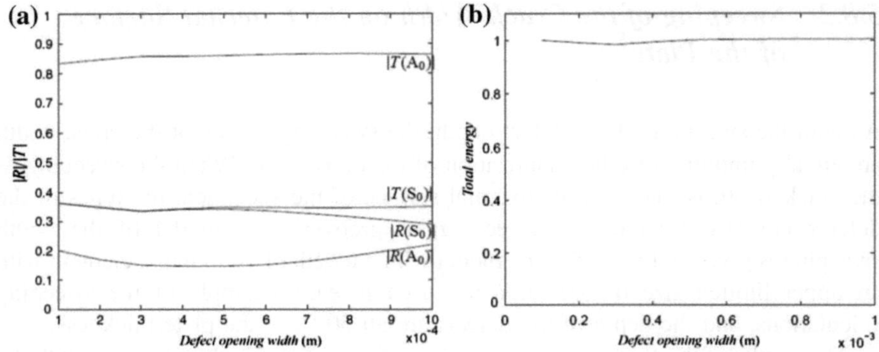

Fig. 5.29 The fd = 1000 Hz × mm A_0 mode incidence for the external triangular defect. **a** Reflection and transmission coefficient. **b** Total energy

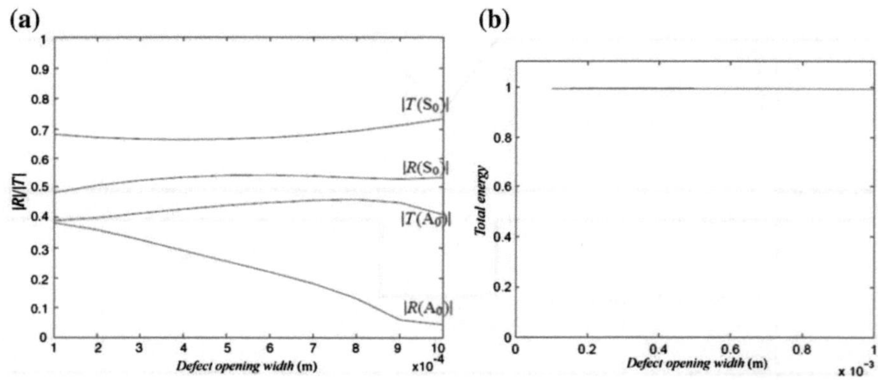

Fig. 5.30 The fd = 1000 Hz × mm S_0 mode incidence for the external rectangular defect. **a** Reflection and transmission coefficient. **b** Total energy

Figs. 5.30 and 5.31. Thus, it is found that the width change has a more serious effect on the coefficients in the specimen with the rectangular defect than in the specimen with the triangular defect.

5.8.3 Sweeping of the Frequency Thickness Product in the Plate with External Defect

Only the triangular groove is considered in this book for the sweeping of the frequency thickness product of the external defect in the plate. The opening width of the defect is 0.2 mm, the length of the model is 10 mm, the maximum size of the element is 0.02 mm, and the depth of the defect is 50 % of the plate thickness. The

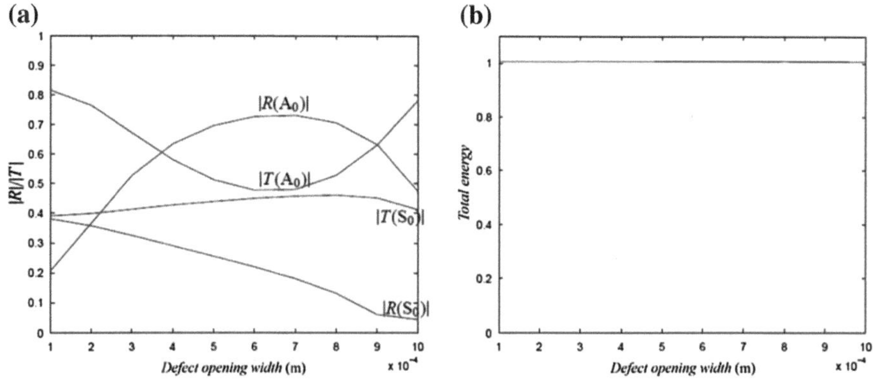

Fig. 5.31 The fd = 1000 Hz × mm A_0 mode incidence for the external rectangular defect. **a** Reflection and transmission coefficient. **b** Total energy

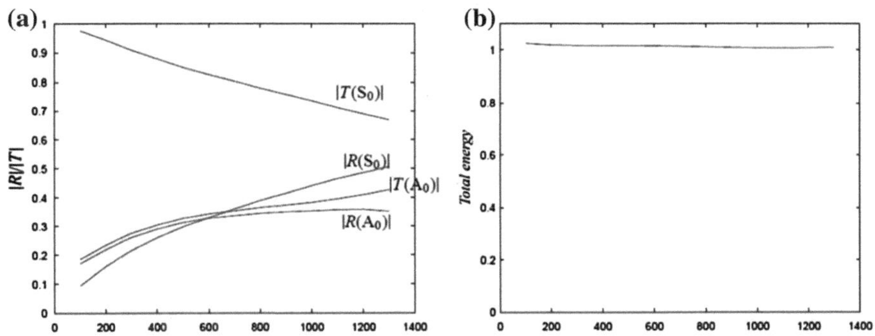

Fig. 5.32 The fd = 1000 Hz × mm S_0 mode incidence for the external rectangular defect. **a** Reflection and transmission coefficient. **b** Total energy

sweeping results of the S_0 and A_0 *Lamb wave* mode incidence are given in Figs. 5.32 and 5.33, respectively. The thickness of the plate is fixed at 1 mm for the sweeping of the frequency thickness product, while the frequency is the only flexible item.

5.9 Model and Numerical Simulation of Internal Defect in the Plate

The reason for studying the interactions of the *Lamb wave* with an internal defect in the plate lies in two facts. First of all, the plate to be tested could possibly contain internal defects, so the results from this study provide guidelines for the detection of

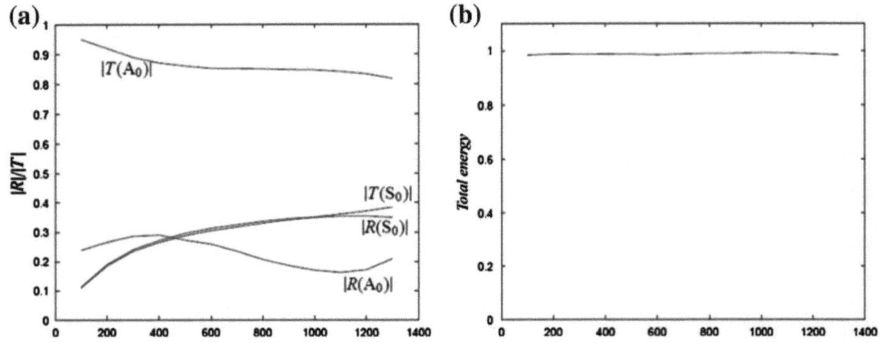

Fig. 5.33 The fd = 1000 Hz × mm A_0 mode incidence for the external rectangular defect. **a** Reflection and transmission coefficient. **b** Total energy

this kind of special defect; secondly, the relationship between the defect detection sensitivity using the *Lamb wave* and the wave structure (or mode shape) of the selected operating/working point of the *Lamb wave* could be researched based on the obtained results.

From the view point of realization by the BEM program, there is a difference in studying the simulation model of the internal defect in the specimen compared with the external defect in the specimen discussed above: the geometrical structure of the internal defect in the specimen is, in fact, a multiply connected domain that has a double-boundary containing both an inner and an outer boundary, which needs special treatment.

A double-boundary model containing an internal opening is shown in Fig. 5.34. It is assumed here that the line connecting point A on the inner boundary Γ_I and point B on the outer boundary Γ_E splits the original double-boundary region from a single-boundary region, so the boundary integration equation is valid in this new single-boundary region. The boundary integration can be written as the following equation:

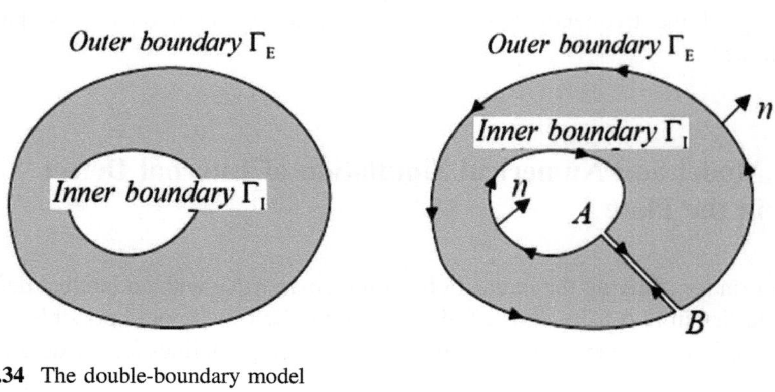

Fig. 5.34 The double-boundary model

$$\int = \int_{\Gamma_E} + \int_{BA} + \int_{\Gamma_I} + \int_{AB}$$

The integral variables are omitted here for simplicity. The integration direction is marked with arrows in Fig. 5.34. For the two integrals on the line AB, there is the following relationship:

$$\int_{BA} = -\int_{AB}$$

So the total boundary integration is listed as follows:

$$\int = \int_{\Gamma_E} + \int_{\Gamma_I}$$

That is, the boundary integration equation is still valid for the original double-boundary model as long as the boundary directions of the inner and outer boundaries follow this rule: When the observation point moves along the boundary direction, the outer normal $\boldsymbol{n}$ always points to the right, so the outer boundary direction is anticlockwise. The inner boundary direction is clockwise. The directions of the outer normal of the two boundaries are also marked in Fig. 5.34. The set of homogenous linear equations to solve is derived from the boundary integration (5.6), so the original BEM process can be applied to the double-boundary model provided that the rule of the selection of the boundary direction is satisfied.

Only internal rectangular slot defects are considered here. Cracks are simulated with slots, and the model is shown in Fig. 5.35. For the external defect model, one end of the opening of the defect is fixed at the surface of the plate, whose variation is relatively simple. Obviously, the internal defect's variation is more complex.

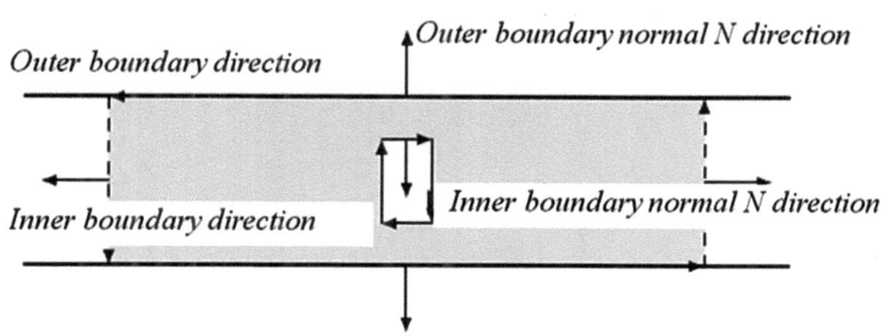

Fig. 5.35 Plate internal defect model with *inner* and *outer* boundaries

5.9.1 *Internal Crack's Height in the Plate*

The depth scanning results of the symmetric rectangular defect with respect to the middle plane of the plate are shown in Figs. 5.36 and 5.37, when the S_0 and A_0 *Lamb* wave modes at fd = 1000 Hz × m are taken as the incident wave.

The length of the model used here is 10 mm, with a plate thickness of 1 mm, so the range of the defect's height corresponds to 10–90 % of the plate thickness. The width of the internal rectangular defect is 0.1 mm, and the sizes of the elements of the non-defect part of the boundary and the defect part of the boundary are limited to 0.02 and 0.01 mm, respectively. In these simulations, the vertical centers of the defect and the plate are always at the same place, or the defect grows in the upward and downward directions symmetrically. Similar to the case of free edge reflection, the symmetry of the defect (and hence the whole model) means that only those reflected and transmitted modes of the same type as the incident mode can exist (the

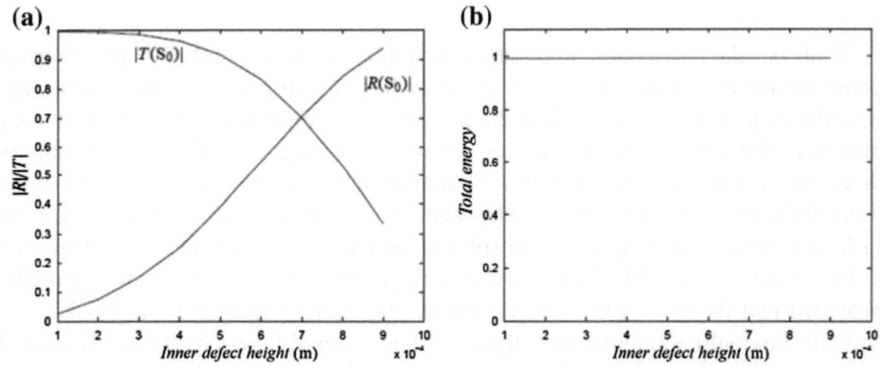

Fig. 5.36 Height sweeping of internal rectangular slot's height, fd = 1 MHz × mm S_0 incident mode. **a** Reflection and transmission coefficient. **b** Total energy

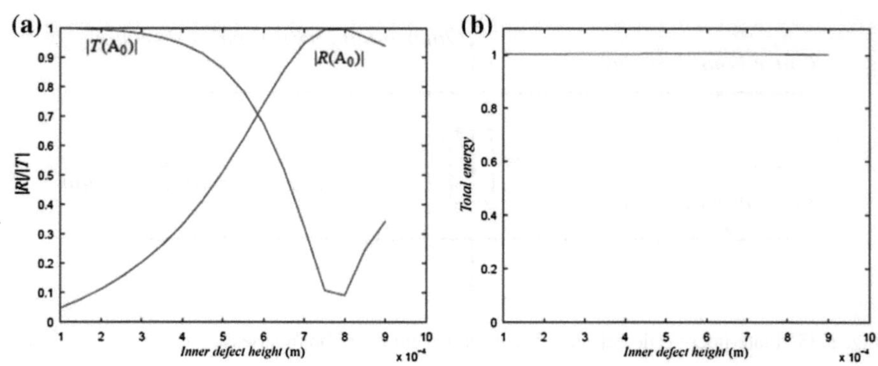

Fig. 5.37 Height sweeping of internal rectangular slot's height, fd = 1 MHz × mm A_0 incident mode. **a** Reflection and transmission coefficient. **b** Total energy

S mode incidence only generates *S* mode reflections and transmissions, and the *A* mode incidence only generates *A* mode reflections and transmissions). From the previous results, it can be learned that the *R* and *T* coefficients change monotonically with increasing defect height for the S_0 mode incidence, while this is not true for the A_0 mode incidence. The normalized total energy is always near 1, so the simulations for the internal defect model are verified. From here onwards, the energy curves will be omitted for simplicity.

5.9.2 Internal Crack's Width in the Plate

The width scanning results of the symmetric rectangular defect at the middle plane of the plate are shown in Figs. 5.38 and 5.39. S_0 and A_0 *Lamb wave* modes at fd = 1000 Hz × m are chosen as the incident wave.

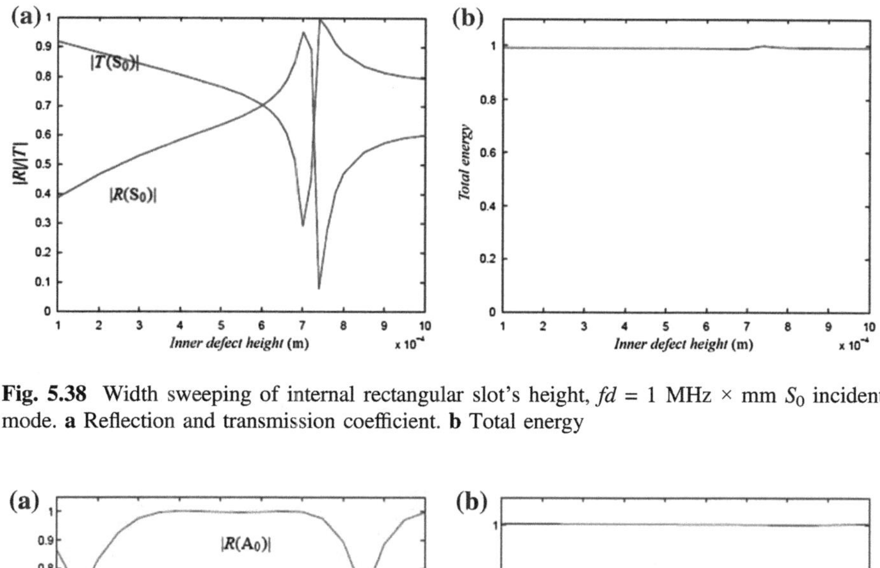

Fig. 5.38 Width sweeping of internal rectangular slot's height, fd = 1 MHz × mm S_0 incident mode. **a** Reflection and transmission coefficient. **b** Total energy

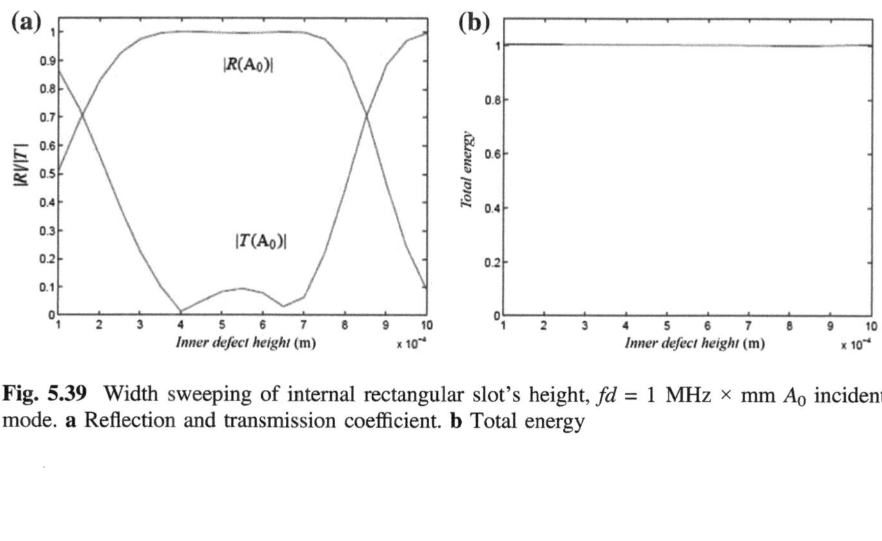

Fig. 5.39 Width sweeping of internal rectangular slot's height, fd = 1 MHz × mm A_0 incident mode. **a** Reflection and transmission coefficient. **b** Total energy

The model's length is 10 mm, the plate thickness is 1 mm, and the height of the internal rectangular defect is 0.5 mm, i.e., 50 % of the plate thickness. The upper limitations of the element sizes of the non-defect and defect parts of the boundary are 0.02 and 0.01 mm, respectively. The symmetry of the defect model means that only reflected and transmitted modes with the same type as the incidence will exist. From the results shown in Figs. 5.38 and 5.39, it can be observed that when the S_0 *Lamb wave* mode is taken as the incident wave, neither $|R(S_0)|$ nor $|T(S_0)|$ changes monotonically with increasing slot width, and they change abruptly near a width of 0.7 mm. The coefficients for A_0 mode incidence do not change monotonically either. The energy curve shows that the normalized total energy is always near 1, so the simulations can be verified.

Figures 5.40 and 5.41 show the results for the R and T coefficients' variations with a non-symmetrical internal rectangular slot width for fd = 1000 Hz × mm S_0 and A_0 incident modes, respectively.

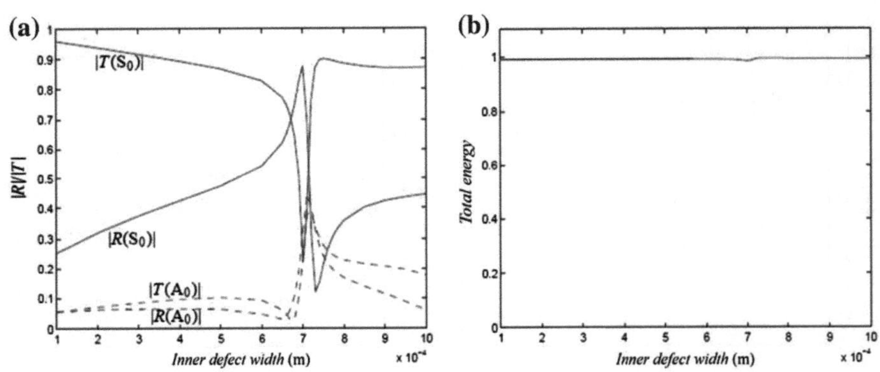

Fig. 5.40 Width sweeping of non-symmetrical internal rectangular slot's width, *fd* = 1 MHz.mm S_0 incident mode. **a** Reflection and transmission coefficient. **b** Total energy

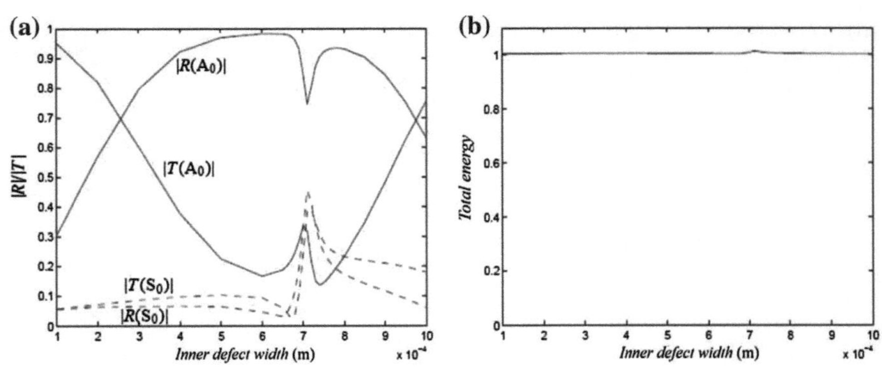

Fig. 5.41 Width sweeping of non-symmetrical internal rectangular slot's width, *fd* = 1 MHz × mm A_0 incident mode. **a** Reflection and transmission coefficient. **b** Total energy

In order to avoid confusion, the curves for the mode-converted coefficients are drawn using dashed lines. A 10 mm model length and 1 mm plate thickness are used. The height of the internal defect is 0.4 mm and the vertical center of the defect is 0.1 mm above the middle plane of the plate. The limits of the element sizes are 0.02 and 0.01 mm, respectively. From these figures, it can be observed that with the S_0 *Lamb wave* mode as incidence, neither $|R(S_0)|$ nor $|T(S_0)|$ changes monotonically. They change abruptly near a 0.7 mm width. This is similar to the A_0 mode incidence. The total energy obtained from the calculation is still near 1 all of the time, so the accuracy of the calculation can be verified.

5.9.3 Frequency Thickness Product of Internal Crack in the Plate

The scanning results of the frequency thickness product of the symmetric rectangular defect at the middle plane of the plate are shown in Figs. 5.42 and 5.43. S_0 and A_0 *Lamb wave* modes are selected as the incident wave.

The model's length is 10 mm and the plate thickness is 1 mm. The rectangular slot at the vertical center of the plate has a width of 0.1 mm and a height of 0.5 mm, i.e., 50 % of the plate thickness. The limits for the element sizes of the non-defect and defect parts of the boundary are 0.02 and 0.01 mm, respectively. Because the whole model is symmetric with respect to the middle plane of the plate, no mode conversion occurs.

In Figs. 5.44 and 5.45, the *Lamb waves'* interactions with the internal non-symmetrical rectangular defects are investigated. The reflection and transmission coefficients' variations with *fd* for the S_0 and A_0 *Lamb wave* modes were selected as the incident waves.

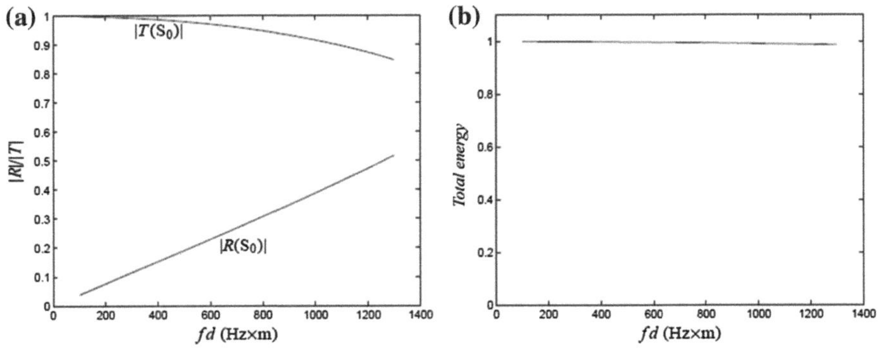

Fig. 5.42 The *fd* sweeping of symmetrical internal rectangular defect, S_0 incident mode. **a** Reflection and transmission coefficient. **b** Total energy

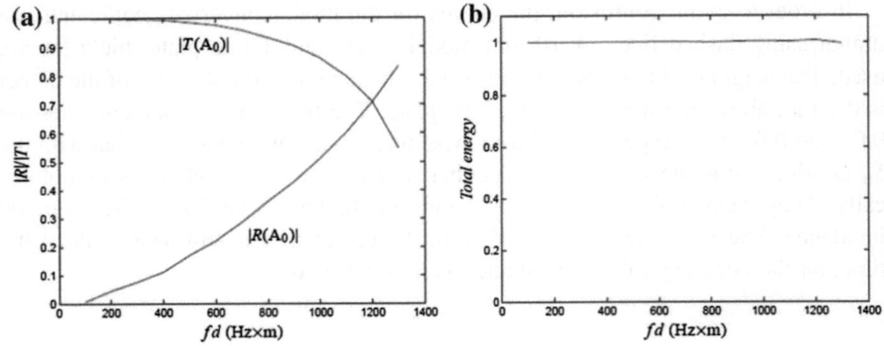

Fig. 5.43 The *fd* sweeping of symmetrical internal rectangular defect, A_0 incident mode. **a** Reflection and transmission coefficient. **b** Total energy

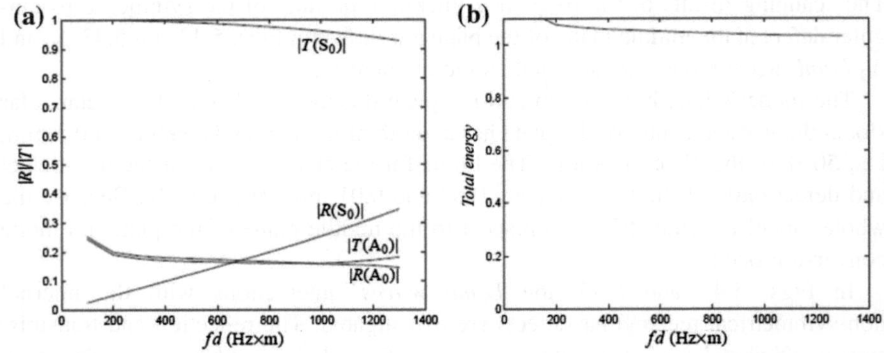

Fig. 5.44 The *fd* sweeping of non-symmetrical internal rectangular defect, S_0 incident mode. **a** Reflection and transmission coefficient. **b** Total energy

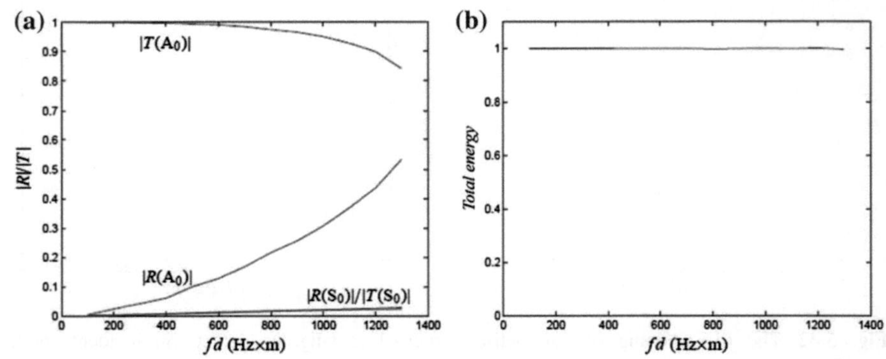

Fig. 5.45 The *fd* sweeping of non-symmetrical internal rectangular defect, A_0 incident mode. **a** Reflection and transmission coefficient. **b** Total energy

The model's length is 10 mm and the plate thickness is 1 mm. The height of the internal defect is 0.4 mm and its vertical center is 0.1 mm above the middle plane of the plate. The width of the defect is 0.1 mm. The limits for the element sizes of the non-defect and defect parts of the boundary are 0.02 and 0.01 mm, respectively. Because the defect investigated here is no longer symmetric with respect to the middle plane, there exist mode-converted reflected and transmitted waves. Under the chosen parameters, the coefficients for the mode of the same type as the incidence have similar trends of variation compared with the case of the symmetrical defect.

5.9.4 *Internal Crack's Movement Along the Vertical Direction*

The variations in reflection and transmission coefficients with the internal crack's movement along the vertical direction are shown in Fig. 5.46. The internal defect is 0.1 mm in width and 0.3 mm in height. The model's length is 10 mm and the plate thickness is 1 mm. The limits for the element sizes of the non-defect and defect parts of the boundary are 0.02 and 0.005 mm, respectively. The horizontal axis in the figure stands for the position of the defect's vertical center with respect to the middle plane. The range of the horizontal axis is from 0 to some place near the upper surface of the plate, so the result shown in the figure actually corresponds to a half plate thickness simulation and the calculated coefficient curves for the other half are symmetrically the same as the drawn curves. It is shown in Fig. 5.46 that the position of the internal defect hardly influences the S_0 mode coefficients, while the A_0 mode coefficients will increase with the defect's increasing distance from the middle plane.

Figure 5.47 shows the case when the A_0 *Lamb wave* mode is taken as the incident wave. Other parameters are the same as for the previous simulation example. From Fig. 5.47, it can be observed that the position of the internal defect has some influence on the A_0 mode reflection coefficient.

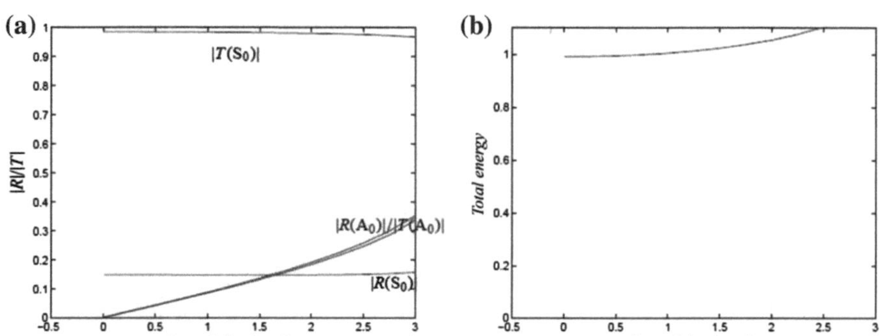

Fig. 5.46 Internal crack's movement along the vertical direction, S_0 incident mode. **a** Reflection and transmission coefficient. **b** Total energy

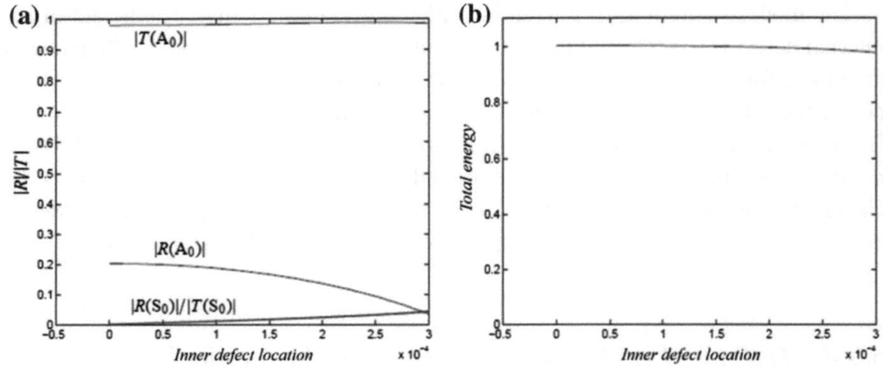

Fig. 5.47 Internal crack's movement along the vertical direction, A_0 incident mode. **a** Reflection and transmission coefficient. **b** Total energy

5.10 Quantitative Crack Detection by Electromagnetic Ultrasonic Guided Waves

The main purpose in detecting cracks using electromagnetic ultrasonic guided waves is to locate and quantitatively size the cracks in the pipe. The information on the crack's location can be determined from the time of the appearance of the characteristic wave signal, but it is relatively difficult to quantitatively size the crack. The characteristics of the reflected and transmitted signals of the ultrasonic wave are dependent on the geometric composition of the crack, which is a positive process. But the process of crack quantification is an inverse process of derivation of the geometric construction of the crack by the waveform of the reflected and transmitted waves. In general, it is very difficult to solve the inverse process, and the results may be not unique. In most cases, the obtained result is only equivalent to the standard defect.

At present, there are studies carried out rarely on crack quantification by guided waves. One of the reasons is that the inverse problems of guided wave propagation and scattering are quite complex, and it generally has to take the positive simulated results as a reference. The linear relationship between the defect depth and the frequency thickness product that corresponds to the minimum point of the transmission coefficient curve in the process frequency thickness product scanning is studied by Rose. However, the scanning of the frequency thickness is very difficult; for instance, if we want to make a scan for the piezoelectric transducer with the oblique incidence longitudinal wave for a certain wave mode, the incident frequency should be changed constantly, and the incident angle calculated by Snell's law should also be flexible. Even though the operating angle of the angle beam transducer is adjustable, it is also difficult to operate and time-consuming. It can be even more difficult using the electromagnetic ultrasonic transducer since the wave length of the guided wave should be changed with the inspection frequency.

The period of the transducer coil used to excite the guided wave should also be changed, and this change of structure is almost impossible to realize, which makes it more difficult to be applied for online testing [2].

Considering the current research status and problems of crack quantification by guided wave, this book is proposing two new methods for crack quantification using BEM simulations.

The crack depth quantification firstly investigates the rectangular groove with a given width, specifically, the functions between the crack depth and reflection, using transmission coefficients by the method of curve fitting. Considering the non-monotonic change of the reflection and transmission coefficients with the crack depth, it makes the solution of the inverse problem more difficult; thus, the range of working points where the coefficients change monotonically is selected to detect the cracks.

As an example, the S_0 *Lamb wave* mode at $fd = 1000$ Hz × m is selected as the incident wave to detect a rectangular crack with an opening width of 0.1 mm in the specimen. The simulated calculation shows that the changes of reflection and transmission coefficients change monotonically with the crack depth in this situation. The results of the BEM simulation are indicated by the discrete points in the coordinate system, where the horizontal coordinate axis represents the reflection or transmission coefficient and the vertical coordinate axis represents the percentage of the crack depth. The simulation process involves exchanging the coordinate axes; expressing the crack depth as the function of the reflection and transmission coefficients; making the curve fit the discrete points, and then studying the corresponding law. For the relationship between the depth and the reflection coefficient, we tried 1–6 times polynomial fitting and 1 and 2 terms exponential fitting. From the fitting results in the fitting schemes, it can be found that a 6-order polynomial fitting is relatively the best choice. The analytical equation obtained from the fitting can be expressed as follows:

$$y = -752.7x^6 + 2478x^5 - 3268x^4 + 2199x^3 - 770.3x^2 + 210.3x + 2.842$$

In the above equation, x is the reflection coefficient and y is the depth percentage of the rectangular groove crack.

The curve of the quantified equation obtained from the 6 order polynomial fitting is shown in Fig. 5.48 as well as the discrete points obtained from the BEM-simulated calculation. It is shown in the figure that the fitting result is a good expression of the relationship between the crack depth and the reflection coefficient of the guided wave; thus, the fitting result can be taken as a foundation of the quantification of the crack depth.

Using a similar method, the relationship between the crack depth and the transmission coefficient of the guided wave is built by the 6 order polynomial fitting. We get the following expression.

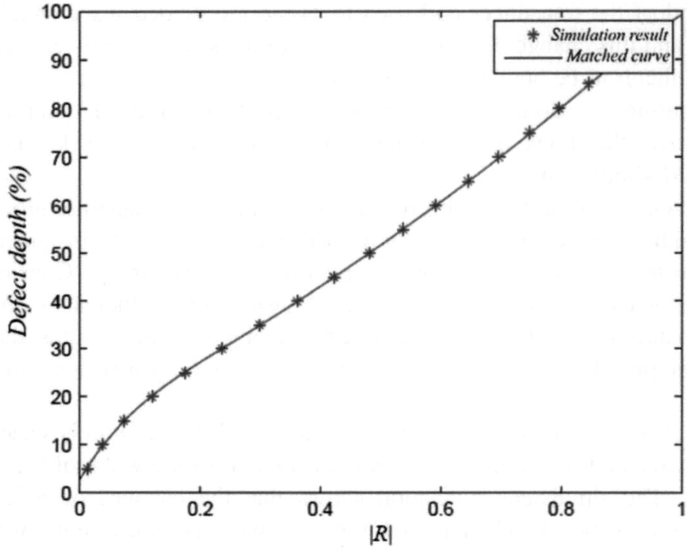

Fig. 5.48 The fitting result of the reflection coefficient-crack depth, S_0 mode at $fd = 1000$ Hz × mm of the rectangular groove crack with a width of 0.1 mm in the specimen

$$y = -5940x^6 + 18850x^5 - 23640x^4 + 14860x^3 - 4881x^2 + 700.7x + 59.7$$

In the above equation, x is the transmission coefficient and y is the depth percentage of the rectangular groove crack. The fitting curve and the discrete point obtained from the simulation are shown in Fig. 5.49. Since the curves of the reflection and transmission coefficients of the tested S_0 mode are monotonous and the crack width is already given, the fitting curves of the guided wave reflection and transmission coefficients can be used to quantify the crack depth of the rectangular groove.

Another example is the symmetric rectangular groove defect in the specimen. The width of the defect is fixed at 0.1 mm and the S_0 *Lamb wave* mode at $fd = 1000$ Hz × mm is taken as the incident wave. The derived fitting equation used to express the relationship between the depth and the reflection coefficient is written as follows:

$$y = -549x^6 + 2056x^5 - 2961x^4 + 2174x^4 - 897.4x^2 + 270.8x + 3.766$$

In the above equation, x is the reflection coefficient. The fitting curve is shown in Fig. 5.50.

Based on the above-quantitative solution of the fitting curve, the defect is assumed as the standard rectangular groove defect and the crack width is given.

In the following section, the quantification of the crack width and the depth is considered in parallel. The S_0 *Lamb wave* mode at $fd = 1000$ Hz × mm is taken as

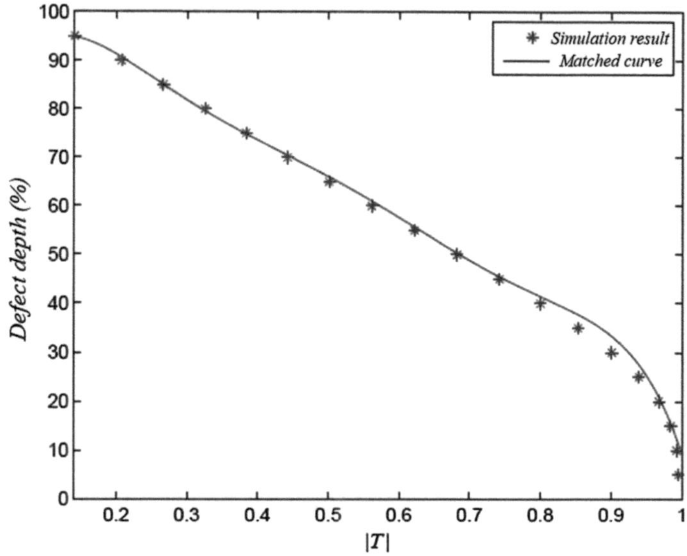

Fig. 5.49 The fitting result of the transmission coefficient-crack depth, S_0 mode at fd = 1000 Hz × mm of the rectangular groove crack with a width of 0.1 mm in the specimen

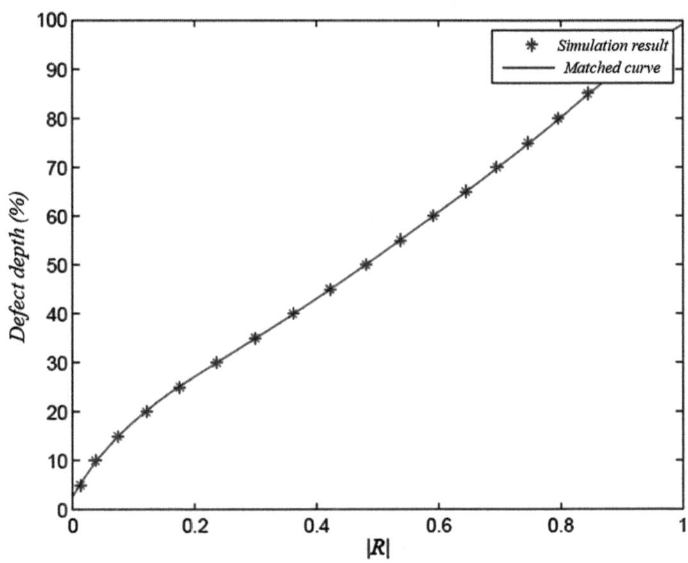

Fig. 5.50 The fitting result of the reflection coefficient-crack depth, S_0 mode at fd = 1000 Hz × mm of the symmetric rectangular groove crack with a width of 0.1 mm in the specimen

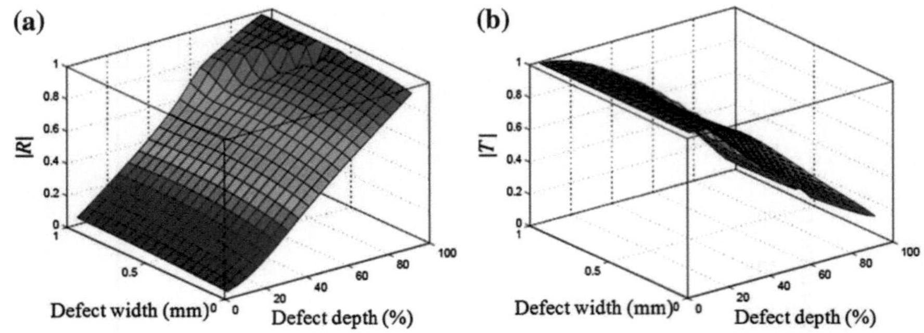

Fig. 5.51 The curves of reflection and transmission coefficients for the rectangular groove defect, when the S_0 *Lamb wave* mode at fd = 1000 Hz × mm is the incident wave. **a** The curve of the reflection coefficient. **b** The curve of the transmission coefficient

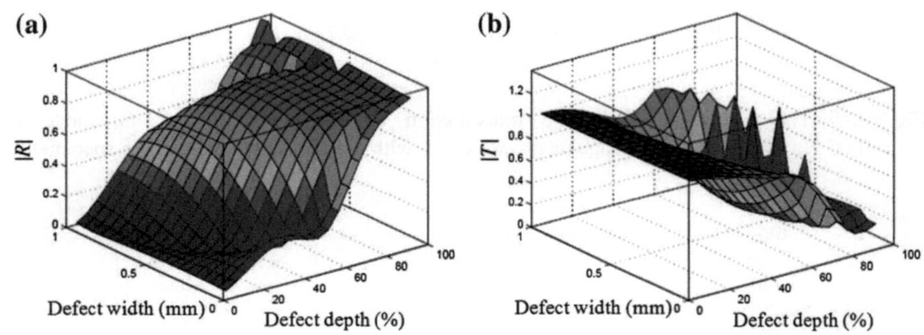

Fig. 5.52 The curves of reflection and transmission coefficients for the rectangular groove defect, when the A_0 *Lamb wave* mode at fd = 1000 Hz × mm is the incident wave. **a** The curve of the reflection coefficient. **b** The curve of the transmission coefficient

the incident wave and the rectangular groove defect on the plate surface is taken as the sample. The percentage of the defect depth is taken as the x-coordinate, the defect width is taken as the y-coordinate, and the reflection and transmission coefficients are taken as the z-coordinate. Based on the data obtained from the simulations, the curves of the reflection and transmission coefficients are provided in Fig. 5.51. It is shown in the figures that these two coefficients of the S_0 *Lamb wave* mode are all monotonous in the depth direction (Fig. 5.52).

Based on the curves shown in Fig. 5.51, if the reflection and transmission coefficients are given, the algorithm that can be used for the quantification of the rectangular groove width and depth in parallel is proposed as follows:

(1) Make the plane correspond to the given reflection coefficient in the reflection coefficient curve, and the subpoint of the intersection line being between the plane and the curve satisfies the requirement of the reflection coefficient;

(2) Make the plane correspond to the given transmission coefficient in the transmission coefficient curve, and the subpoint of the intersection line being between the plane and the curve satisfies the requirement of the transmission coefficient;
(3) The horizontal and vertical coordinates of the intersection point of the two projection lines represent the quantized percentile depth and width of the rectangular groove crack corresponding to the given reflection and transmission coefficients.

For example, in the case of the rectangular groove defect with a width of 0.26 mm and a 40 % of depth, its reflection and transmission coefficients are 0.3975 and 0.7848, respectively. The projection line of the intersection line between the reflection and transmission curves and the plane composed of the given coefficients on the surface of the percentile depth and width is shown in Fig. 5.53. The vertical and horizontal coordinates of its intersection point are the quantitative depth percentage 39.9995 % and the quantitative width percentage 0.2614 mm, respectively.

A set of quantitatively calculated examples of the rectangular groove cracks are provided in Table 5.1. It is shown in Table 5.1 that the quantitative depth of the rectangular groove crack has a better accuracy than the quantitative width of the rectangular groove crack, for the reason that the variation in the reflection and transmission coefficients is relative small as the crack width changes.

A condition for the application of the above algorithm is that the curves of the reflection and transmission coefficients should be monotonous within the examined

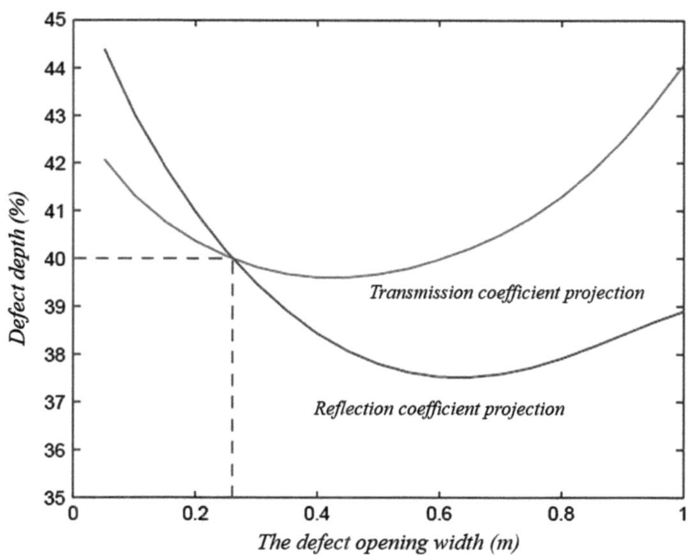

Fig. 5.53 The projection of the intersection line between the coefficient curve and the plane of the reflection and transmission coefficient

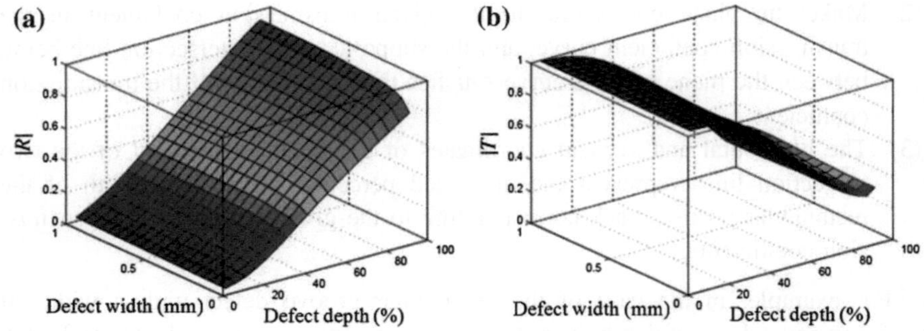

Fig. 5.54 The curves of reflection and transmission coefficients for the triangular groove defect, when the S_0 *Lamb wave* mode at $fd = 1000$ Hz × mm is the incident wave. **a** Reflection coefficient curve. **b** Transmission coefficient curve

scope of the defect size, for instance, for the condition of the S_0 *Lamb wave* mode at $fd = 1000$ Hz × mm being taken as the incident wave.

The quantitative scheme based on the curves of rectangular groove reflection and transmission coefficients can only be used for the quantification of the rectangular groove defect. The corresponding results of the triangular defect will be checked in the following section. The curves of the reflection and transmission coefficients for the condition of the S_0 *Lamb wave* mode as the incident wave in the triangular groove defect are shown in Fig. 5.54. It is shown in the figure that the change in the coefficient curves of the S_0 *Lamb wave* mode is monotonous in the depth direction, which is similar to that of the rectangular groove (Fig. 5.55).

Using the curves of the reflection and transmission coefficients, the quantitative research for the opening width and depth of the triangular groove defect is conducted using the same method as that for the quantification of the rectangular groove defect. A set of examples and their results are given in Table 5.2. It can be

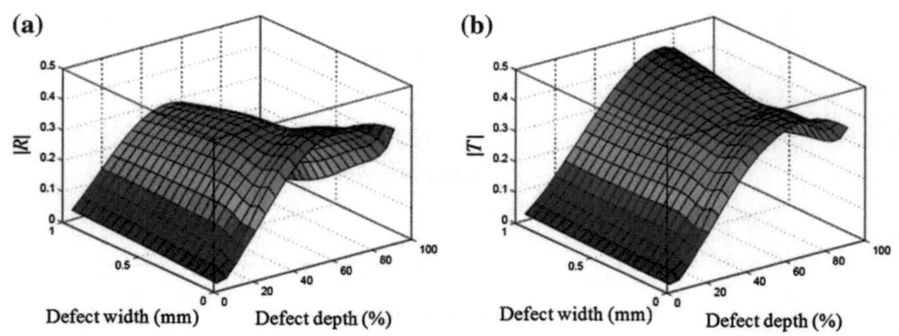

Fig. 5.55 The curves of reflection and transmission coefficients for the triangular groove defect, when the A_0 *Lamb wave* mode at $fd = 1000$ Hz × mm is the incident wave. **a** Reflection coefficient curve. **b** Transmission coefficient curve

Table 5.2 The quantitative calculated examples of the triangular groove cracks

Width (mm)	Depth (%)	Reflection *R*	Transmission *T*	Width (mm)	Depth (%)
0.26	40	0.33396	0.84023	0.2820	39.8454
0.45	23	0.1405	0.9626	0.5159	22.6980
0.15	55	0.4904	0.6748	0.1491	55.0176
0.56	22	0.1342	0.9664	0.6346	21.6318
0.67	46	0.4146	0.7833	0.6768	46.0067

seen that the accuracy of the depth quantification for the triangular groove crack is better than that of the width quantification.

Based on the same principle, elliptical defects and other types of standard defects can also be inspected. A systematical database is composed of the reflection and transmission coefficient curves of all the standard defects, which is beneficial to the quantification of crack-like defects whose shapes are unknown.

The quantitative method given in this section is mainly targeted at the standard defect, that is, the model of an artificial defect with a normalized shape; thus, it has certain limitations. The establishment and design of the new algorithm which is more applicable is the research orientation of the intensive study of the inverse problem of the defect quantification.

References

1. Wang, S., Huang, S., Zhao, W.: BEM simulation of Lamb wave propagation in plate with crack. China Mech. Eng. **20**(8), 887–891 (2009). (In Chinese)
2. Wang, S., Huang, S., Zhao, W., et al.: Approach to Lamb wave lateral crack quantification in elastic plate based on reflection and transmission coefficients surfaces. Res. Nondestr. Eval. **21** (4), 213–223 (2010)

Chapter 6
Finite Element Simulation of Ultrasonic Guided Waves

The commercial software *Abaqus* is used in this chapter to simulate the *Lamb wave* propagation in the steel plate, the circumferential *Lamb wave* in the pipe, and the *L*- and *T*-type guided waves in the axial direction of the pipe, as well as their interactions with the transverse defects. The "*transverse*" means that the extensional direction of the defect is perpendicular to the propagation direction of the guided waves. As a classic example, a stress corrosion crack on a gas transmission pipeline will generally stretch along the axial direction of the pipe. When the circumferential guided wave in the pipe is used, it is possible that the propagation path is perpendicular to the stretching direction of the tested defect.

6.1 The Explicit Integration Finite Element Method

The corresponding simulation of the ultrasonic guided wave is based on the explicit time integration of the finite element method. The simulation principle of this method is given.

The selection of explicit time integration determines that the *Abaqus/Explicit* should be automatically taken as the finite element solver. The central difference method is used in the application of the *Abaqus/Explicit* to carry out the explicit time integration for the kinematic equation. It is to use the kinestate of a time increment to calculate the kinestate of another time increment.

At the beginning of the one-time increment, an expression of the dynamic balance with the following form is established: $\mathbf{M}\ddot{u} = \mathbf{P} - \mathbf{I}$, where $\mathbf{M}$ is the node mass matrix, $\ddot{u}$ is the node acceleration, $\mathbf{P}$ is the applied external force, and $\mathbf{I}$ is the acting force of the internal elements.

S. Huang et al., *Electromagnetic Ultrasonic Guided Waves*, Springer Series in Measurement Science and Technology, DOI 10.1007/978-981-10-0564-0_6

At the starting point of the time increment, that is, time t, the acceleration is $\ddot{u}_t = (\mathbf{M})^{-1}(\mathbf{P} - \mathbf{I})_t$. Then, using the central difference method to create the time integral for the acceleration, the velocity of the present-time increment midpoint $\dot{u}_{t+\frac{\Delta t}{2}}$. is obtained from the velocity of the previous time increment midpoint $\dot{u}_{t-\frac{\Delta t}{2}}$, as shown in the following equation:

$$\dot{\mathbf{u}}_{t+\frac{\Delta t}{2}} = \dot{\mathbf{u}}_{t-\frac{\Delta t}{2}} + \frac{\Delta t_{t+\Delta t} + \Delta t_t}{2} \ddot{u}_t$$

The displacement $\mathbf{u}_{t+\Delta t}$ at the end of the time increment is obtained from the starting point of the time increment $\mathbf{u}_t$, as shown in the following equation:

$$\mathbf{u}_{t+\Delta t} = \mathbf{u}_t + \Delta t_{t+\Delta t} \dot{u}_{t+\frac{\Delta t}{2}}$$

The earlier calculations are all based on nodes. Then, the strain increment, stress, and the internal force of the node $I_{t+\Delta t}$ are all calculable. The whole calculation process will be repeated in the next-time increment.

When the explicit time integration method is used, in order to get accurate simulation results, the time increment should be small enough so that the acceleration within a time increment is approximately constant. The explicit time integration method is very suitable for the high-speed dynamic problems, so that small time increments are needed in this case to get the high-resolution solution.

6.2 Finite Element Simulation of the Lamb Wave in the Plate [1]

6.2.1 Establishment of the Lamb Wave Equation in the Elastic Plate

The coordinate system and the *Lamb wave* displacement component are shown in Fig. 6.1. In Fig. 6.1, O is the origin of the coordinates, $d = 2h$ is the plate thickness, x and y are the coordinate axes, and u and v are the displacement components of the x and y coordinate axes. The extension directions of the plate are along the positive and negative directions of the x-axis.

One basic assumption for the *Lamb wave* propagation in the elastic plate is that it is traction-free on the top and bottom surfaces of the plate, and the displacement vector in the wave propagation process contains only the x-axis (u) and y-axis (v) non-zero components; thus, this is a plane-strain problem.

Here, the procedure to derive the *Lamb wave* equation is listed in the following sections:

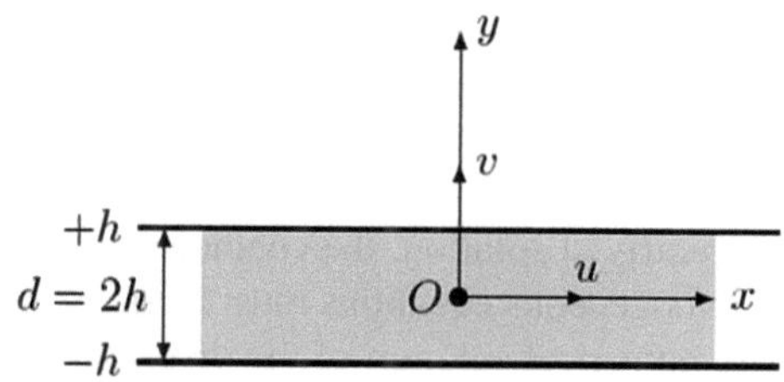

Fig. 6.1 The coordinate system in the plate and the displacement components of the *Lamb wave*

(1) The elastic-dynamic control equation, that is, the *Navier* wave equation, is shown in the following equation:

$$\mu\nabla^2\mathbf{u} + (\lambda + \mu)\nabla\nabla \cdot \mathbf{u} = \rho\frac{\partial^2\mathbf{u}}{\partial t^2}$$

In the above equation, **u** is the displacement vector, λ and μ are *Lame constants*, and ρ is the mass density. The potential function is used to decompose the displacement vector **u** (*Helmholtz decomposition*) so that the displacement vector can be expressed as $\mathbf{u} = \nabla\varphi + \nabla \times \boldsymbol{\psi}$, where φ and $\boldsymbol{\psi}$ are the scalar and vector potential functions, respectively. The potential function is used to express the elastic-dynamic wave equation. As the solution of the wave equation, the potential function has the following form:

$$\varphi = \Phi(y)\mathrm{e}^{i(kx-\omega t)}, \quad \psi = \boldsymbol{\psi}(y)\mathrm{e}^{i(kx-\omega t)}$$

In the above equation, k is the wave number, ω is the angular frequency, and ψ is the Z component of the $\boldsymbol{\psi}$-vector. The expressions of these potential functions describe the traveling wave along the x-direction and the standing wave along the y-direction.
$\Phi(y)$ and $\Psi(y)$ are assumed to be the following forms:

$$\Phi(y) = A_1\sin(py) + A_2\cos(py), \Psi(y) = B_1\sin(qy) + B_2\cos(qy)$$

In the above equation, A_1, A_2, B_1, and B_2 are unknown coefficients, and we have the equation listed as follows:

$$p^2 = \frac{\omega^2}{C_L^2} - k^2, \quad q^2 = \frac{\omega^2}{C_T^2} - k^2$$

In the above equation, C_L is the velocity of the longitudinal wave and C_T is the velocity of the shear wave.

(2) According to the strain–displacement relationship and the stress–strain relationship in elastic theory, the potential functions φ and ψ are used to express the stress where unknown coefficients are involved.

(3) Based on the assumption of a traction-free boundary condition on the surfaces, the equation set generated by the boundary condition is actually a system of linear homogeneous equations, and the coefficient in the stress expression discussed above is now an unknown value.
(4) In order to get the untrivial solution, the coefficient matrix determinant of the system of linear homogeneous equations must be 0; the dispersion equation is thus obtained. The numerical solution of the dispersion equation is expressed as the dispersion curve of the phase velocity (C_p). The dispersion curve of the group velocity is obtained from the frequency dispersion curve of the phase velocity (C_p), as shown in the following equation, where f is the frequency:

$$C_g = \frac{\mathrm{d}\omega}{\mathrm{d}k} = \frac{C_p^2}{C_p - f\frac{\mathrm{d}C_p}{\mathrm{d}f}}$$

According to the symmetry of the displacement components about the middle plane of the plate, the *Lamb wave* can be decomposed into two types: a symmetrical type and an antisymmetrical type.

The dispersion equation of the symmetrical type is listed as follows:

$$\frac{\tan(qh)}{\tan(ph)} = -\frac{4k^2pq}{(k^2 - q^2)^2}$$

The dispersion equation of the antisymmetrical type is listed as follows:

$$\frac{\tan(qh)}{\tan(ph)} = -\frac{(k^2 - q^2)^2}{4k^2pq}$$

The dispersion curve of the *Lamb wave* propagation in the steel plate is shown in Fig. 6.2. The symmetrical-type guided wave is marked as "S_n" and the antisymmetrical-type guided wave is marked as "A_n." The range of the subscript n is from 0 to infinity. The x-axis thickness product (fd) and its unit is "Hz × mm"; the y-axis is the phase velocity c_p or group velocity c_g and its unit is "km/s." The material parameters are: *Young's modulus* $E = 207 \times 10^9$ Pa, *Poisson ratio* $v = 0.3$, mass density $\rho = 7800$ kg/m^3.

6.2.2 *Finite Element Simulation of the Lamb Wave in the Plate*

The simulation based on *Abaqus* includes three main stages: pretreatment, calculation, and post-processing. Each of these stages corresponds to one or several function modules of the *Abaqus/CAE* environment.

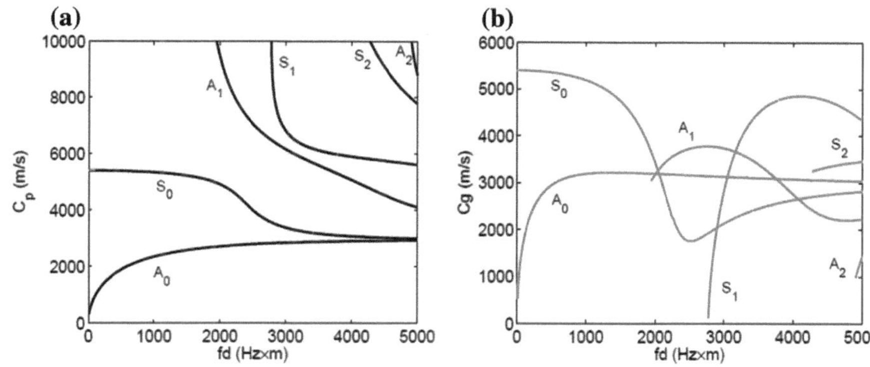

Fig. 6.2 The dispersion curve of the *Lamb wave* in the steel plate, with a plate thickness of 1 mm. **a** Phase velocity dispersion curve. **b** Group velocity dispersion curve

In the preliminary treatment process, in general, the simulation model will be defined by the *Abaqus/CAE* environment. The purpose is to generate the suitable solving input file (.inp file).

(1) The *Part* module. This module allows users to establish the part to correspond with the different parts of the simulated structure in the real world. Each part is a blueprint, which should be implemented as a specific instance first, and then it can be analyzed in *Abaqus*. In the simulation model used in this section, a part composed of a planar waveguide geometric structure and character is drawn in *Sketcher*.
(2) The *Property* module. The material parameter is defined in this module, and then is referenced in the section definition. Then, the defined section is given to the part to complete the specification of the material parameter.
(3) The *Assembly* module. The whole model is an *Assembly*, which is composed of one or several instances, but the instance generated in this module is implemented from the defined part. The instance can be selected from the dependent instance or independent instance. This selection appoints that the grid generated on the part is taken as the blueprint, and then inherits all the instances realized by this part, or generates on the instance itself directly. At this moment, each instance has the grid of itself.
(4) The *Step* module. Except for the internal initial step, the analytical procedure should also be established. In this simulation, the type of the analytical procedure is appointed as the "*dynamic explicit.*" This is the main step to simulate the *Lamb wave* propagation and its interaction with the defect. The time period parameter defined in the step decides the simulation time of duration, so its value should be selected carefully to catch the useful guided wave signal characteristics and avoid the useless characteristics. The *Field Output* and *History* output are also defined in this module. The *Field Output* indicates the following requests sent to the solver: The solver is required to generate the values of some selected variables of the whole model or partial region of

the model on the time points, to establish the carton composed by the isoline map that is used to describe the wave propagation. By contrast, *History Output* requests the solver, under the precise time step, to calculate the time history of the selected variables of the specified space points in the model, mainly used to generate the time waveform of the variables on those points.

(5) The *Load* module. The loading and boundary conditions are all defined in this module. The guided wave excitation source is defined and applied in this module. The narrow-band tone burst signal is taken as the excitation source. This signal is composed of a several cycles sine wave and modulated by the "*Hamming*" window function defined as follows:

$$w(n) = 0.54 - 0.46\cos\left(\frac{2\pi n}{N}\right), \quad n = 0, 1..., N-1$$

In the above equation, N is the point that is used to define the window function. Another usable typical window function is the "*Hanning*" window function, as listed as follows:

$$w(n) = 0.5 - 0.5\cos\left(\frac{2\pi n}{N}\right), \quad n = 0, 1, ..., N-1$$

A key problem of guided wave simulation modeling is how to apply the excitation source to generate the required type of guided wave. The excitation source of different types can be used as shown in Fig. 6.3. For the S-type wave, edge excitation is used, where the applied displacement boundary condition is perpendicular to the edge. For the A-type wave, the excitation method is also edge excitation, where the applied displacement boundary condition is parallel with the edge. The top excitation can also be used; that is, the load is applied on the micro-region of the top surface of the end, where the guided wave is close to the start edge. The generated type (S or A) cannot be controlled any more when the top excitation is used, which means that the S- and A-type waves will be generated at the same time. In the study of the *Lamb wave* propagation, both the edge and the top excitation will be used; for the study of the defect scattering, the only used excitation method is edge excitation to generate a single pure guided wave mode.

The *Mesh* module. The finite element mesh is defined in this module. The finite element model is shown in Fig. 6.4, and the element size of x-direction l_{ex} and

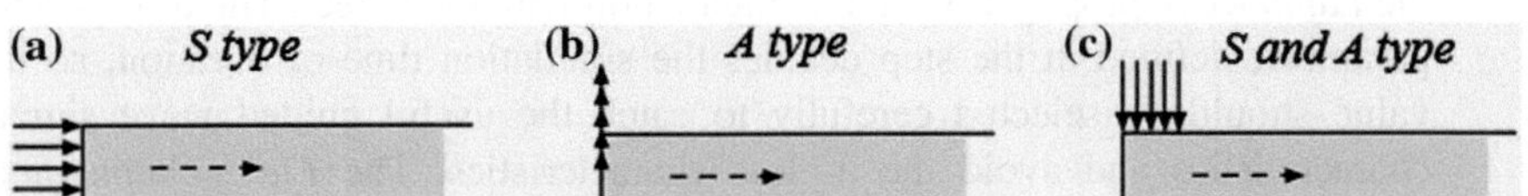

Fig. 6.3 Different types of excitation methods used for *Lamb wave* generation in the plate. **a** Perpendicular to the edge. **b** Parallel to the edge. **c** Top excitation

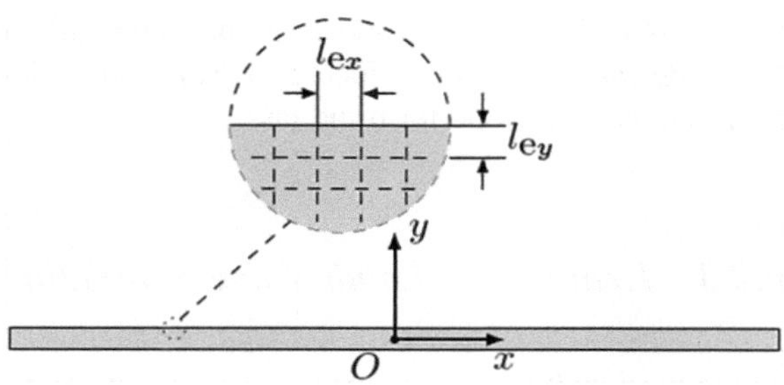

Fig. 6.4 Finite element model of the *Lamb wave* in the plate

y-direction l_{ey} must be selected carefully to guarantee the accuracy of the simulation. The l_{ey} should be small enough to guarantee that the element number of the *y*-direction is big enough, so that the wave structure can be described accurately, that is, the distribution of the displacement, stress, or other physical quantity in the *y*-direction of the plate waveguide (the thickness direction). For the defect-free circumstance, the element number in the *y*-direction is selected as 10; for the circumstance of defects, the element number in the *y*-direction is selected as 100. l_{ex} should also be small enough to guarantee the element number within a wave length is sufficient.

$$\frac{\lambda}{l_{ex}} \geq N$$

In the above equation, $\lambda = C_p/f$ is the wave length of the *Lamb wave*, (C_p is the phase velocity, and f is the frequency, that is, the center frequency of the tone burst signal), and N is the selected integer. The value of N is selected as 20 in most of the simulations except the "*defect area*" in the scattering simulation.

If it is available, the "*Structured*" gridding technique can be used, since the quality of the gridding generated by this technique is easy to be guaranteed. The successful application of this technique requests that the geometric structure of the model should be simple enough, or the model can be divided into several partitions with a simple structure. The plane-strain quadrilateral element *CPE4R* is selected as the type of these elements.

The second stage of the finite element simulation is calculated, which will be finished in the *Job* module. In this module, the generated input file is submitted to the solver selected in the *Step* module of the previous processing stage.

The last stage of the finite element simulation is the post-processing, which will be realized in the *Visualization* module. In this module, the simulated results stored in the output database file (.odb) can be shown, and then processed according to the actual demands.

Most simulated work can be realized through the *GUI* operation in the *Abaqus/CAE* environment, for instance, the special tasks when the tone burst signal is defined by the equally spaced type Amplitude. In the post-processing process, it is also possible to complete some data processing work, which is difficult to realize using *GUI*. For these works, the programmatic interface named as "*Abaqus Scripting Interface*" is provided by *Abaqus*; specifically, the *Python* programming language is used to perform the task.

6.2.3 Example of Lamb Wave Simulation in the Plate

When the *Lamb wave* is propagating freely, it means that there is no defect on the propagation path of the *Lamb wave*, although it is possible for the existence of edge reflection. Then, the interaction between the *Lamb wave* and the transverse defect will be studied.

The model of the *Lamb wave* propagation in the free steel plate is shown in Fig. 6.5, where O is the origin of the coordinates, d is the plate thickness, and $l = 0.25$ m is the length of the plate. P1 [coordinate $(-0.075, d/2)$] and P2 [coordinate $(0.075, d/2)$] are the observation points. The waveform of these observation points is recorded in the simulation. The "*starting edge*" in Fig. 3.7 is the applied position of the excitation source, as well as being the initial position of the *Lamb wave* propagation. The figure also shows the occurring locations of the edge reflections of the second time, the fourth time, etc.

The tone burst signal is modulated by the *Hamming window* function, it has 5.5 sinusoidal periods. The plate thickness is selected as 1 mm and the material parameters are: *Young's modulus* $E = 207 \times 10^9$ Pa, *Poisson's ratio* $v = 0.3$, and the mass density $\rho = 7800$ kg/m^3. From these material parameters, the bulk wave velocities can be calculated as: longitudinal wave velocity $C_L = 5977.03$ m/s, shear wave velocity $C_T = 5977.03$ m/s.

Some typical "*working points*" can be obtained from the dispersion curve of the phase and group velocity, as shown in Table 6.1. These data can be used in the latter simulated examples.

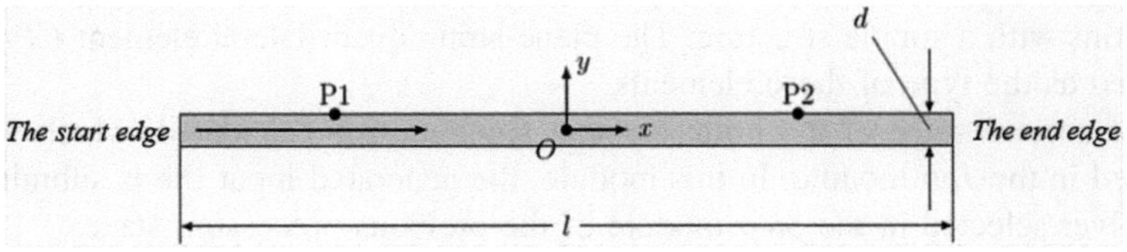

Fig. 6.5 The model of the *Lamb wave* propagation in the sample without defect

Table 6.1 The typical "working points" and the corresponding velocities of wave propagation

Working point number	Mode	fd (Hz × m)	C_p (m/s)	C_g (m/s)
1	S_0	1000	5335.55	5189.70
2	A_0	1000	2347.40	3191.47
3	A_0	2500	2793.65	3161.44
4	A_1	2500	7049.18	3728.66

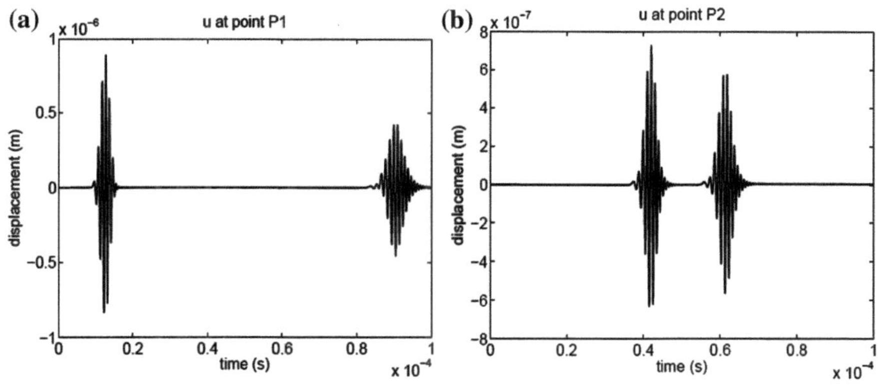

Fig. 6.6 The simulation of the S_0 mode by edge excitation. **a** *u* at P1 point. **b** *u* at P2 point

From the distance of the guided wave propagation (that is the plate length l = 0.25 m) and the calculated group velocity of the guided wave propagation, which is listed in Table 6.1, the propagation time of the S_0 mode, fd = 1000 Hz × m, from the starting edge to the end edge is 4.82×10^{-5} s. The propagation time of the A_0 mode is 6.70×10^{-5} s. The time period parameter in the step definition of the finite element simulation can be determined from these time durations.

Under the edge excitation in the simulation, the waveform of the displacement components *u* at the input mode points P1 and P2 is shown in Fig. 6.6. The corresponding waveform of *v* components is not shown here, due to the similarity between the *v* and *u* waveforms. Both of the two waveforms in Fig. 6.6 contain two larger waveform characteristics; the first characteristic is the signals for P1 and P2 and the second is the reflected signal from the end edge. The signal amplitude attenuation and time dilation caused by the dispersion and material damping can be observed from the figure. The recorded signal characteristic of propagation velocity C_g' from points P1 to P2 is calculated using the following equation:

$$C_g' = \frac{\Delta d}{\Delta t}$$

In the above equation, $\Delta d (=0.15\,m)$ is the propagation distance of the guided wave and Δt is the propagating time of the guided wave, which is solved through the written *Python function*. The value of C_g' should be very close to the theoretically calculated value of C_g of the guided wave mode.

As shown in Table 3.1, the group velocity C_g of the S_0 mode at fd = 1000 Hz × m is 5189.70 m/s. The C_g' obtained from the calculation of the signal from the P1 to the P2 points in Fig. 6.6 is 5126.77 m/s. The error between the propagation velocity C_g' of the signal and the theoretical group velocity C_g is around 1 %, which demonstrates that the expected guided wave mode is generated in the simulation.

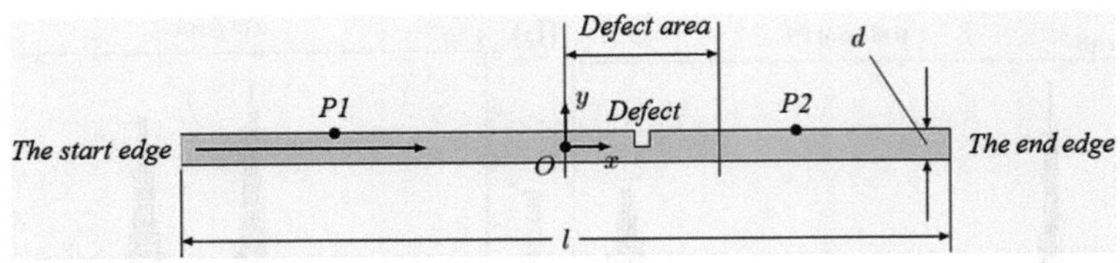

Fig. 6.7 The model of *Lamb wave* propagation in the sample with defect

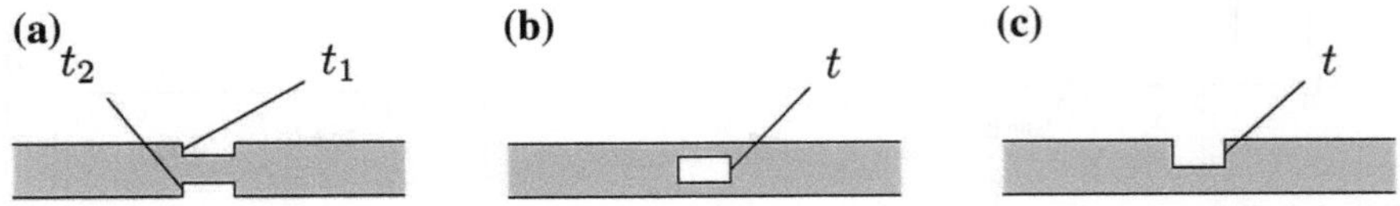

Fig. 6.8 The different types of defects for the *Lamb wave* scattering simulation

In the following section, the interactions between the *Lamb wave* and the defects on the propagation path are studied. The model for this study is shown in Fig. 6.7. In the figure, the plate length is assumed to be $l = 0.25$ m, plate thickness $d = 1$ mm, and the excitation frequency $f = 1$ MHz, thus $fd = 1000$ Hz × m. The location of the defect in the x-direction is in the range of 0–0.025 m. For the other part of the model, a more detailed mesh is used in the defect area to accurately catch the scattering phenomenon of the guided wave in the defect area.

Three types of defects in the cross section of the plate are inspected as shown in Fig. 6.8, and they are listed as follows:

The external symmetric defect of the plate, Fig. 6.8a;
The inner symmetric defect of the plate, Fig. 6.8b;
The external antisymmetric defect of the plate, with the opening located on the surface of the plate as shown in Fig. 6.8c.

The "*symmetry*" and "*asymmetry*" are all defined relative to the middle of the plate. "*Outside*" and "*inside*" refer to the locations on the cross section of the plate. For all the simulations, the width of the simulated rectangular defect is 1 mm and the size of the defect in the y-direction/plate thickness direction t (in Fig. 6.8a $t_1 = t_2$, $t_1 + t_2 = t$) is 50 % of the plate thickness.

The scattering circumstance of the external symmetric defect is inspected first and the defect shape is shown in Fig. 6.8a. For the incident S_0 mode, the u waveforms of points P1 and P2 are shown in Fig. 6.9. In the u waveform (Fig. 6.9a) of point P1, the characteristics of waveforms of the propagation, reflection from the defect, and reflection from the edge of the S_0 mode can be seen. In the u waveform (Fig. 6.9b) of point P2, only the characteristics of the waveforms of transmission from the defect and reflection from the edge of the S_0 mode can be found. The simulated result shows that there is no existence of the A_0 mode in the reflection or transmission wave when the S_0 mode is taken as the incident wave.

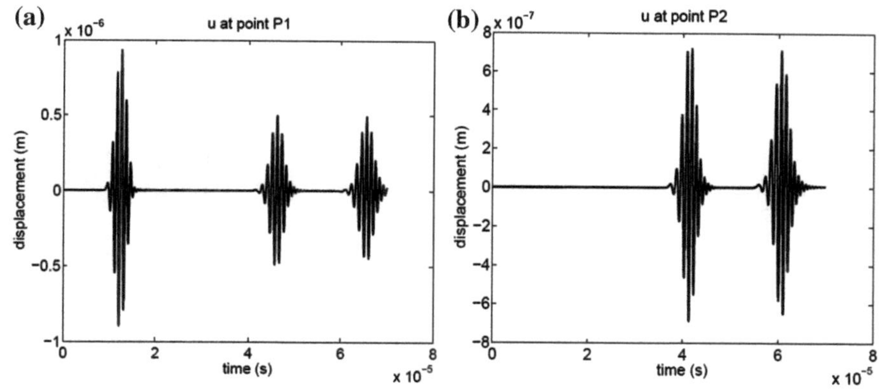

Fig. 6.9 The S_0 mode incident into the sample with an external symmetric defect. **a** u at point P1. **b** u at point P2

Then, the scattering condition of the internal symmetric defect is checked. The defect shape is shown in Fig. 6.8b. The defect is actually located inside the plate. When the S_0 mode is taken as the incident wave, the u waveforms recorded at points P1 and P2 are shown in Fig. 6.10. These results are almost the same as the simulated results of the external symmetric defect shown in Fig. 6.9.

In the following section, the scattering situation of the external antisymmetric defect in the plate will be discussed. The opening of the defect is located on the surface of the plate and its shape is shown in Fig. 6.8c. The defect is no longer symmetric of the middle plane of the plate.

When the S_0 mode is taken as the incident wave, the u waveforms recorded at points P1 and P2 are shown in Fig. 6.11. In Fig. 6.11a, the signal around the time $t = 1.26 \times 10^{-5}$ s is the propagating signal of the S_0 mode at *point* P1; after the time $t = 4.0 \times 10^{-5}$ s, the first waveform corresponds to the end edge reflection S_0 mode

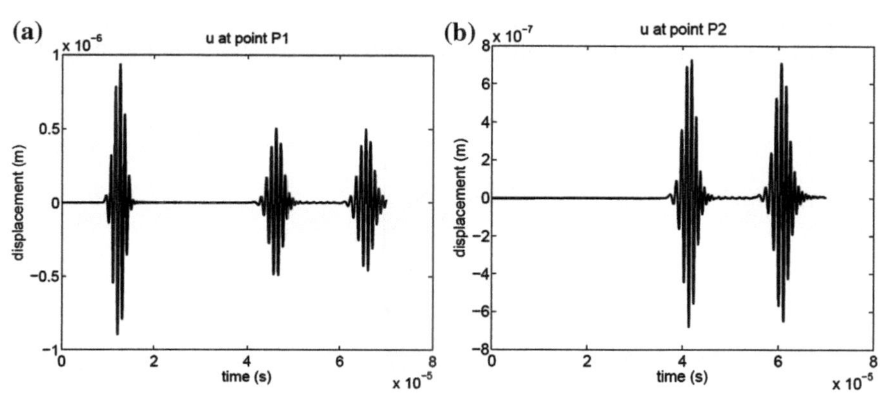

Fig. 6.10 The S_0 mode incident into the sample with an internal symmetric defect. **a** u at point P1. **b** u at point P2

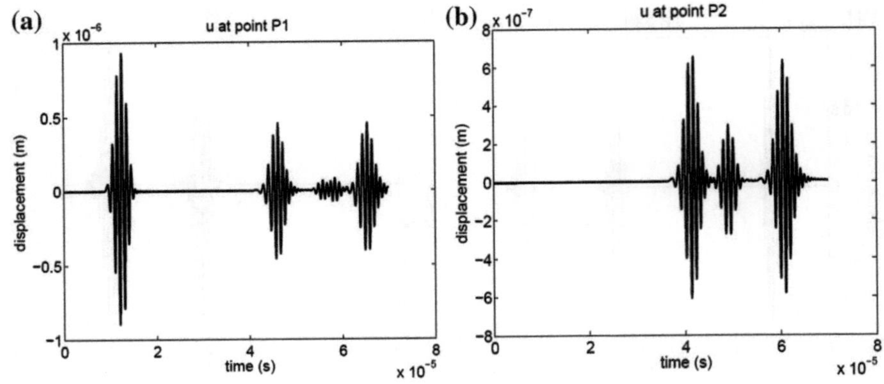

Fig. 6.11 The S_0 mode incident into the external antisymmetric defect. **a** u at point P1. **b** u at point P2

signal. Another waveform is the end edge reflection A_0 mode, where the last characteristic is the starting edge reflection S_0 mode. In Fig. 6.11b, there are three waveforms after the time $t = 4.0 \times 10^{-5}$ s, corresponding to the S_0 mode, which has the fastest propagation velocity, the A_0 mode, which has the slower propagation velocity, and the end edge reflection of the transmission S_0 mode.

6.3 Finite Element Simulation of the Circumferential Lamb Wave in Pipe [2]

6.3.1 Establishment of the Dispersion Equation of Circumferential Lamb Waves

The definition of the circumferential coordinate system and the displacement components of the pipeline circumferential *Lamb wave* are shown in Fig. 6.12. *O* is

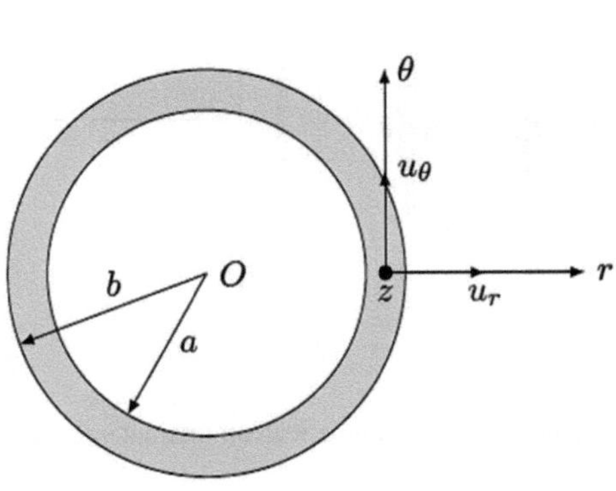

Fig. 6.12 The model used for the circumferential coordinate system and the displacement components of the circumferential *Lamb wave*

the original point, a is the internal radius of the pipeline, b is the external radius of the pipeline, r, θ, and z are coordinates, and u_θ and u_r are the displacement components on the θ and r axes.

For the circumferential *Lamb wave*, u_θ and u_r are non-zero terms and are functions of coordinates r and θ. The displacement u_z perpendicular to the cross section is equal to 0.

$$u_\theta = u_\theta(r,\theta), \quad u_r = u_r(r,\theta), \quad u_z = 0 \tag{6.1}$$

The above assumptions indicate the plane-strain problem of the *Lamb wave*.

When the potential functions are used to express the displacement components, then the *Helmholtz decomposition expression* is $\mathbf{u} = \nabla\varphi + \nabla\times\boldsymbol{\psi}$. In the cylindrical coordinates, this decomposition is expressed as in Eq. (6.2).

$$\begin{cases} u_r = \frac{\partial\varphi}{\partial r} + \frac{1}{r}\frac{\partial\psi_z}{\partial\theta} - \frac{\partial\psi_\theta}{\partial z} \\ u_\theta = \frac{1}{r}\frac{\partial\varphi}{\partial\theta} + \frac{\partial\psi_r}{\partial z} - \frac{\partial\psi_z}{\partial r} \\ u_z = \frac{\partial\varphi}{\partial z} + \frac{1}{r}\frac{\partial(r\psi_\theta)}{\partial r} - \frac{1}{r}\frac{\partial\psi_r}{\partial\theta} \end{cases} \tag{6.2}$$

In the above equation, ψ_z, ψ_θ and ψ_r are the components of the three coordinate axes, and the vector $\boldsymbol{\psi}$ and scalar φ are the displacement potential functions.

Considering (6.1), the *Helmholtz decomposition* (6.2) can be simplified, as shown in the following equation:

$$u_r = \frac{\partial\varphi}{\partial r} + \frac{1}{r}\frac{\partial\psi_z}{\partial\theta}, \quad u_\theta = \frac{1}{r}\frac{\partial\varphi}{\partial\theta} - \frac{\partial\psi_z}{\partial r}, \quad u_z = 0$$

If the subscript z in ψ_z is ignored, we have (6.3).

$$u_r = \frac{\partial\varphi}{\partial r} + \frac{1}{r}\frac{\partial\psi}{\partial\theta}, \quad u_\theta = \frac{1}{r}\frac{\partial\varphi}{\partial\theta} - \frac{\partial\psi}{\partial r} \tag{6.3}$$

The wave equation is listed as follows, where C_L is the longitudinal wave velocity and the C_T is the shear wave velocity:

$$\nabla^2\varphi = \frac{1}{C_L^2}\cdot\frac{\partial^2\varphi}{\partial t^2}, \quad \nabla^2\psi_z = \frac{1}{C_T^2}\frac{\partial^2\psi_z}{\partial t^2}$$

In the cylindrical coordinate system, we have the following equation:

$$\nabla^2\varphi = \frac{1}{r}\frac{\partial}{\partial r}\left(r\frac{\partial\varphi}{\partial r}\right) + \frac{1}{r^2}\frac{\partial^2\varphi}{\partial\theta^2} + \frac{\partial^2\varphi}{\partial z^2}$$

For the time harmonic wave, we have the following equation:

$$\frac{\partial^2 \varphi}{\partial t^2} = -\omega^2 \varphi, \quad \frac{\partial^2 \psi_z}{\partial t^2} = -\omega^2 \psi_z$$

If the subscript z in ψ_z is ignored, the wave equation can be written as shown in (6.4).

$$\begin{cases} \frac{1}{r}\frac{\partial \varphi}{\partial r} + \frac{\partial^2 \varphi}{\partial r^2} + \frac{1}{r^2}\frac{\partial^2 \varphi}{\partial \theta^2} = \frac{1}{C_L^2}(-\omega^2 \varphi) \\ \frac{1}{r}\frac{\partial \psi}{\partial r} + \frac{\partial^2 \psi}{\partial r^2} + \frac{1}{r^2}\frac{\partial^2 \psi}{\partial \theta^2} = \frac{1}{C_T^2}(-\omega^2 \psi) \end{cases} \tag{6.4}$$

The solutions of the potential function φ and ψ are assumed to be as shown in (6.5).

$$\begin{cases} \varphi = \Phi(r)\mathrm{e}^{i\left(k\frac{a+b}{2}\theta - \omega t\right)} \\ \psi = \Psi(r)\mathrm{e}^{i\left(k\frac{a+b}{2}\theta - \omega t\right)} \end{cases} \tag{6.5}$$

The distribution of the potential function exists in the wall thickness direction of the pipeline. ω is the angular frequency of the particle vibration of the guided wave propagating along the circumferential direction, k is the wave number, and $(a+b)\theta/2$ is the arc length in the propagation direction, which means that the middle surface of the pipeline wall is taken as the guided wave propagation path under normal conditions.

The solutions of these equations are *Bessel functions*, as shown in the following equation:

$$\begin{cases} \Phi(r) = A_1 J_M\left(\frac{\omega r}{C_L}\right) + A_2 Y_M\left(\frac{\omega r}{C_L}\right) \\ \Psi(r) = B_1 J_M\left(\frac{\omega r}{C_T}\right) + B_2 Y_M\left(\frac{\omega r}{C_T}\right) \end{cases}$$

In the above equation, J is the first-order *Bessel function*, Y is the second-order *Bessel function*, $M = k(a+b)/2$ is the order of the *Bessel function*, and A_1, A_2, B_1, and B_2 are all unknown coefficients.

From the recurrence relation of the *Bessel function*, we have the following equations:

$$\frac{\mathrm{d}\left(J_M\left(\frac{\omega r}{C_L}\right)\right)}{\mathrm{d}r} = \frac{\omega}{C_L} J'_M\left(\frac{\omega r}{C_L}\right) = \frac{\omega}{2C_L}\left(J_{M-1}\left(\frac{\omega r}{C_L}\right) - J_{M+1}\left(\frac{\omega r}{C_L}\right)\right)$$

$$\frac{\mathrm{d}^2\left(J_M\left(\frac{\omega r}{C_L}\right)\right)}{\mathrm{d}r^2} = \frac{\omega^2}{4C_L^2}\left(J_{M-2}\left(\frac{\omega r}{C_L}\right) - 2J_M\left(\frac{\omega r}{C_L}\right) + J_{M+2}\left(\frac{\omega r}{C_L}\right)\right)$$

So, we have the following equation, where the parameter of the *Bessel function* parameter omitted in the first three equations is $\omega r/C_L$ and the *Bessel function* parameter omitted in the latter three equations is $\omega r/C_T$.

$$\begin{cases} \Phi(r) = J_M A_1 + Y_M A_2 \\ \Phi'(r) = \frac{\omega}{2C_L}(J_{M-1} - J_{M+1})A_1 + \frac{\omega}{2C_L}(Y_{M-1} - Y_{M+1})A_2 \\ \Phi''(r) = \left(\frac{\omega}{2C_L}\right)^2 (J_{M-2} - 2J_M + J_{M+2})A_1 + \left(\frac{\omega}{2C_L}\right)^2 (Y_{M-2} - 2Y_M + Y_{M+2})A_2 \\ \Psi(r) = J_M B_1 + Y_M B_2 \\ \Psi'(r) = \frac{\omega}{2C_T}(J_{M-1} - J_{M+1})B_1 + \frac{\omega}{2C_T}(Y_{M-1} - Y_{M+1})B_2 \\ \Psi''(r) = \left(\frac{\omega}{2C_T}\right)^2 (J_{M-2} - 2J_M + J_{M+2})B_1 + \left(\frac{\omega}{2C_T}\right)^2 (Y_{M-2} - 2Y_M + Y_{M+2})B_2 \end{cases}$$

Based on the plane-strain assumptions for the circumferential *Lamb wave*, the relationship between the displacement and stress in the cylindrical coordinate system can be simplified into the following equation, where λ and μ are *Lame constants*:

$$\begin{cases} \sigma_r = \lambda\left(\frac{\partial u_r}{\partial r} + \frac{u_r}{r} + \frac{1}{r}\frac{\partial u_\theta}{\partial \theta}\right) + 2\mu\frac{\partial u_r}{\partial r} \\ \tau_{r\theta} = \mu\left(\frac{\partial u_\theta}{\partial r} - \frac{u_\theta}{r} + \frac{1}{r}\frac{\partial u_r}{\partial \theta}\right) \end{cases}$$

Substituting (6.5) into (6.3), we get (6.6).

$$\begin{cases} u_r = \left[\frac{\partial \Phi(r)}{\partial r} + \frac{ik\frac{a+b}{2}}{r}\Psi(r)\right] e^{i\left(k\frac{a+b}{2}\theta - \omega t\right)} \\ u_\theta = \left[\frac{ik\frac{a+b}{2}}{r}\Phi(r) - \frac{\partial \Psi(r)}{\partial r}\right] e^{i\left(k\frac{a+b}{2}\theta - \omega t\right)} \end{cases} \tag{6.6}$$

After the $\Phi(r)$ and $\Psi(r)$ are obtained, the wave structure of the displacement can be calculated by the above equations.

$$\begin{cases} \sigma_r = \frac{\mu}{r^2}\left[\chi^2 r^2 \Phi''(r) + (\chi^2 - 2)\left(r\Phi'(r) - k^2\left(\frac{a+b}{2}\right)^2 \Phi(r)\right) + 2ik\frac{a+b}{2}\left(r\Psi'(r) - \Psi(r)\right)\right] \\ \tau_{r\theta} = \frac{\mu}{r^2}\left[-r^2\Psi''(r) + r\Psi'(r) - k^2\left(\frac{a+b}{2}\right)^2 \Psi(r) + 2ik\frac{a+b}{2}\left(r\Phi'(r) - \Phi(r)\right)\right] \end{cases} \tag{6.7}$$

In (6.7), $\chi = C_L/C_T$.

Based on the traction-free boundary condition, the stress on both the inside and the outside of the surfaces is equal to 0, as shown in (6.8).

$$\sigma_r|_{r=a,b} = 0, \quad \tau_{r\theta}|_{r=a,b} = 0 \tag{6.8}$$

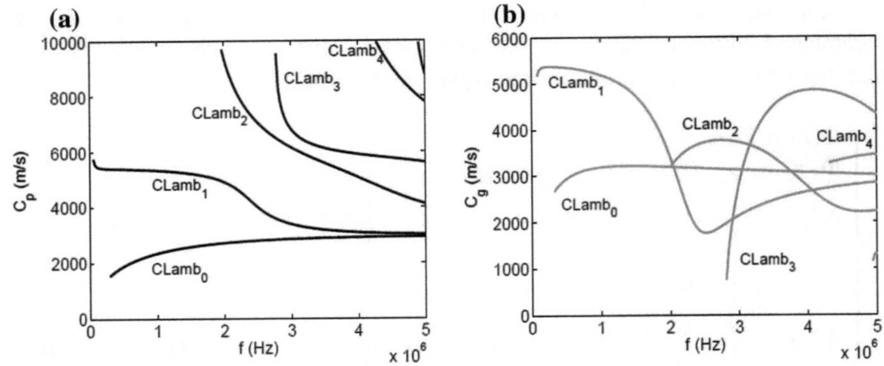

Fig. 6.13 The circumferential *Lamb wave* dispersion wave in the pipeline. **a** Phase velocity dispersion curve. **b** Group velocity dispersion curve

From the boundary conditions (6.8) and (6.7), we can get the following equation:

$$\begin{cases} \chi^2 r^2 \Phi''(r) + (\chi^2 - 2)\left(r\Phi'(r) - k^2\left(\frac{a+b}{2}\right)^2\Phi(r)\right) + 2ik\frac{a+b}{2}\left(r\Psi'(r) - \Psi(r)\right)\Big|_{r=a,b} = 0 \\ -r^2\Psi''(r) + r\Psi'(r) - k^2\left(\frac{a+b}{2}\right)^2\Psi(r) + 2ik\frac{a+b}{2}\left(r\Phi'(r) - \Phi(r)\right)\Big|_{r=a,b} = 0 \end{cases}$$

There are actually *four* equations included in the above equation set: two for $r = a$ and two for $r = b$. A_1, A_2, B_1, and B_2 are the unknown numbers of the solutions.

Now the determinant of the matrix composed by the coefficients of these unknown numbers is 0; thus, the dispersion equation of the circumference *Lamb wave* can be obtained. This equation can only be solved using a numerical method. The numerical solution is expressed as the dispersion curve of the phase velocity; the dispersion curve of the group velocity can be calculated from that of the phase velocity.

The dispersion curves of the *Lamb waves* in the circumferential direction of the pipeline with an inner and outer radius of 49 and 50 mm are shown in Fig. 6.13. The "$CLamb_n$" form is used as the guided wave mode, where the integer subscript n increases from 0 to infinity. The material parameters are *Young's modulus* $E = 207 \times 10^9$ Pa, *Poisson's* ratio $\nu = 0.3$, and mass density $\rho = 7800$ kg/m^3.

6.3.2 Finite Element Simulation of the Circumferential Lamb Wave in the Pipe

The simulation process of the circumferential *Lamb wave* in the pipe is similar to that of the *Lamb wave* in the plate. In this chapter, only the differences are provided.

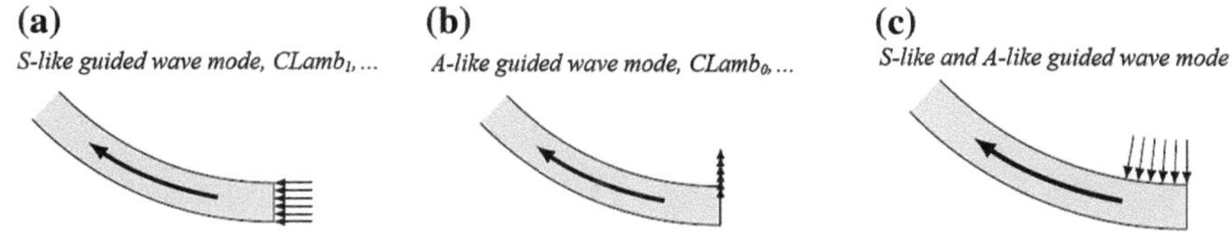

Fig. 6.14 The loadings of the excitation used to generate different types circumferential *Lamb waves*. **a** Excitation perpendicular to edge. **b** Excitation parallel to edge. **c** Top excitation

In the *Load* module of the preprocessing stage, there are still three different types of application methods of the excitation, as shown in Fig. 6.14. For the *S*-like modes such as $\mathrm{CLamb_1}$, $\mathrm{CLamb_3}$, and $\mathrm{CLamb_4}$, edge excitation is used and the applied displacement boundary condition is perpendicular to the edge. For the *A*-like mode that includes the $\mathrm{Clamb_0}$ and $\mathrm{CLamb_2}$ modes, the applied displacement boundary condition is parallel to the edge. The top excitation can also be used; in this case, the pressure loading is applied to the micro-region on the top surface, where the waveguide is near to the starting edge. In this situation, the *S*-like and *A*-like modes will be generated at the same time. For studies of the *Lamb wave* free-propagation and edge reflection, the edge excitation and top excitation will be examined. For the study of the defect scattering, the method used to generate a simple single input mode is the edge excitation.

6.3.3 *Simulation of the Circumferential Lamb Wave in the Pipe*

The situation of free-propagation should be checked first, and then the interaction between the circumferential *Lamb wave* and the defect will be studied.

The model that is used to study the *Lamb wave* free-propagation and edge reflection is shown in Fig. 6.15. *O* is the original point, *a* and *b* are the inner and outer radius of the pipeline, respectively, and P1 and P2 are the observation points of the waveform that is expected to be recorded.

In the given simulated example, the inner radius and outer radius are 49 and 50 mm, respectively. The parameters of the tone burst signal, material parameters, and bulk wave velocities are all the same as in the simulation of the *Lamb wave* in the plate. The dispersion curve is calculated as shown in Fig. 6.13. Based on the calculated *Lamb wave* dispersion curve, the phase and group velocities of the different guided wave modes and frequencies are already selected. As a result, for the "S_0"/$\mathrm{CLamb_1}$ mode at 1 MHz, the phase velocity c_p = 5335.52 m/s and group velocity c_g = 5187.68 m/s, and for the "A_0"/$\mathrm{CLamb_1}$ mode at the same frequency, c_p = 2347.23 m/s and c_g = 3191.88 m/s.

Under the edge excitation and using "S_0"/$\mathrm{CLamb_1}$ as the incident mode, the simulation u_r waveforms at points P1 and P2 are shown in Fig. 6.16. The

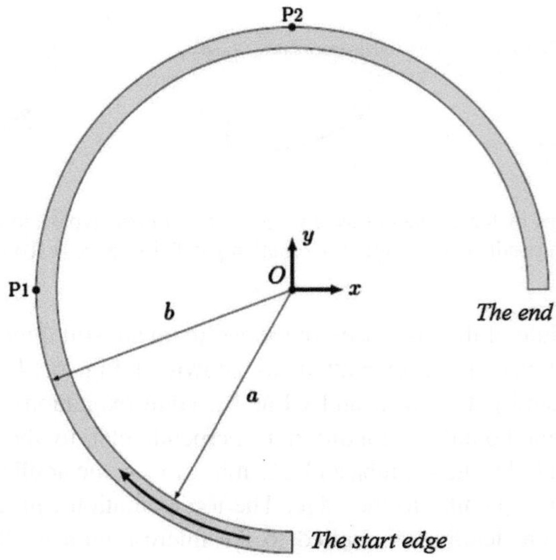

Fig. 6.15 The model used to simulate the defect-free propagation of the circumferential *Lamb wave*

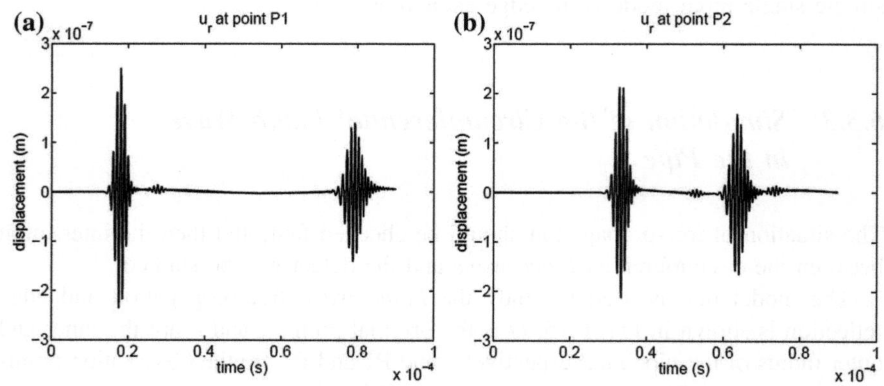

Fig. 6.16 The simulated results of the "S_0"/CLamb$_0$ mode. **a** u_r at point P1. **b** u_r at point P2

corresponding u_θ waveforms at points P1 and P2 are shown in Fig. 6.17. Two big waveforms are included in these four waveforms, where the first characteristic is the propagating signal at points P1 and P2 and the second characteristic is the reflection signals from the end edge. The propagation velocity (C'_g) of the recorded waveform from points P1 to P2 along the middle of the pipeline can be calculated, as shown in the following equation:

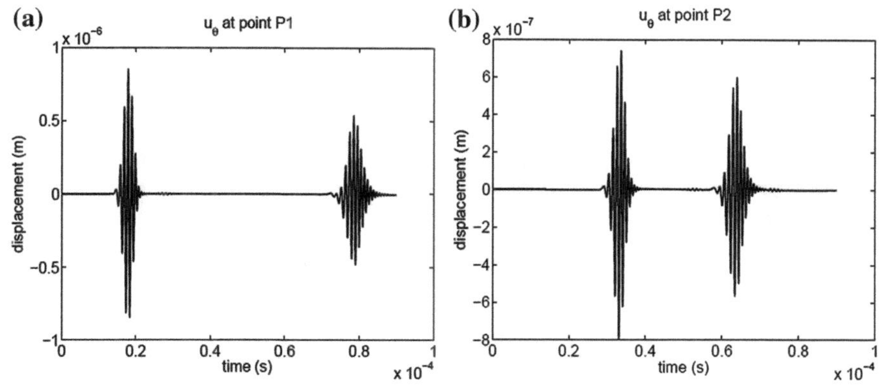

Fig. 6.17 The simulated results of the "S_0"/CLamb$_1$ mode. **a** u_θ at point P1. **b** u_θ at point P2

$$C'_g = \frac{\Delta d}{\Delta t} \times \frac{\frac{a+b}{2}}{b}$$

$\Delta d = 1/4 \times 2\pi b$ is the propagation distance of the characteristic from points P1 to P2 on the external surface of the pipeline and Δt is the propagation time of the recorded waveform characteristics.

As discussed above, for the "S_0"/CLamb$_1$ mode, c_g = 5187.68 m/s. In Fig. 6.16, the value of C'_g calculated from the propagation characteristic from points P1 to P2 is 5132.38 m/s, and from the calculation of the propagation characteristic in Fig. 6.17, $C'_g = 5132.39\,\text{m/s}$. The error between the simulated velocity C'_g and the theoretical group velocity C_g is around 1 %.

In the next section, the propagation of the A mode will be investigated and parallel edge excitation will be used. Under edge excitation, the incident mode is "A_0"/CLamb$_0$. The u_r waveforms at points P1 and P2 are shown in Fig. 6.18. The

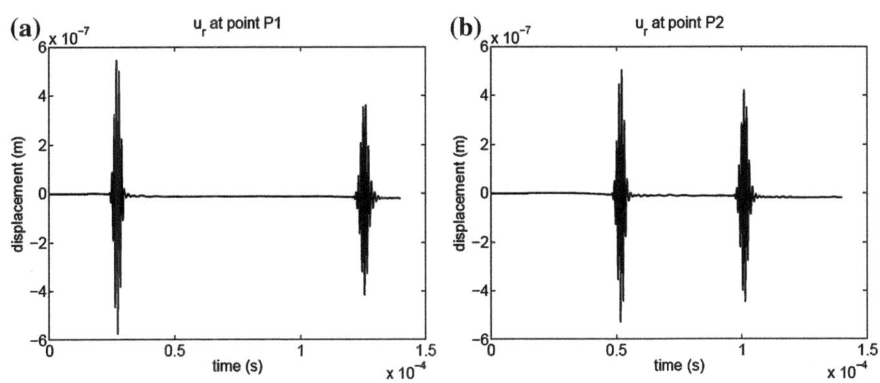

Fig. 6.18 The simulated results of the "A_0"/CLamb$_0$ mode. **a** u_r at point P1. **b** u_r at point P2

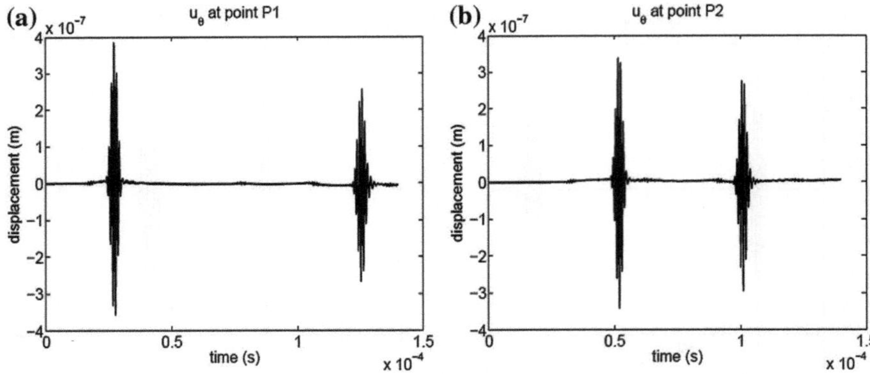

Fig. 6.19 The simulated results of the "A_0"/CLamb$_0$ mode. **a** u_θ at point P1. **b** u_θ at point P2

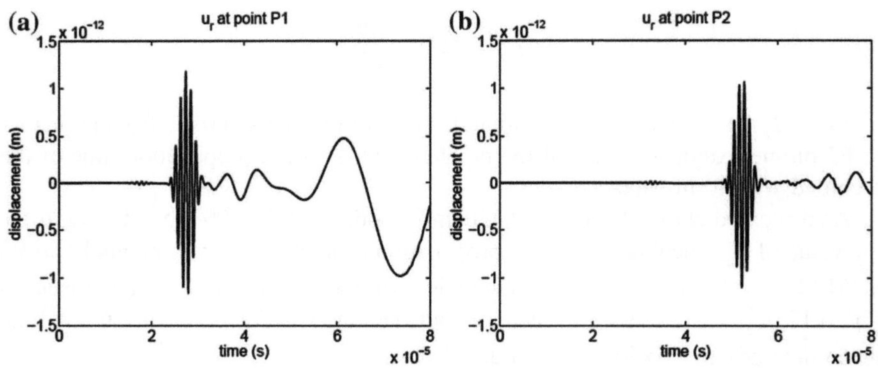

Fig. 6.20 The simulated results of the top excitation. **a** u_r at point P1. **b** u_r at point P2

corresponding u_θ waveforms at points P1 and P2 are shown in Fig. 6.19. For the "A_0"/CLamb$_0$ mode, $C_g = 3191.88\,\text{m/s}$. Based on the propagating signal at points P1 and P2 in Fig. 6.18, it can be calculated that $C_g' = 3212.34\,\text{m/s}$, and it can be calculated from Fig. 6.19 that $C_g' = 3148.25\,\text{m/s}$. The error between the simulated velocity C_g' and the theoretical group velocity C_g is also around 1 %.

The simulated result of the top excitation is shown in Fig. 6.20. The circumferential width of the top excitation source is 2°. It is shown in the figure that, for the "S_0"/CLamb$_1$ mode characteristic near the time 1.8×10^{-5} s, u_θ is much bigger than u_r, but for the "A_0"/CLamb$_0$ mode characteristic near the time 2.7×10^{-5} s, u_r is much bigger than u_θ.

After the discussions of the free-propagation and edge reflection of the circumferential *Lamb waves*, the most important thing is to study their interaction with the defects on the propagation path and the scattering of the defects. The model used to study this problem is shown in Fig. 6.21. The circumferential width of the

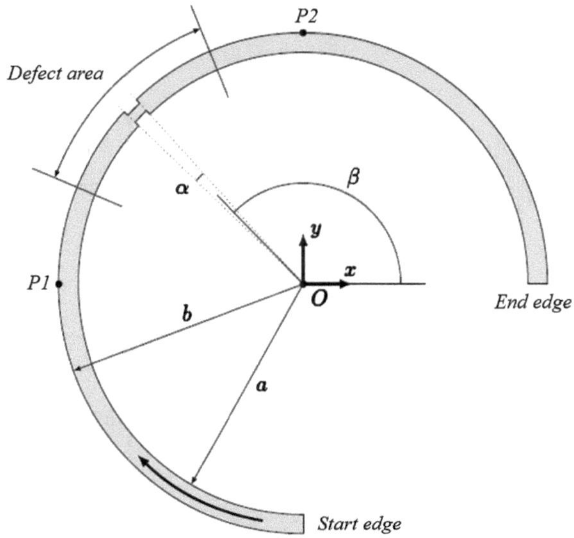

Fig. 6.21 The model of the interaction between the circumferential *Lamb wave* and the defect

defect is defined by angle α. The circumferential position of the defect is defined by the counterclockwise angle β, from the perspective of the x-axis.

The "*defect area*" is also drawn out in the model, which happens within an angle range of 112.5°–157.5°, that is, 45° or one-eighth of the whole circle. Compared to the rest of the model, a more detailed mesh generation is used in the defect area.

Four kinds of defects are investigated: the external symmetric defect as shown in Fig. 6.22a; the internal symmetric defect, where the defect is located inside the pipeline, as shown in Fig. 6.22b; the external asymmetric defect, the opening of which is located on the external surface of the pipeline, as shown in Fig. 6.22c; and the internal symmetric defect, the opening of which is located on the inner surface of the pipeline, as shown in Fig. 6.22d.

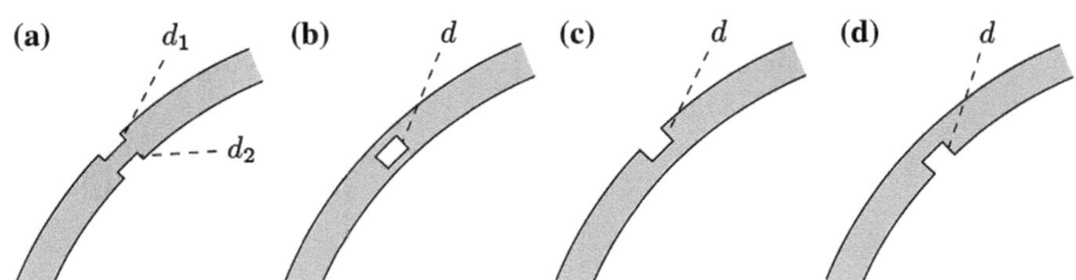

Fig. 6.22 Different defects used in the scattering simulation of the circumferential *Lamb wave* in the pipe

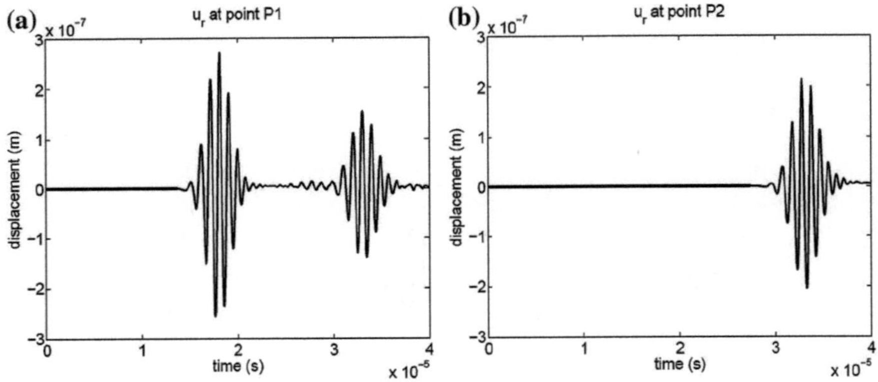

Fig. 6.23 The simulated results when the "S_0"/CLamb$_1$ mode is incident into the sample with the external symmetric defect. **a** u_r at point P1. **b** u_r at point P2

Here, the "*symmetry*" and "*asymmetry*" are in respect to the middle surface of the pipeline. The "*external*" and "*inner*" are used to describe the location of the defect in the pipeline. For all the simulated examples of the defect's scattering, $\alpha = 1°$ and $\beta = 135°$. The depth of the rectangular cross-sectional defect d (in Fig. 6.22a, $d_1 = d_2$, $d_1 + d_2 = d$) is equal to 50 % of the wall thickness.

When the model for the external symmetric defect case, with the "S_0"/CLamb$_1$ as the incident wave, is studied, the waveforms of u_r and u_θ recorded at points P1 and P2 are shown in Figs. 6.23 and 6.24. In the waveforms recorded at point P1, the propagation and defect reflection characteristics of the "S_0"/CLamb$_1$ model are very easy to observe. In the waveforms recorded at point P2, only the defect transmission "S_0"/CLamb$_1$ model characteristic exists.

The limitations of the simulation time mean that the edge reflection is not recorded. The simulated results also prove that there is no mode conversion from

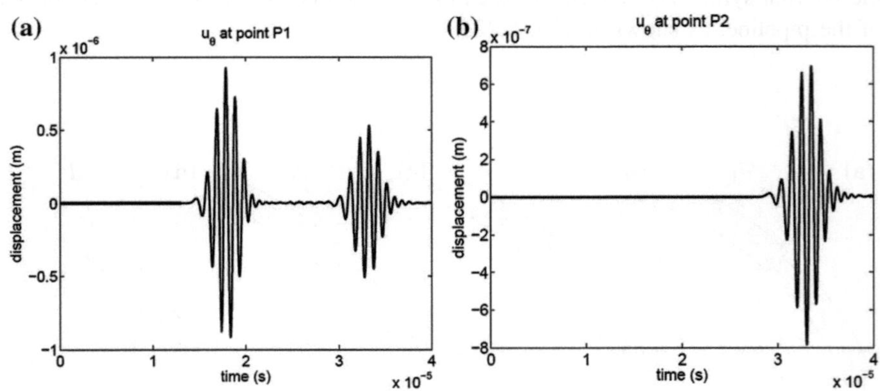

Fig. 6.24 The simulated results when the "S_0"/CLamb$_1$ mode is incident into the sample with the external symmetric defect. **a** u_θ at point P1. **b** u_θ at point P2

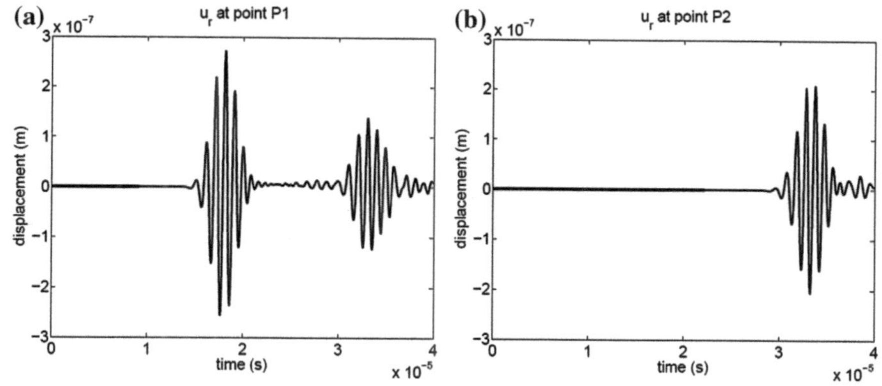

Fig. 6.25 The simulated results when the "S_0"/CLamb$_1$ mode is incident into the sample with the internal symmetric defect. **a** u_r at point P1. **b** u_r at point P2

the incident "S_0"/CLamb$_1$ mode to the reflection or transmission mode of "A_0"/CLamb$_0$.

The situation with the internal symmetric defect will be investigated in the following section; the defect is actually located inside the pipeline. The "S_0"/CLamb$_1$ mode is the incident wave; the u_r and u_θ waveforms recorded at points P1 and P2 are shown in Figs. 6.25 and 6.26. These results are almost the same as the simulated results with the external symmetric defect, as shown in Figs. 6.23 and 6.24.

Then the situation of the external asymmetric defect on the outside surface of the pipeline is discussed. The defect is not symmetric of the middle plane of the pipeline, so the mode conversion from the input "S_0"/CLamb$_1$ mode to the reflection and transmission "A_0"/CLamb$_0$ mode can be observed.

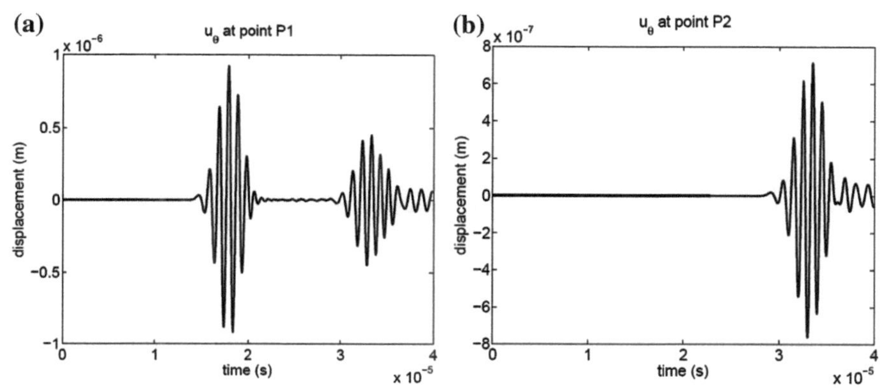

Fig. 6.26 The simulated results when the "S_0"/CLamb$_1$ mode is incident into the sample with the internal symmetric defect. **a** u_θ at point P1. **b** u_θ at point P2

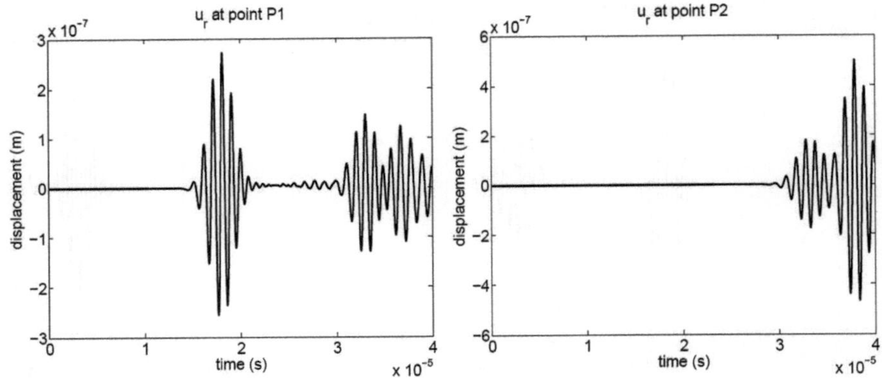

Fig. 6.27 The simulated results when the "S_0"/CLamb$_1$ mode is incident into the sample with the external asymmetric defect. **a** u_r at point P1. **b** u_r at point P2

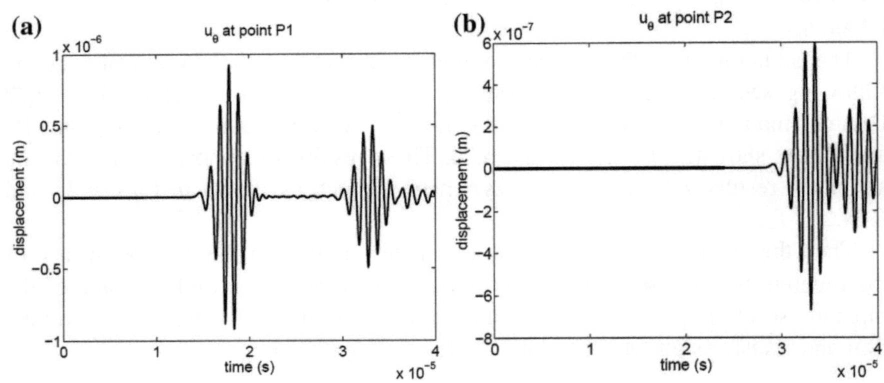

Fig. 6.28 The simulated results when the "S_0"/CLamb$_1$ mode is incident into the sample with the external asymmetric defect. **a** u_θ at point P1. **b** u_θ at point P2

For the "S_0"/CLamb$_1$ mode as the incident wave, the u_r and u_θ waveforms recorded at points P1 and P2 are shown in Figs. 6.27 and 6.28. In Figs. 6.27a and 6.28a, the characteristic near the time 1.85×10^{-5} s is the nonstop characteristic of the incident "S_0"/CLamb$_1$ mode at the point P1; and after the time 3×10^{-5} s, it is the end reflection mode of "S_0"/CLamb$_1$ and then the end reflection mode of "A_0"/CLamb$_0$.

In Figs. 6.27b and 6.28b, the two characteristics after the time 3×10^{-5} s are the faster "S_0"/CLamb$_1$ mode and the slower "A_0"/CLamb$_0$ mode. The mode conversion from the input "S_0"/CLamb$_1$ mode to the reflection and transmission "A_0"/CLamb$_0$ mode can be observed clearly.

The isoline of the reflection and transmission wave after the defect's scattering is shown in Fig. 6.29. The drawn variables are the displacement amplitude.

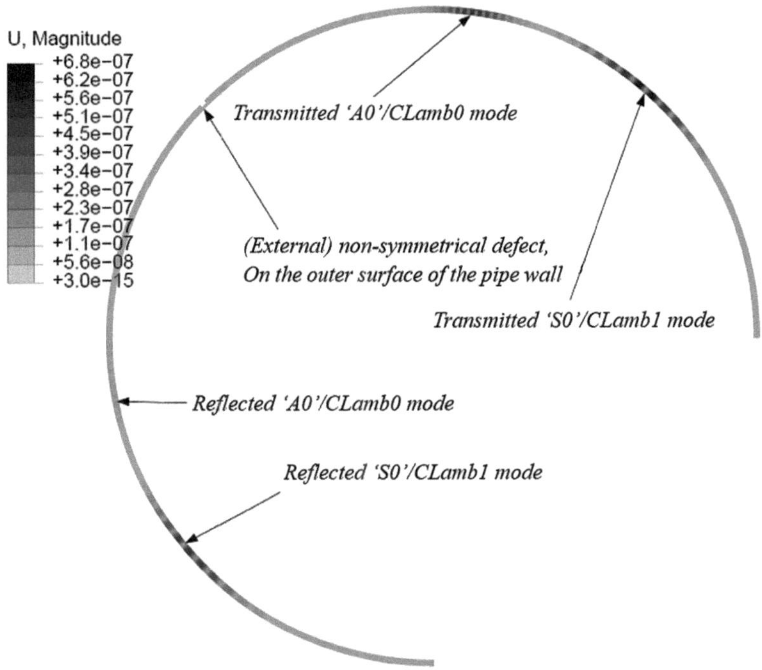

Fig. 6.29 The isoline of the reflection and transmission wave after the scattering from the external antisymmetric defect

Finally, the antisymmetric defect on the inner surface of the pipeline is inspected with the "S_0"/CLamb$_1$ taken as the incident mode; the u_r and u_θ waveforms recorded at points P1 and P2 are shown in Figs. 6.30 and 6.31.

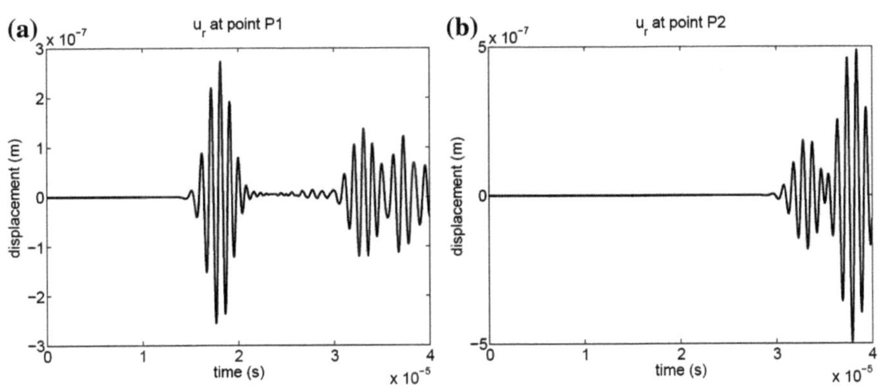

Fig. 6.30 "S_0"/CLamb$_1$ mode is incident into the sample with external antisymmetric defect on the inner surface of the pipeline. **a** u_r at point P1. **b** u_r at point P2

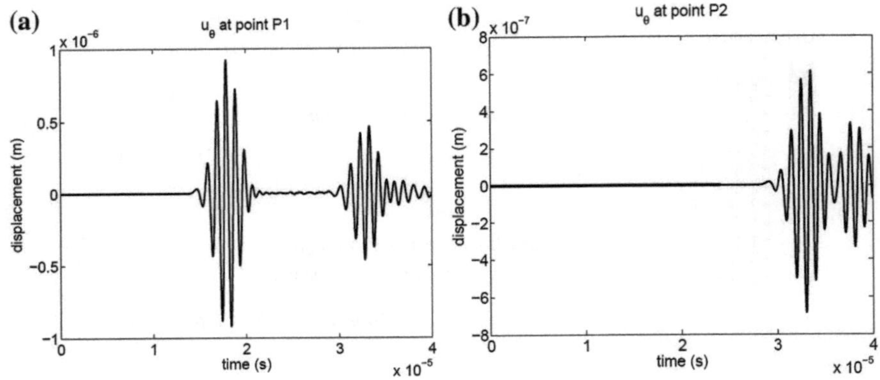

Fig. 6.31 "S_0"/CLamb$_1$ mode is incident into the sample with the external antisymmetric defect on the inner surface of the pipeline. **a** u_θ at point P1. **b** u_θ at point P2

6.4 Finite Element Simulation of the *L*-Type Guided Wave Along the Axial Direction of the Pipeline

The guided wave along the axial direction of the pipeline can be divided into three types: *L* mode (longitudinal modes), *T* mode (torsional modes), and *F* mode (flexural modes), where the *L*- and *T*-type guided wave modes are axisymmetric, which means that the displacement vibration of the guided wave propagation is symmetric about the axial of pipeline. The structure of its displacement is relatively simple. In practical applications, usually the detecting system and guided wave transducer are controlled intentionally to generate the axisymmetric guided wave. The *F*-type guided wave is a non-axisymmetric guided wave and its vibration displacement has three components: axial direction, circumferential direction, and radial direction. Its constitution is very complex and in practical applications the positive generation of the *F*-type guided wave along the axial direction of the pipe is usually to be avoided. Although the guided wave transducer can generate the axisymmetric *L*- or *T*-type guided waves, if defects in the pipeline are non-axisymmetric, the generation of the *F*-type guided wave is inevitable.

In this section, the *L*-type guided wave simulation in the case of guided waves and defects are all axial symmetrically distributed. Initially, the dispersion curve of the *L*-type guided waves in the pipeline is as shown in Fig. 6.31. The phase velocity is the propagation velocity of the steady-state guided wave. In practical inspections, its value is used to design the transducer parameter. In the simulation, it is used to decide the finite element size along the guided wave propagation direction, that is, the axial direction in the pipeline. The group velocity is the propagation velocity of the short-time impulse; its value is used to verify the generation of the expected guided wave mode, and it also confirms the length of the finite element simulation as well as the guided wave propagation path length.

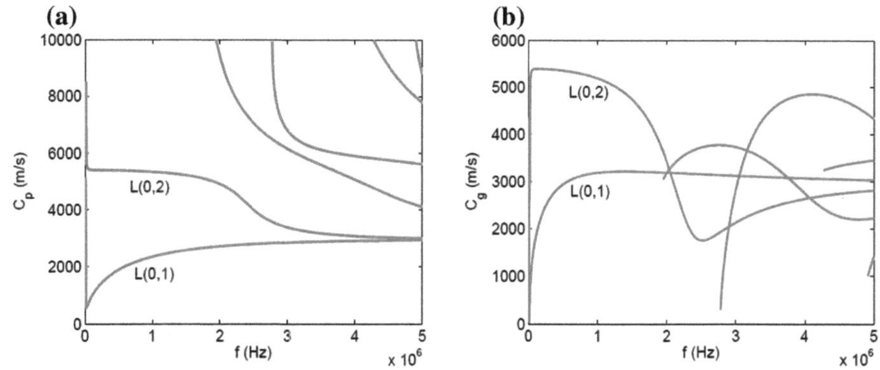

Fig. 6.32 The dispersion curve of the *L*-type guided wave in the steel pipeline. **a** Phase velocity dispersion curve. **b** Group velocity dispersion curve

The material parameters are *Young's modulus* $E = 207 \times 10^9$ Pa, *Poisson's ratio* $\nu = 0.3$, and mass density $\rho = 7800\ \text{kg/m}^3$. The inner radius of the pipe $a = 49$ mm and the external radius $b = 50$ mm.

In the dispersion of the *L*-type guided wave, the modes are numbered as *L*(0,*n*) in the dispersion curve, where *n* is a positive integer. It can be seen from the dispersion curve that the *L*(0,1) mode is similar to the A_0 *Lamb wave* mode in the plate; here it is called "A_0*-like mode.*" The *L*(0,2) mode is similar to the S_0 *Lamb wave* mode in the plate; here, it is called "S_0*-like mode*" (Fig. 6.32).

The center frequency of the used tone burst signal is 250 kHz; at this frequency, the phase velocity of the *L*(0,1)mode/A_0-like mode is 1440.51 m/s and the group velocity $C_g = 2478.84$ m/s. The phase velocity of the *L*(0,2) mode/S_0-like mode is 5397.97 m/s and the group velocity $C_g = 5388.5$ m/s.

The 3D pipe model is generated using the rotating method, which means that the two-dimensional cross section and rotation axle of the 3D pipe are drawn first. Then, the rotation angle of the two-dimensional cross section around the rotation axle is appointed as 360° and a complete 3D model can be generated. In all the simulation studies of the *L*-type guided wave, the length of the pipeline is 0.4 m.

Similar to the simulation of *Lamb wave* in the plate, the excitation on the edge of the waveguide starting terminal is used to generate the *L*-type *S-like* and *A-like* guided wave modes. At the frequency of 250 kHz, the *S-like* mode is the *L*(0,2) mode and the *A-like* mode is the *L*(0,1) mode. In the 3D model of the pipeline, the edge excitation that is parallel to the edge is actually along the radial direction; thus, the auxiliary cylindrical coordinates are established on the end edge of the pipeline and so the edge excitation along the radial direction is very easy to apply (Fig. 6.33).

Firstly, the guided wave propagation in the defect-free pipeline is simulated. The circumstance of the free-propagation of the *L*-type guided wave in the steel pipeline is shown in Fig. 6.34. Figure 6.34a corresponds to the *L*(0,2) mode/S_0-like mode and Fig. 6.34b corresponds to the *L*(0,1) mode/A_0-like mode. The variable drawn in

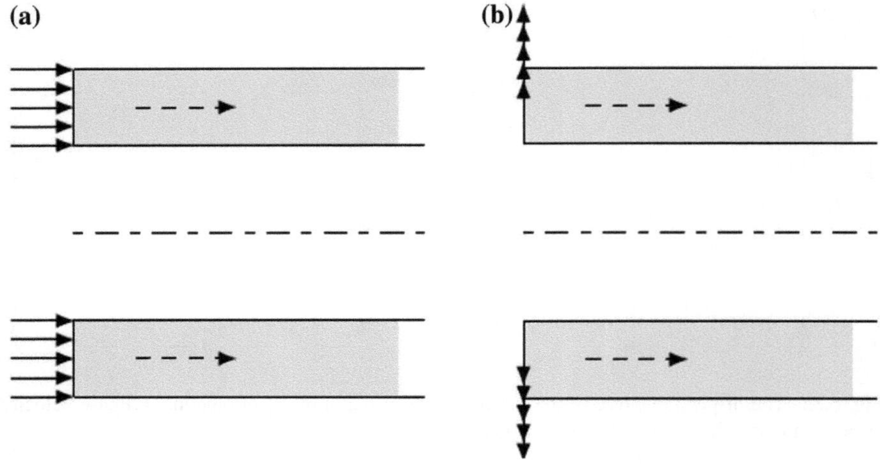

Fig. 6.33 The different types of excitation used to generate the *L*-type guided wave along the axial direction in the pipe. **a** Edge vertical excitation. **b** Edge parallel excitation

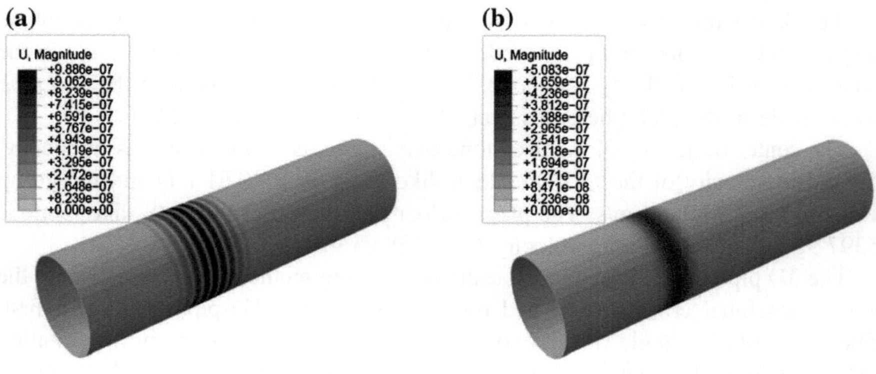

Fig. 6.34 The simulation of the *L*-type guided wave free-propagation in the steel pipeline. **a** *L* (0,2) mode. **b** *L*(0,1) mode

the figure is the displacement amplitude, the axial direction ultrasonic guided wave propagating from the lower left side of the pipeline to the upper right side of the pipeline. According to the guided wave mode phase velocity and excitation frequency, the wavelength of the *L*(0,1) mode/A_0-like mode can be calculated as $\lambda = C_p/f = 5.8$ mm, and the wave length of the *L*(0,2) mode/S_0-like mode is $\lambda = 21.6$ mm.

Now, the axial symmetric defect with a circumferential distribution existing in the pipeline will be discussed. Here, the interactions between the *L*-type ultrasonic guided wave and *four* types of axisymmetric transversal defects are studied. The defect's axial cross section includes the external axisymmetric defect, as shown in

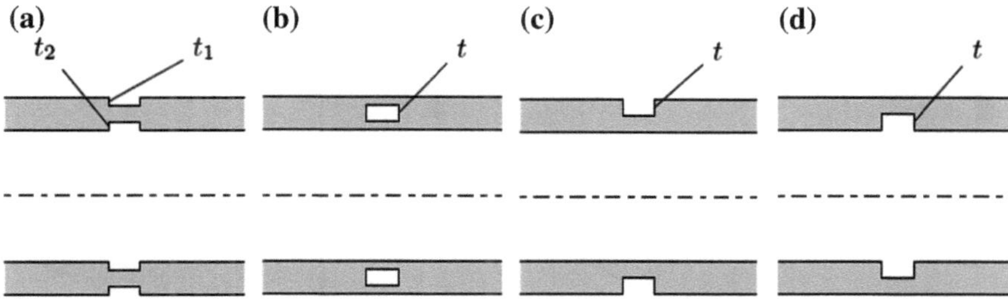

Fig. 6.35 Different types of transverse defects in the simulation of the *L*-type guided wave along the axial direction of the pipeline. **a** External axisymmetric defect. **b** Internal symmetric defect. **c** Outside surface asymmetric defect. **d** Inside surface asymmetric defect

Fig. 6.35a; the internal symmetric defect, where the defect is located inside the pipeline, as shown in Fig. 6.35b; the external axisymmetric defect, the opening of which is located on the outside surface of the pipe wall, as shown in Fig. 6.35c; and the external axisymmetric defect, the opening of which is located on the inside surface of the pipeline, as shown in Fig. 6.35d.

Here, the "*symmetry*" and "*asymmetry*" are all in respect to the middle surface of the pipeline. "*Outside*" and "*inside*" are used to describe the location of the defect inside the pipeline. Because of the axial symmetry of the simulated defect, the scattering signal of the incident *L*-type guided wave mode is still an *L*-type guided wave mode.

The simulated results between the *L*-type ultrasonic guided wave and defect are given in the following section. The axial width of the transverse/circumferential cracks in the following example are all 1 mm, and they are all located in the middle position of the pipe axial direction in the model. The radial dimension (t) of the defect with a rectangular axle cross section is 50 % of the pipe wall thickness.

In Fig. 6.36, subgraph (a) shows the circumstance after the scattering of the $L(0,2)$ mode/S_0-*like* mode guided waves at the external symmetric defect in the steel

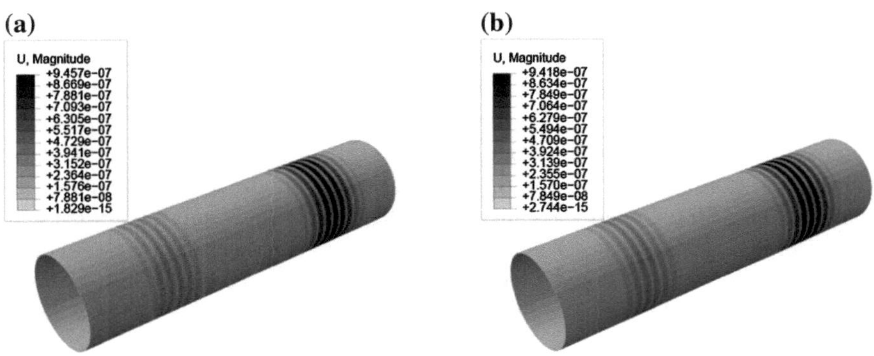

Fig. 6.36 The scattering of the $L(0,2)$ mode guided wave at the defect that is symmetric about the middle surface of the pipe wall in the steel pipe. **a** External symmetric defect. **b** Internal symmetric defect

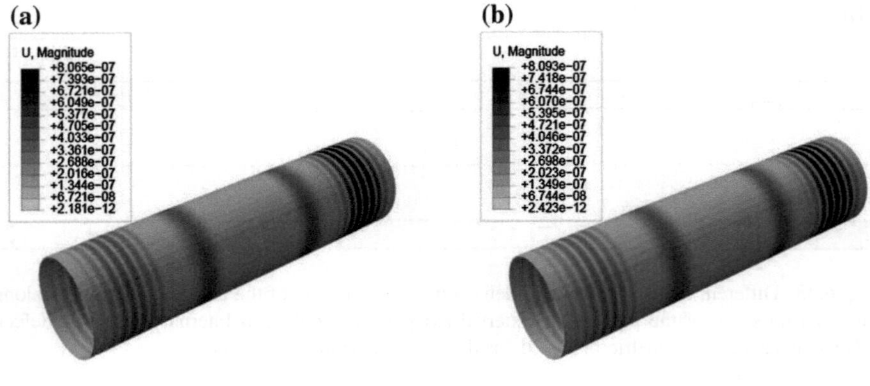

Fig. 6.37 The scattering of the *L*(0,2) mode guided wave about the defect that is asymmetric about the middle surface of the pipe wall inside the steel pipe. **a** Axisymmetric defect on the outside surface. **b** Axisymmetric defect on the inside surface

pipeline. Because of the symmetry of the symmetric defect of the middle surface of the pipeline, there is only the *L*(0,2) mode, which is the same as the incident mode that exists of the reflection and transmission guided wave. This means there is no existence of the reflection and transmission *L*(0,1) mode/A_0-*like* mode generated from the mode conversion.

In Fig. 6.37, subgraph (a) shows the circumstance after the scattering of the *L*(0,2) mode/S_0-*like* mode guided wave at the axisymmetric defect on the outside surface of the steel pipeline, while subgraph (b) shows the circumstance after the scattering at the axisymmetric defect on the internal surface. Because the defect is not symmetric at the pipe wall's middle surface any more, at this time the mode conversion from the incident *L*(0,2) guided wave mode to the reflection and transmission *L*(0,1) mode guided wave will happen. In the scattering guided wave shown in the figure, the faster guided wave characteristic near the two ends of the pipeline is the reflection and transmission *L*(0,2) mode, while the slower guided wave characteristic near the pipe central defect is the reflection and transmission *L*(0,1) mode generated by the mode conversion.

In Fig. 6.38, subgraph (a) shows the circumstance after the scattering of the *L*(0,1) mode/A_0-like mode guided wave at the asymmetric defect on the outside surface of the steel pipeline, while subgraph (b) shows the circumstance after the scattering at the asymmetric defect on the internal surface. Because the defect is not symmetric of the pipe wall's middle surface any more, at this time the mode conversion from the incident *L*(0,1) guided wave mode to the reflection and transmission *L*(0,2) guided wave modes will happen. In the scattering guided wave shown in the figure, the faster guided wave characteristic near the two ends of the pipeline is the reflection and transmission *L*(0,2) mode generated by the mode conversion, while the slower guided wave characteristic near the central defect in the pipe is the reflection and transmission *L*(0,1) mode.

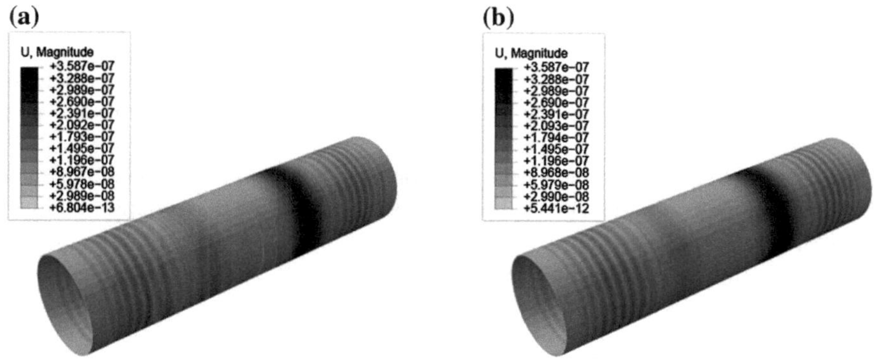

Fig. 6.38 The scattering of the $L(0,1)$ mode guided wave about the defect that is asymmetric about the middle surface of the pipe wall inside the steel pipe. **a** Axisymmetric defect on the outside surface. **b** Axisymmetric defect on the internal surface

6.5 Finite Element Simulation of the *T*-type Guided Wave Along the Axial Direction in the Pipeline

In this section, the ideal circumstance of the *T*-type guided wave simulation, if the guided waves and defects are all symmetrically distributed along the axial direction in the pipeline, will be investigated. The target steel pipeline has an inner radius and outer radius of 49 and 50 mm, respectively. The phase and group velocity dispersion for the *T*-type guided wave in the pipeline is shown in Fig. 6.39.

The material parameters used in the dispersion curve calculation are *Young's modulus* $E = 207 \times 10^9$ Pa, *Poisson's ratio* $v = 0.3$, and mass density $\rho = 7800$ kg/m^3. The inner radius of the pipeline in the model is 49 mm and the outer radius is 50 mm. In the dispersion curve of the *T*-type guided wave, $T(0,n)$

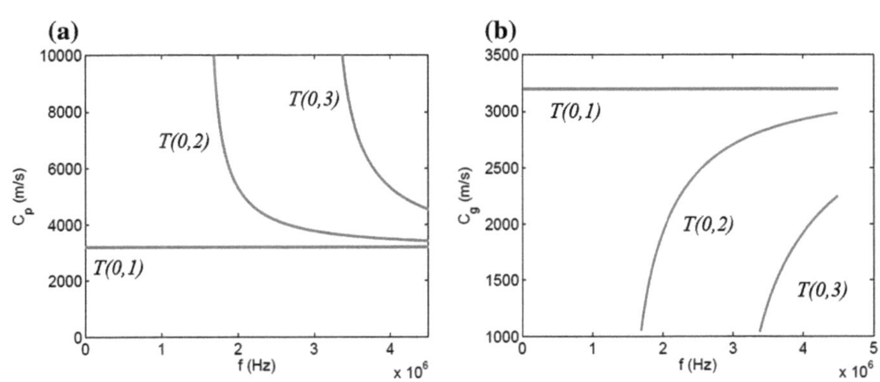

Fig. 6.39 The dispersion curves of the *T*-type guided wave in the steel pipeline. **a** Phase velocity dispersion curve. **b** Group velocity dispersion curve

stands for the *n*th torsional mode. It can be seen from the dispersion curve that each mode of the *T*-type guided wave is similar to that of the SH guided wave in the plate. Similar to the SH_0 mode in the plate, the phase velocity and group velocity of the *T*(0,1) mode is a straight line and there is no existence of the dispersive phenomenon of this mode.

The center frequency of the used excitation signal is 250 kHz. At this frequency, the phase velocity of the *T*(0,1) mode is 194.86 m/s and the group velocity is 3194.86 m/s. The 3D model is also used in the simulation for the *T*-type guided wave along the axial direction in the pipeline. In all the *T*-type guided wave simulations, the pipe length is 0.4 m.

Considering the particle vibration in the propagation process of the *T*-type guided wave mode, the tangential excitation is applied on the edge of the starting terminal in the pipeline. The application method of the excitation is shown in Fig. 6.40. In order to make the assignment of the tangential excitation more convenient, the auxiliary cylindrical coordinates are established at the starting end surface in the pipeline. It can be seen from the dispersion curve that there is only the *T*(0,1) guided wave mode existing at the used frequency of 250 kHz; thus, there is only the *T*(0,1) mode generated by the tangential excitation, as shown in Fig. 6.40.

Firstly, the circumstance of the guided wave propagation in the defect-free pipeline is simulated. The free-propagation of the *T*-type guided wave inside the steel pipeline is shown in Fig. 6.41, where the guided wave mode is *T*(0,1). The variable drawn in the figure is the displacement amplitude, the axial ultrasonic guided wave propagating from the lower left side of the pipeline to the upper right side of the pipeline.

Then, the simulated results of the interactions between the *T*-type ultrasonic guided wave and the defects are provided. The axial width of the transverse/circumferential crack in the following examples are all 1 mm, and they are all located at the middle position of the axial direction of the pipeline in the model. The radial dimension of the defect with a rectangular axial cross section is 50 % of the pipe wall thickness.

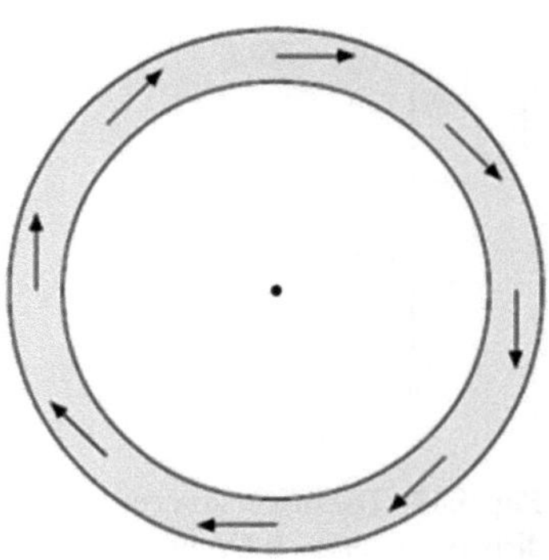

Fig. 6.40 The excitation method that is used to generate the *T*-type guided wave along the pipeline axial direction

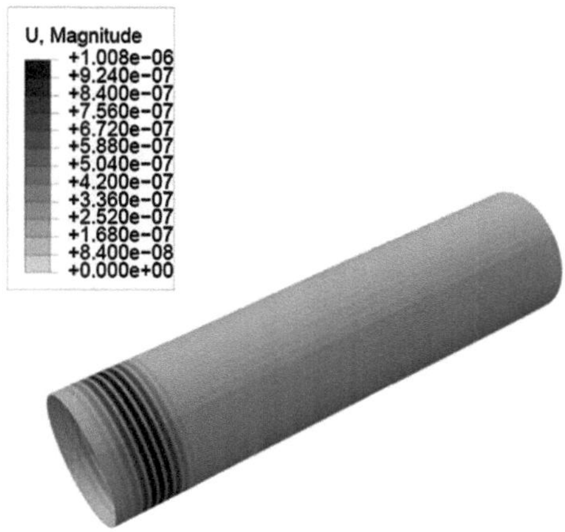

Fig. 6.41 The free-propagation of the *T*(0,1) guided wave mode in the steel pipeline

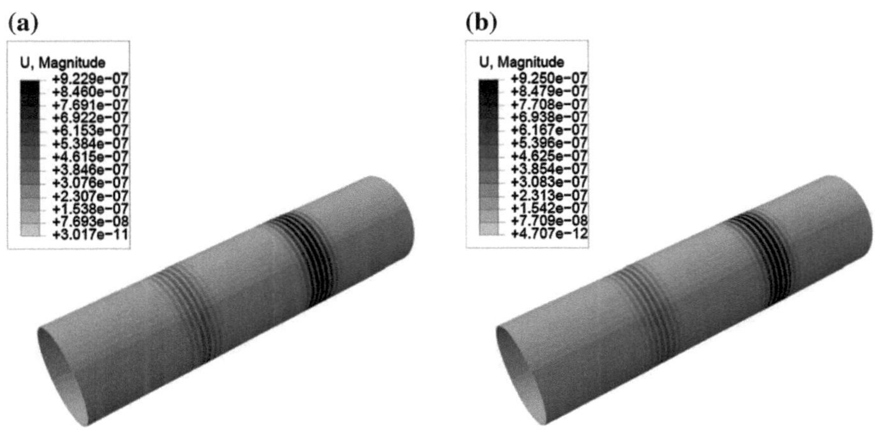

Fig. 6.42 The scattering of the *T*(0,1) guided wave at the symmetric defect inside the steel pipeline. **a** External symmetric defect. **b** Internal symmetric defect

In Fig. 6.42, subgraph (a) shows the circumstance of the *T*(0,1) mode guided wave at the external symmetric defect inside the steel pipeline, while subgraph (b) shows the circumstance after the scattering at the inside symmetric defect.

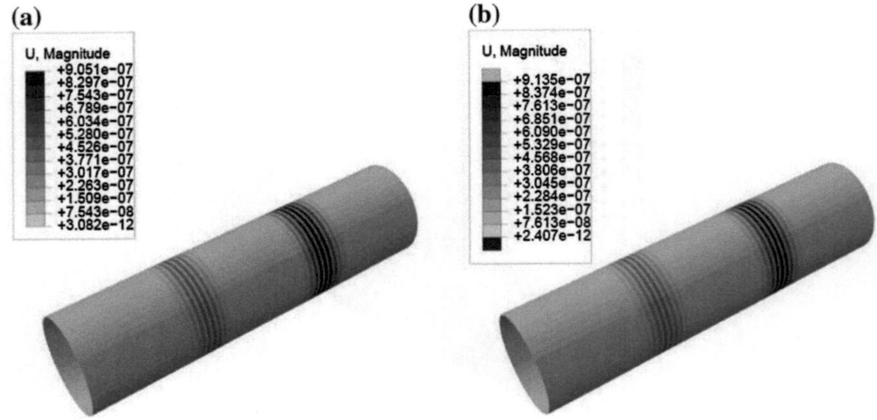

Fig. 6.43 The scattering of the *T*(0,1) guided wave at the asymmetric defect inside the steel pipeline. **a** Asymmetric defect on the outside surface. **b** Asymmetric defect on the inside surface

In Fig. 6.43, subgraph (a) shows the circumstance of the *T*(0,1) mode guided wave at the external asymmetric defect inside the steel pipeline, while subgraph (b) shows the circumstance after the scattering in the inside asymmetric defect.

References

1. Wang, S., Huang, S., Zhao, W.: Simulation of Lamb waves interactions with transverse internal defects in an elastic plate. Ultrasonics **51**(4), 432–440 (2011)
2. Wang, S.: Crack detection by electromagnetic ultrasonic guided waves in natural gas pipeline. Tsinghua University, Beijing (2008). (in Chinese)

Chapter 7
Applications of the Electromagnetic Ultrasonic Guided Wave Technique

7.1 Thickness Measurement by Electromagnetic Ultrasonics

7.1.1 Principle of the Thickness Measurement by Electromagnetic Ultrasonics

There are two principles of the thickness measurement by electromagnetic ultrasonics: One is based on *Lorentz force*, which is suitable for conductor materials, and the other is based on magnetostrictive force, which is suitable for ferromagnetic conductive materials. In the process of thickness measurement, the ferromagnetic conductive materials will be affected by two different forces.

The pulse echo method of the straight beam is used for the electromagnetic ultrasonic thickness measurement, while for the fast scanning and high-resolution detection, or detection of thin material, the pitch–catch approach of the oblique beam can also be used.

In the propagation process of the ultrasonic wave, the reflection and transmission phenomenon will happen when it meets the interface of the two kinds of medium. In the bounce-back-and-forth process of the vertical incident wave between the top and bottom surfaces of the specimen, the reflection and transmission phenomenon will happen at both the upper and lower interfaces of the specimen. Assuming the acoustic impedance between the specimen (metallic material) and air are Z_1 and Z_2, respectively, for the circumstance of the vertical incidence of the ultrasonic wave, the reflection coefficient R and transmission coefficient T can be expressed, respectively, as shown in (7.1).

S. Huang et al., *Electromagnetic Ultrasonic Guided Waves*, Springer Series in Measurement Science and Technology, DOI 10.1007/978-981-10-0564-0_7

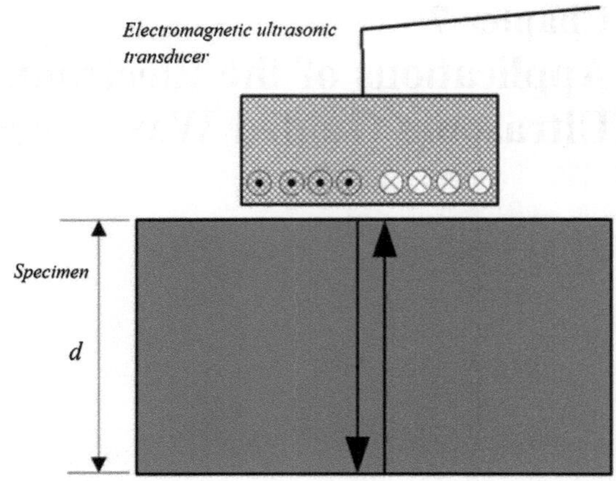

Fig. 7.1 A sketch map of the electromagnetic ultrasonic thickness measurement

$$R = \left(\frac{Z_1 - Z_2}{Z_1 + Z_2}\right)^2 \times 100\ \%$$
$$T = \frac{4Z_1Z_2}{(Z_1 + Z_2)} \times 100\ \% \tag{7.1}$$

Since the acoustic impedance of the air Z_2 is much less than that of the specimen Z_1—thus, R is close to 1, but T is close to 0. The total reflection of the ultrasonic wave will happen at the interface, and many times, the reflections will occur between the top and bottom surfaces of the specimen. Then, the thickness of the tested specimen can be calculated by (7.2).

$$d = \frac{CT_n}{2} \tag{7.2}$$

In the above equation, d is the thickness of the tested specimen and T_n is the time difference between the two echo signals, which is the propagation time of the ultrasonic wave between the top and bottom surfaces of the detected specimen. C is the wave velocity of the electromagnetic ultrasonics in the specimen. The principle of the thickness measurement is shown in Fig. 7.1. Since under different temperatures, the propagation velocity of the ultrasonic wave in the specimen can be different, sometimes the temperature compensation should be conducted for the modification of the ultrasonic wave velocity.

7.1.2 Setup of the Electromagnetic Ultrasonic Thickness Measurement [1]

The electromagnetic ultrasonic thickness measurement system mainly consists of 4 parts: the electromagnetic ultrasonic transducer, match circuit, ultrasonic

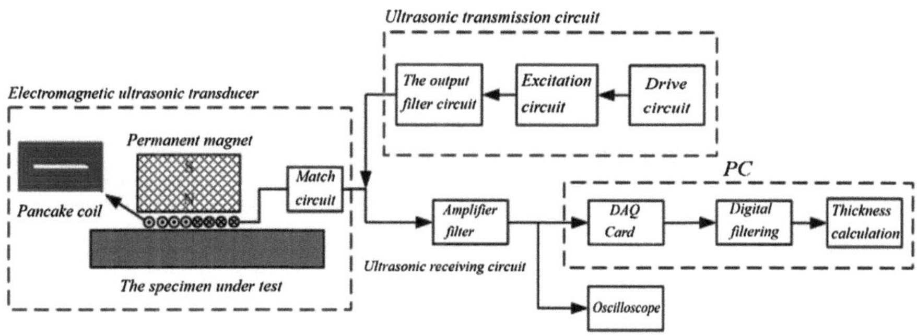

Fig. 7.2 The setup of the electromagnetic ultrasonic thickness measurement

transmission circuit, signal-receiving amplifier circuit, as well as the data acquisition and processing circuit. The permanent magnet can provide a constant bias magnetic field. The match circuit can improve the transfer efficiency of the energy converter. The tested specimen is ferromagnetic material or conductor. The ultrasonic transmission circuit includes the power amplifier, pulse-generating circuit, and the frequency regulator. The pulse-generating circuit is charged by the power amplifier. The frequency regulator is used to modulate the pulse width of the generated pulse and the frequency of the sine wave. The core component of this circuit is programmable logic component FPGA. The waveform of any frequency can be modified by the online program. After the ultrasonic echo signal is received by the electromagnetic ultrasonic transducer through amplification, smoothing, and data sampling, and then incident into PC through the digital signal filter, the value of the thickness is calculated. The waveform after the amplification and smoothing can also be observed in the oscilloscope. The testing system is shown in Fig. 7.2.

7.1.3 Hardware of the Electromagnetic Ultrasonic Thickness Measurement [2]

Among the electromagnetic ultrasonic thickness measurement devices, the key parts are the signal reception and amplification, which have a significant effect on the accuracy of the measurement. The signals received from the electromagnetic ultrasonic transducer require amplification and filtering, as shown in Fig. 7.3.

The echo signals are received by the electromagnetic ultrasonic transducer and transferred to the filter amplifier circuit; the preamplifier is usually a broadband, low-gain device. The ultrasonic signal in the whole amplitude range can be amplified linearly, modifying the signal to the level which is big enough to overcome the noise of the subsequent amplifier stage and filtering stage, which can maximize the signal-to-noise ratio. The amplifier modifies the ultrasonic echo signal amplitude to the level that will enable the signal processing circuit to work

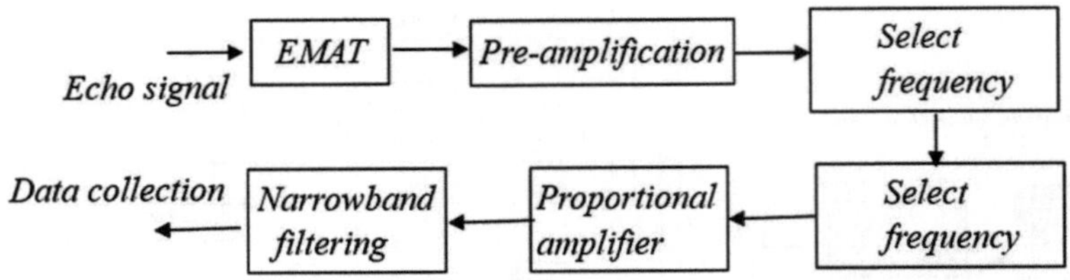

Fig. 7.3 The receiving and processing process of the electromagnetic ultrasonic signal

normally; the filter is narrowband, which can eliminate the unnecessary frequency response of EMAT, reducing the interference of the power. The frequency characteristic of the signal is determined by the band-pass filter, controlling the frequency band of the amplified signal, which will contribute to improving the signal-to-noise ratio. The signal after the amplification and filtering is incident into the acquisition system for the data collection.

Usually, when the circuit reaches a good matching status and the current peak of the excitation coil reaches 50 A, the voltage of the receiving coil is only dozens of microvolt. At the same time, the high-power excitation circuit will generate strong electromagnetic interference during its working period, which induces the serious acoustic noise in the EMAT system. Thus, in order to obtain a signal with a high signal-to-noise ratio, the receiving circuit of EMAT must have the ability to detect a weak signal under the background of strong noise. In order to reduce the interference of the external environment, a coaxial cable and BNC joint are used in the connections of the receiving circuit. Moreover, it is necessary to avoid the influence of the high-voltage excitation on the receiving circuit. A block diagram of the protection of the amplification circuit and frequency-selective amplification signal processing, etc., is shown in Fig. 7.4.

The coil excitation needs a high-voltage and high-power excitation circuit in order to provide excitation to the receiving circuit, so the excitation circuit generates a high-frequency sinusoidal signal or pulse signal. According to the current technology and the request of the electromagnetic ultrasonic thickness measurement, a full-bridge inverter amplifier circuit is used to generate sine waves. A single switch pipeline is used for the charge–discharge of the capacitance to generate the high-voltage narrow pulse.

The principle of the full-bridge inverter power amplification excitation circuit is shown in Fig. 7.5. 220 V power frequency AC main is connected to a rectifier bridge

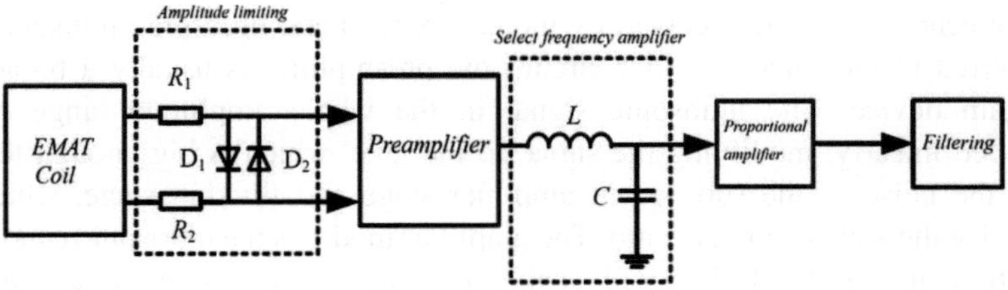

Fig. 7.4 The amplitude limitation and amplification processes of the receiving circuit

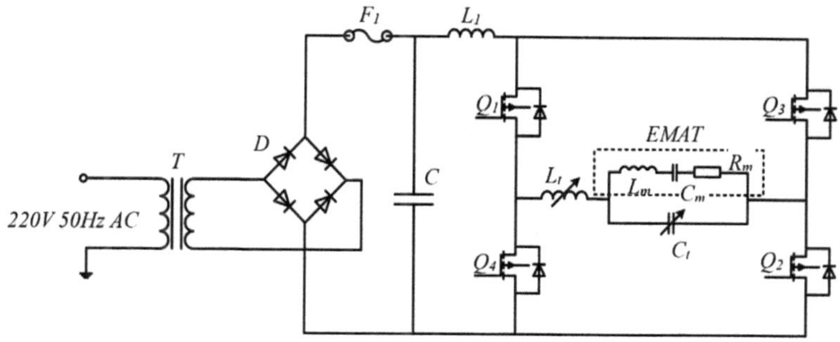

Fig. 7.5 The principle of the full-bridge inversion of the excitation circuit

through an adjustable transformer. The AC main is transferred into the DC current through the full-bridge rectifier, providing the input voltage for the inverter circuit. At the end of the AC input, the alternating voltage is transferred into the direct voltage, the maximum value of which can attain 400 V through the full-bridge rectifier composed of *four* diodes. The peak voltage output by the full-bridge inverter circuit is close to the direct voltage at the input terminal, but the value of the output voltage is only half of the direct voltage after the rectification of the front-end rectifier. Compared to the half-bridge inverter circuit, the full-bridge converter has a higher efficiency. The inductance L_1 is the smoothing section after the rectification, in which the high-frequency spike pulse generated from the rectifying process can be filtered. The capacitance at the back-end is the high-voltage charging capacitance with a large capacity, which can form a stable DC voltage at the input terminal. F_1 is the fuse wire which is used to control the electric current in the switching pipeline, so that the average current in the switching pipeline can be restricted. The full-bridge inverter mainly includes the driver module, inverter module, and tunable filter module. The drive capability of the pulse signal generated by the signal-generating module is limited, and the driving for the power switch pipeline should be placed after the amplification of the driver module. The power electronics switch with high frequency, high voltage, and high power should be used in the inverter module through the control of the connection and disconnection of the switch, outputting the symmetric square pulse. The snubber circuit is used to absorb the sharp pulse and noise generated in the conduction and turnoff process.

7.1.4 Analysis and Processing of the Echo Signal in the Electromagnetic Ultrasonic Thickness Measurement [3]

The electromagnetic ultrasonic thickness measurement is the same as that of the traditional piezoelectric ultrasonic approach, where the impulse reflection process is used. The ultrasonic wave reflection and propagation have strict repeatability, so the waveform characteristic is obvious. The time difference corresponding to the peak values is used to calculate the thickness. Since pulse energy is mostly used as the excitation of the thickness measurement, such as the pulse-modulated sine wave excitation transducer, its echo signal has the same amplitude as the modulated signal and its frequency is the same as that of the excitation pulse. The echoed signal is usually highly contaminated by noise and the signal-to-noise ratio is relatively low. The superposition of the noise makes the output signal unsteady, where the amplitude, phase position, and some other parameters of the sinusoidal components of each frequency are time-varying parameters. However, the analog filter circuit has a scarce capacity when the random noise of the unsteady signal is modified, and thus, it is necessary to carry out some digital filtering and processing of the signal.

Time-domain analysis is used to extract the time-varying sine signal submerged in noise, which can improve the signal quality. An averaging technique is used for the randomicity characteristic by making multiple samples in the time domain and then averaging them to improve the signal-to-noise ratio. Now, it becomes a general method for dealing with the signal received by EMAT. For the recurring signal received by EMAT, the waveforms within the error range are basically the same every time, but the noise of the signal is random. After a lot of signal acquisition, the noise level of the useful signal can be significantly reduced, and the echo signals are strengthened through the summation and averaging of the corresponding sampling points.

As shown in Fig. 7.6, the ultrasonic echo signal generated by the sinusoidal burst signal with a center frequency 2 MHz and periodicity 12 in the aluminum plate with a thickness of 29.2 mm is provided. Figure (a) is the original signal, and figure (b) is the waveform after averaging 128 times. It can be seen from the figure that the

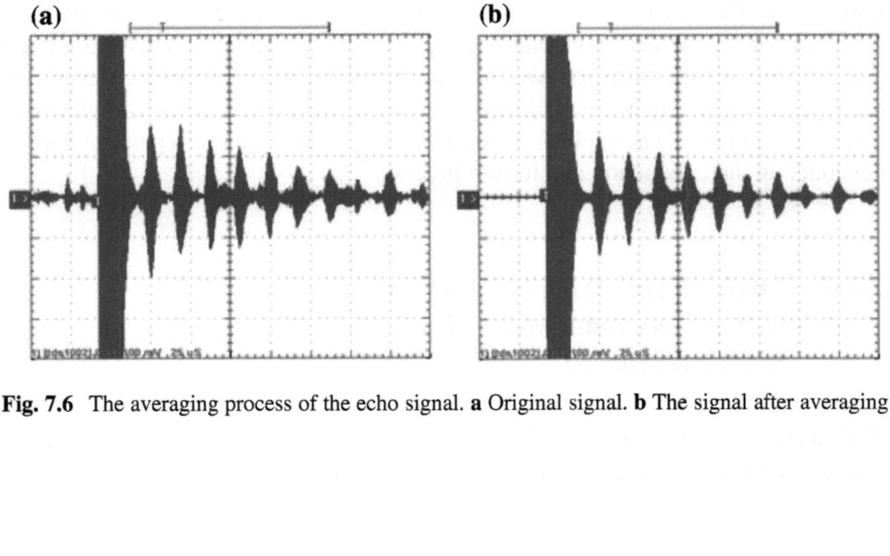

Fig. 7.6 The averaging process of the echo signal. **a** Original signal. **b** The signal after averaging

random noise is suppressed effectively, and as there is no obvious attenuation of the useful signal, the signal-to-noise ratio is improved.

In order to measure the time difference of the echo signal to calculate the thickness of the tested specimen, the envelope of the detected signal should be obtained first. The envelopes are the up and down lines of the parcel signal generated by connecting the signal maximum and minimum points sequentially.

Common methods that are used to extract the signal envelope are listed as follows: the polyphonic method, the full-wave rectification method, the detection filtering method, and the *Hilbert* transform method (*Hilbert*). Those modulation methods are generally needed to work with the filter, so that it can be effectively applied in practice. The *Hilbert* is a simple and effective method and has wide applications in the engineering field.

After the extraction of the ultrasonic echo signal envelope, in order to achieve the automatic thickness measurement, the time difference between the two adjacent echo peaks should be checked. The thickness of the specimen is calculated from the calculation equation of the electromagnetic ultrasonic thickness. However, the difficulty of the method is how to check the locations of the two wave peaks. In order to perform further analysis and processing of the ultrasonic echo signal, two functions used to reflect the echo signal characteristic are defined: the energy distribution representing the energy distribution of the echo signal and the comparability distribution representing the level of similarity between the *n*th waveform and the 1st waveform of the ultrasonic signal. Initially, the tone burst signal in the envelope is removed. A rectangular window with a width W_{d} is constructed as shown in Fig. 7.7. Moving the rectangular window along the time axis, within the scope of the rectangular window, the second norm of the vectors composed of all the data points is the value of the energy distribution function corresponding to the central point of the rectangular window. Its function waveform is shown in Fig. 7.8a. The horizontal ordinate corresponding to the maximum value of the energy distribution function is the *FirstPeak* at the time corresponding to the peak of the first echo. Setting this moment as the center position of the rectangular window, at this time the vector composed of the energy distribution function value within this rectangular window is $P_1(W_d)$. Moving the rectangular window along the time axis on the waveform of the energy distribution function, when it moves to any position, the vector composed of the values of the energy distribution function within the rectangular window is $P_n(W_d)$. The comparability distribution function can be expressed by (7.3).

$$f(n) = \exp\left[-\|P_1(W_{\mathrm{d}}) - P_n(W_{\mathrm{d}})\|^2\right] \tag{7.3}$$

In (7.3), $\|P_1(W_{\mathrm{d}}) - P_n(W_{\mathrm{d}})\|^2$ represents the second norm of the two vectors, when the waveform of the *FirstPeak* corresponding to the first echo peak is most similar to itself. The corresponding $f(n)$ is the largest, followed by the waveform of the *SecondPeak* corresponding to the second echo peak and then by the waveforms corresponding to the time of the *n*th echo peak. The waveform of this function is

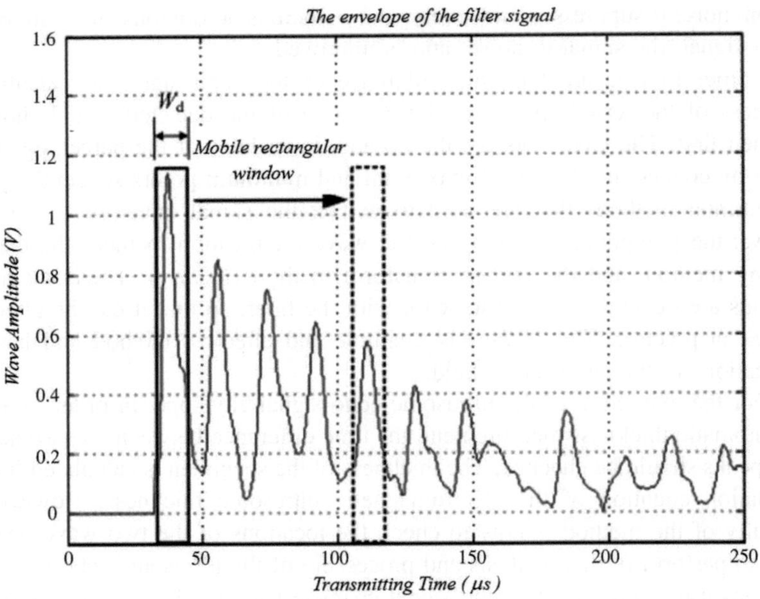

Fig. 7.7 Establishment of the characteristic function of the ultrasonic echo signal envelope

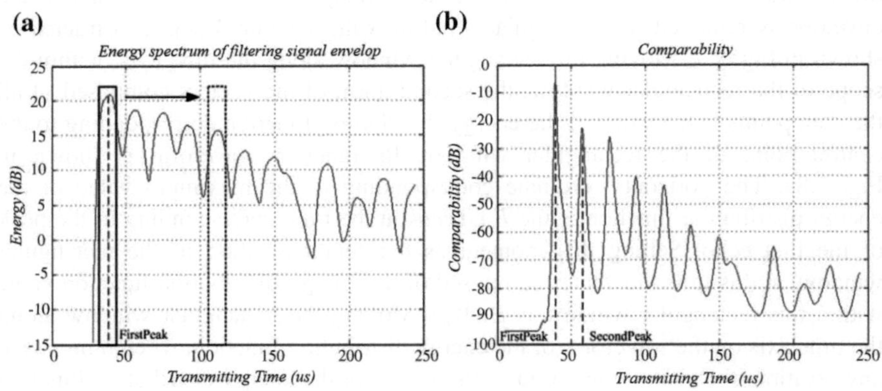

Fig. 7.8 The characteristic function of the ultrasonic echo signal. **a** Energy distribution function. **b** Waveform similarity function

listed in Fig. 7.8b. From the *FirstPeak*, the horizontal ordinate corresponding to the maximum value of the function *f*(*n*) is the *SecondPeak* corresponding to the second echo peak. The time difference between these two peaks is the propagation time T_n of the ultrasonic wave bouncing back and forth once inside the specimen. If the propagation velocity *C* of the ultrasonic wave in the specimen is already known, then the thickness of the tested specimen can be measured by the electromagnetic ultrasonic thickness calculation (7.2).

The energy distribution function represents the energy distribution of the echo signal. Compared to the method by which the first wave peak time is obtained directly from the echo signal envelope after comparison with the echo signal amplitude, this method is more accurate and reliable for the reason that the first echo peak is obtained from the energy distribution function. The energy distribution function can eliminate the singular value contained in the echo signal envelope and the misjudgment caused by the high-frequency noise spike. The width of the rectangular window depends on the frequency of the ultrasonic wave, the data sampling rate, and the shape of the echo peak; thus, the selection of the width of the rectangular window depends on various circumstances.

7.2 Electromagnetic Ultrasonic Guided Wave Test Along the Axial Direction of the Pipeline

7.2.1 Electromagnetic Ultrasonic Transducers

The structure of the longitudinal mode guided wave EMAT for the pipeline is shown in Fig. 7.9. Two kinds of schemes of single cluster coil and multiple cluster coils are used separately. Varnished wire is used to unwind the electromagnet coil. After the DC current is input, a static bias magnetic field will be generated along the pipeline in the axial direction. The EMAT coil is unwound along the pipeline in the circumferential direction, where a dynamic magnetic field will be generated when high-frequency alternating current is introduced. The direction of the dynamic magnetic field is parallel to that of the static bias magnetic field.

From the phase velocity and group velocity dispersion curves of the longitudinal mode, within a frequency range of 0–1 MHz, there are four guided wave modes, from $L(0,1)$ to $L(0,4)$. Since the $L(0,2)$ guided wave mode is not dispersive, the $L(0,2)$ mode is selected for the test.

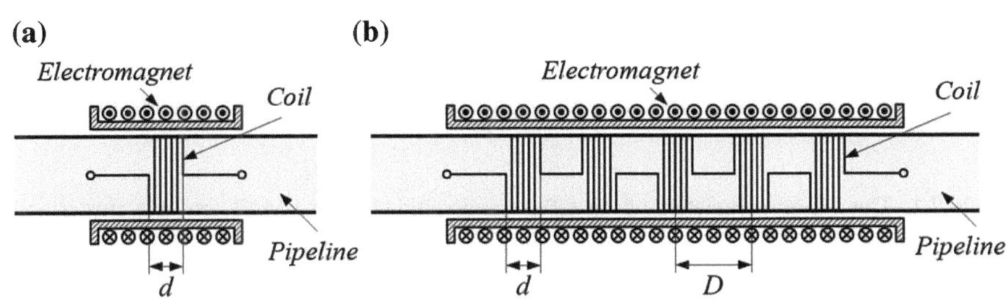

Fig. 7.9 The structure diagram of the longitudinal mode guided wave EMAT. **a** Single cluster coil. **b** Multiple cluster coils

In the design of the multiple cluster coils, the space D between adjacent cluster coils is half the wavelength of the longitudinal guided wave mode, as shown in (7.4).

$$D = \frac{\lambda}{2} = \frac{c_{\mathrm{p}}}{2f} \tag{7.4}$$

In the above equation, λ is the wavelength of the guided wave mode, c_{p} is the phase velocity of the guided wave, and f is the frequency. Thus, the space between adjacent cluster coils corresponds to the selected guided wave frequency. At the same time, adjacent cluster coils are unwound along the clockwise and anticlockwise directions, respectively, making the direction of the current in adjacent cluster coils opposite to each other to generate a dynamic magnetic field with different directions. Based on this design, when the propagation distance of the particle vibration caused by a bunch of coils is $\lambda/2$ (the position of the adjacent cluster coil) at phase velocity of c_{p}, it will produce constructive interference in the phase with the particle vibration caused by the adjacent cluster coil, which will enhance the vibration intensity. For the guided waves of other modes, if they do not satisfy the corresponding relationship expressed in (7.4), the particle vibration caused by the adjacent cluster coil cannot generate the constructive interference, so adjacent cluster coils might offset each other. Thus, if multiple cluster coils are used, they can both improve the vibration intensity of the guided wave mode and restrain the generation of other guided wave modes.

The design parameters of the EMAT multiple cluster coils that can excite the $L(0,2)$ guided wave mode are given in Table 7.1. From the phase velocity dispersion curve of the longitudinal guided waves in the pipe, at the 250 kHz frequency point, the phase velocity of the $L(0,1)$ mode is 2169 m/s and the wavelength is 8.68 mm, while the phase velocity of the $L(0,2)$ mode is 5364 m/s and the wavelength is 21.46 mm. In this way, when the spacing between adjacent cluster coils is designed as 10.73 mm, that is, half of the wavelength of the $L(0,2)$ mode, the $L(0,2)$ mode can be excited, while at the same time, the generation of the $L(0,1)$ mode guided wave is restrained.

The width d of the single cluster coil should be small or equal to half the wavelength of the expected guided wave mode; otherwise, the vibration caused by the wire at the end of the coil will offset each other, reducing the vibration intensity of the expected guided wave mode. When designing multiple cluster coils, the

Table 7.1 The design parameters of the $L(0,2)$ mode guided wave EMAT multiple cluster coils

Parameter	Value			
Guided wave frequency f (kHz)	100	150	200	250
Guided wave phase velocity c_{p} (m/s)	5412	5392	5378	5364
Distance between the adjacent cluster coil D (mm)	27.06	17.97	13.45	10.73

Fig. 7.10 Coils made by the flexible printed circuit board

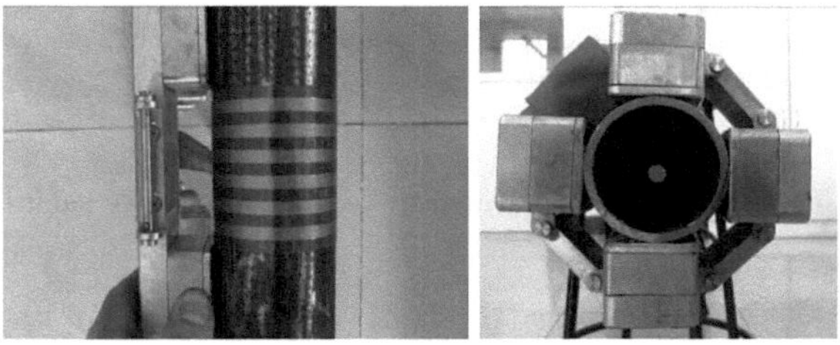

Fig. 7.11 A magnet made by the neodymium–iron–boron permanent magnet

width of each cluster coil also needs to meet this requirement. Although the principle of interferometry cannot be used to enhance the vibration intensity of the expected guided wave mode, when the single cluster coil is used, the longitudinal guided wave of the different frequencies can be generated by adjusting the coil width with great flexibility.

In the above discussion, the electromagnet and coil are both entwined by varnished wire, so it is easy to adjust the intensity of the bias magnetic field, the working frequency of EMAT, the thickness and turns of the coil wire, and so on, which is convenient for use in the laboratory. However, multiple cluster coils used in practical engineering are always made by a flexible printed circuit board, while the magnet is made by the neodymium–iron–boron permanent magnet, as shown in Figs. 7.10 and 7.11.

A printed circuit board is used to make the EMAT coil. The width of the conductor line in the coil and the spacing between the adjacent coils can be controlled with high accuracy, which benefits the miniaturization of the EMAT so that its energy conversion efficiency can be improved. The magnet can be divided into four parts. Each set of magnets is composed of two neodymium–iron–boron permanent magnets and an iron yoke. The different magnet sets are connected by a hinge. When it is used, the magnet can be placed on the outer surface of the pipeline axisymmetrically in order to provide a static bias magnetic field along the axial direction of the pipeline for the coils.

The structure of the torsional mode guided wave EMAT is shown in Fig. 7.12. The single cluster coil and multiple cluster coils are used, respectively. A premagnetized nickel belt is wrapped around the outer surface of the pipeline to generate the static bias magnetic field along the pipeline in the circumferential direction. The coil is wrapped along the circumferential direction in the pipeline to generate the dynamic magnetic field, the direction of which is perpendicular to the static bias magnetic field. Similar to the longitudinal guided wave EMAT, the width

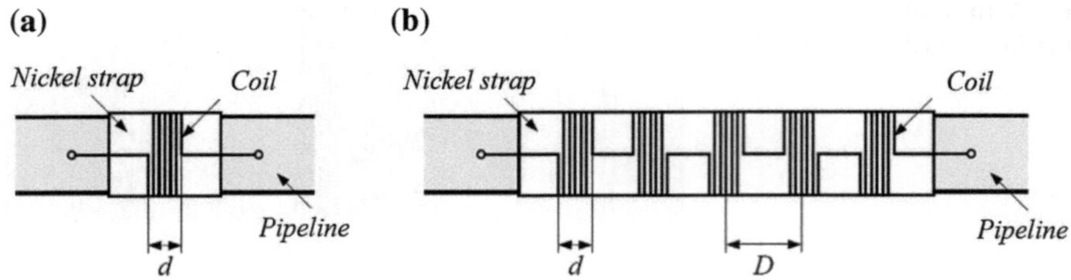

Fig. 7.12 The structure diagram of the torsional mode guided wave EMAT. **a** Single cluster coil. **b** Multiple cluster coils

Table 7.2 The design parameters of the $T(0,1)$ mode guided wave EMAT multiple cluster coils

Parameter	Value		
Guided wave frequency f (kHz)	100	150	190
Guided wave phase velocity c_p (m/s)	3200	3200	3200
Distance between adjacent cluster coils D (mm)	16.00	10.67	8.42

of single cluster coil, distance between the adjacent coils, etc., should meet the matching relationship of the selected guided wave mode.

The design parameters of the EMAT multiple cluster coils that can excite the $T(0,1)$ mode guided wave are given in Table 7.2. From the phase velocity dispersion curve of the torsional mode in the pipe, at the 190 kHz frequency point, the phase velocity of the $T(0,1)$ guided wave mode is 3200 m/s and the wavelength is 1684 mm. In this situation, when the spacing between adjacent cluster coils is designed as 8.42 mm, that is, half the wavelength of the $T(0,1)$ mode, the $T(0,1)$ guided wave mode can be excited.

The premagnetized nickel strap can be fixed on the pipeline surface using either epoxy resin glue or a mechanical clamping ring, and the coil can be made by the varnished wire or the flexible printed circuit board. Since the EMAT with this kind of structure will first generate vibration inside the ferromagnetic nickel strap and then couple the vibration into the tested pipeline, it can also be used to detect a pipeline made by non-conducting and non-ferromagnetic material.

7.2.2 *Electromagnetic Ultrasonic Excitation Source and the Filter Amplifier [4]*

The electromagnetic ultrasonic excitation source is mainly composed of a control circuit, a drive circuit, a rectifier and filter circuit, an inverter circuit, and a current foldback circuit. Its principle structure is shown in Fig. 7.13. Initially, the control circuit generates a control signal, either through the computer or using the parameters set by the hand knob. Then, the control signal is amplified by the drive

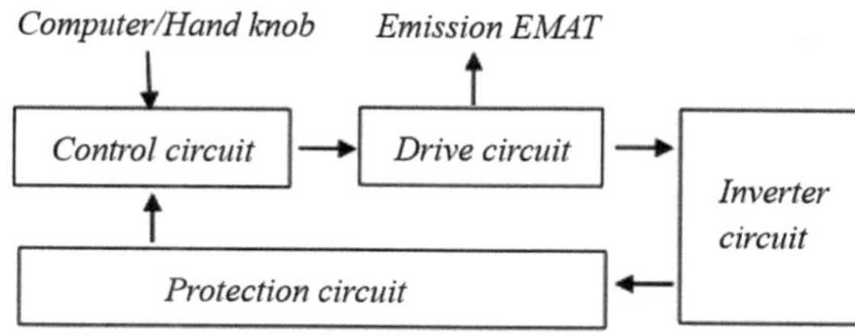

Fig. 7.13 A structure diagram of the electromagnetic ultrasonic drive source

circuit and drives the power MOSFET in the inverter circuit. Finally, the inverter circuit converts the direct voltage generated by the rectifier and filter circuit into the high-power pulse signal and outputs it to the excitation EMAT. At the same time, the output current is measured by the current foldback circuit composed of a *Hall* sensor and an analog-to-digital converter, and it is fed back to the control circuit to realize the overcurrent protection.

The programming language VHDL is used to write the control circuit program. It mainly includes a clock divider, trigger, signal from control and overcurrent protection, and some other modules. Its structure diagram is shown in Fig. 7.14. The clock divider module is based on the phase-locked coil implanted into the FPGA, on the outside of which there is a global clock at a frequency of 200 MHz with a multiplication frequency of 50 MHz generated by the source oscillator, with all the signals formatted from the edge-triggered count of the global clock. In the trigger module, the counter is defined as X. When the value of X is summed as the global clock number according to the adjacent time interval, the trigger module will output a trigger signal, the time width of which is a global clock cycle. Then, the counter X is reset and counting restarts. The rising edge of the trigger signal

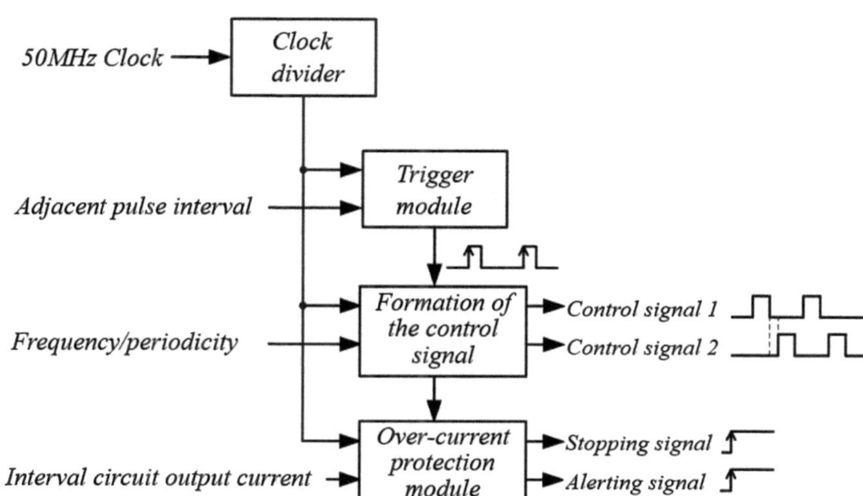

Fig. 7.14 A structure diagram of the control circuit program

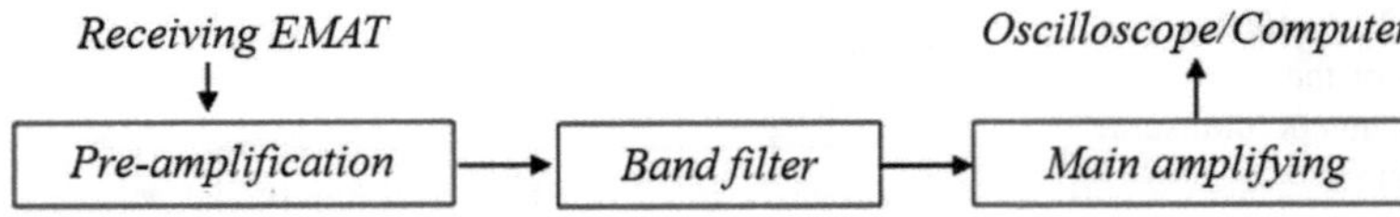

Fig. 7.15 A structure diagram of the electromagnetic ultrasonic filter amplifier

prompts the formation of the control signal and the counters *Y* and *Z* begin counting, while at the same time judging the numerical range of *Y* and *Z* using a branch structure. Controlling the switching element in the inverter circuit to establish the excitation signal frequency and cycle number generates the double-channel control signal. The output current is monitored by the overcurrent protection module through the overcurrent protection circuit of the excitation source. If the peak value of the output current exceeds the established threshold value, the alerting signal and stopping signal are reset from a low level to a high level through the overcurrent protection module, which will cause a buzzer to ring. At the same time, the output of the excitation signal is stopped to make sure that the whole excitation source is working within the security scope.

The electromagnetic ultrasonic filter amplifier is mainly composed of a preamplifier, a band filter circuit, and the main amplifying circuit. Its structure is shown in Fig. 7.15. Initially, the voltage signal received by the receiving EMAT is amplified preliminarily; then, the band-pass filtering is made by the band filter circuit, in which the guided wave frequency used in the actual application is taken as the center frequency; then, the signal after filtering is amplified further. Finally, an oscilloscope and computer are used to collect and display the signal.

7.2.3 *Experiment of the Electromagnetic Ultrasonic Guided Wave Test and the Factors*

Firstly, longitudinal mode and torsional mode ultrasonic guided waves are generated by EMAT, and the correctness of excitation of the modes is verified through the measured wave velocity. Then, the influences of the number of cycles of the signal, static bias magnetic field and dynamic magnetic field on the guided wave signal amplitude are studied. Finally, for the artificial defect on the outer wall of the pipeline, the experiment using the electromagnetic ultrasonic guided wave in the axial direction of the pipeline for defect detection is conducted.

The experimental system of the electromagnetic ultrasonic guided wave excitation and defect detection in the pipeline is provided in Fig. 7.16. It is mainly composed of the electromagnetic ultrasonic source, the transmitting EMAT, the receiving EMAT, a filter amplifier, computer, direct current main, and the experimental pipeline. In the experiment, the high-power pulse excitation signal is generated by the electromagnetic ultrasonic source and output to the excitation EMAT. The excitation EMAT is based on the magnetostrictive effect, where the ultrasonic

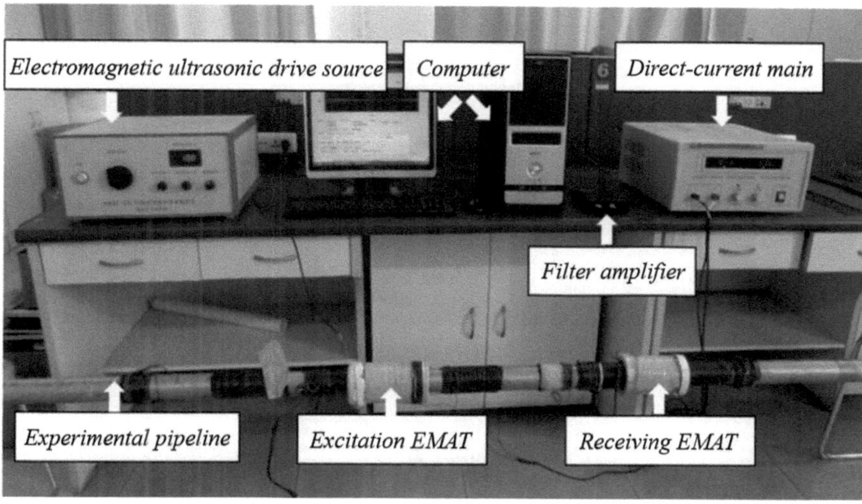

Fig. 7.16 The experimental system of the electromagnetic ultrasonic guided wave excitation and defect detection

guided wave propagating along the axial direction of pipeline is incident on the pipeline. The receiving EMAT is based on the inverse effect of magnetostriction. The ultrasonic guided wave signal in the pipeline is converted into voltage signal. The voltage signal detected by the receiving EMAT is amplified and filtered virtually and then collected and processed by the computer. The direct current inside the electromagnet of the EMAT is provided by direct current main to generate the static bias magnetic field. The parameters of the tested pipeline are as follows: external diameter = 59 mm, inner diameter = 53 mm, and length of the pipe = 3 m.

A PCI-9812 data acquisition card is assembled in the computer in the experimental system to collect the ultrasonic wave signal and convert it into a digital signal. The data acquisition card has 4-way independent *A*/*D* channels, a 12-bit *A*/*D* resolution ratio, and a synchronous sampling rate of up to 20 MS/s, which is good enough to meet the requirements of the acquisition and conversion of the ultrasonic guided wave signal. The PCI-9812 data acquisition card has very high flexibility, and it can be used either in the circumstance that the ultrasonic guided wave is taken as the trigger signal or in the circumstance where the external digital signal is taken as the trigger signal. Data acquisition and analysis software is also installed in the computer. Its main interface is shown in Fig. 7.17. Using this software, we can set the trigger type, trigger level, sampling rate, and sampling point, and the collected data can also be processed and analyzed by the software, such as for band-pass filtering, times averaging, and calculating the envelope.

The ultrasonic guided wave has multimode characteristics. Multiple modes can exist at a certain frequency point. It is the basic precondition of the pipeline detection to verify the correctness of the excited guided wave mode. In practice,

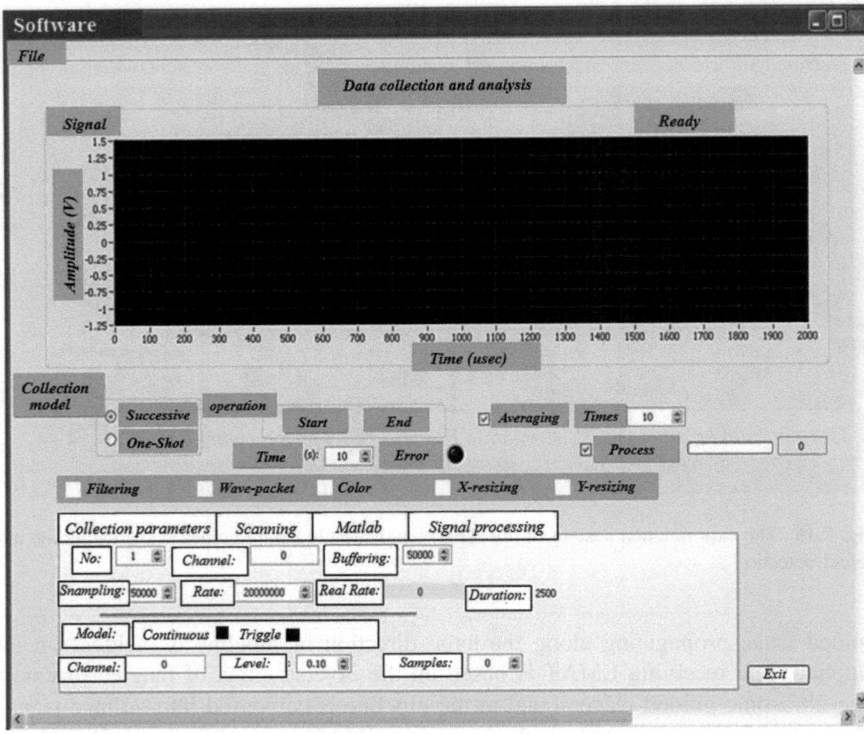

Fig. 7.17 The main interface of the EMAT data collection and analysis software

comparison of the measured wave velocity and theoretical wave velocity is usually used to verify the excited guided wave mode.

When we verify the longitudinal guided wave mode, the experimental setup shown in Fig. 7.18 is used. The distance between the central position of the *excitation* EMAT and the left margin of the pipeline is 0.5 m. Copper varnished wire with a diameter of 1 mm is used in these coils. The wire is divided into 2 layers, each layer having 25 turns, and the total width is 25 mm. The distance between the central position of the *receiving* EMAT and the left margin of the pipeline is 2.0 m. Copper varnished wire with a diameter of 0.31 mm is used as these coils. This wire is divided into 4 layers, each layer having 70 turns, and the total width is 21.7 mm.

The excitation voltage amplitude of the excitation EMAT is set at 50 V, the excitation frequency is 100 kHz, the cycle number is 6, and the magnetizing current of the electromagnet is 6 A. A computer is used to collect and filter the voltage signal detected by the receiving EMAT, and the obtained longitudinal guided wave signal is shown in Fig. 7.19.

After the generated signal is incident on the excitation EMAT, the longitudinal guided wave excited by the EMAT will propagate to left and right directions. The guided wave propagation path is shown in Fig. 7.18, and the temporal relation of each signal is shown in Fig. 7.19. It is known that *signal* (1), *signal* (3), and *signal*

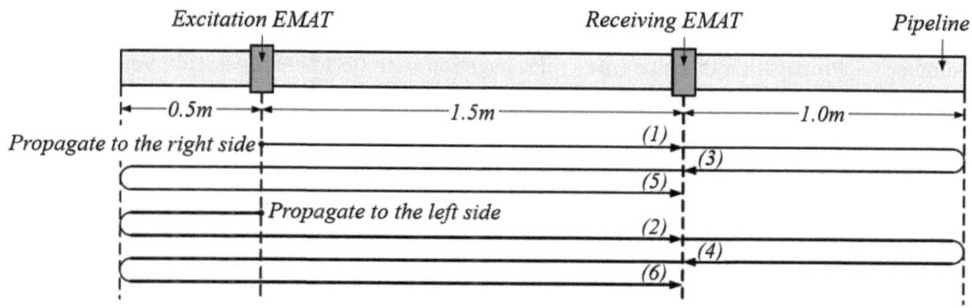

Fig. 7.18 The experimental setup for the mode verification of the longitudinal guided wave

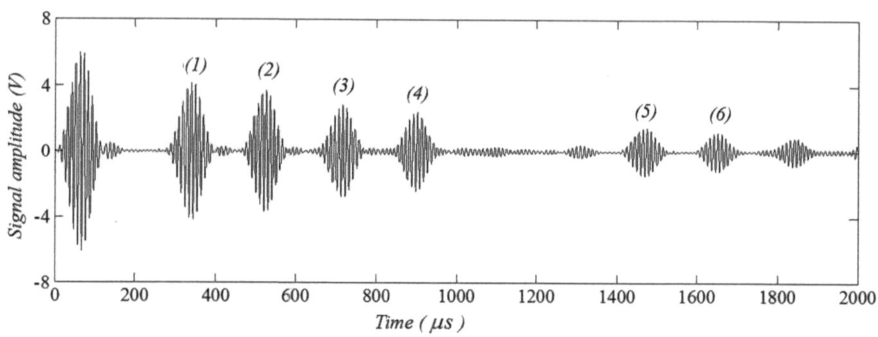

Fig. 7.19 The longitudinal guided wave signal

(5) are the direct signal propagating to the right side and the reflection wave signals from the right margin and the left margin, respectively. *Signal* (2), *signal* (4), and *signal* (6) are the first reflection wave signal propagating to the left side, the reflection wave signal from the right margin, and the second reflection wave signal from the left margin, respectively. The unnumbered signals on the left side of *signal* (1) in Fig. 7.19 are the initial excitation signals sensed by the receiving EMAT through the space coupling, when the excitation signal is applied on the excitation EMAT.

If *signal* (1) is taken as the reference, the difference in distance between other signals and *signal* (1) can be obtained from Fig. 7.18. The delay of each signal compared with *signal* (1) can be obtained from Fig. 7.19, and then, the propagation velocity of the excited longitudinal guided wave can be calculated, with the results shown in Table 7.3.

From the group velocity dispersion curve of the longitudinal guided waves in the pipeline, at the frequency of 100 kHz, the theoretical value of the $L(0,2)$ guided wave group velocity is 5349 m/s. As shown in Table 7.3, the average value of the longitudinal guided wave propagation velocity is 5365 m/s. The error between the theoretical value and experimental value of the $L(0,2)$ mode guided wave group

Table 7.3 Experimental results of the longitudinal guided wave propagation velocity

Number	Propagation distance (m)	Propagation time (μs)	Propagation velocity (m/s)
(2)	1.0	183	5464
(3)	2.0	375	5333
(4)	3.0	559	5366
(5)	6.0	1128	5319
(6)	7.0	1310	5344

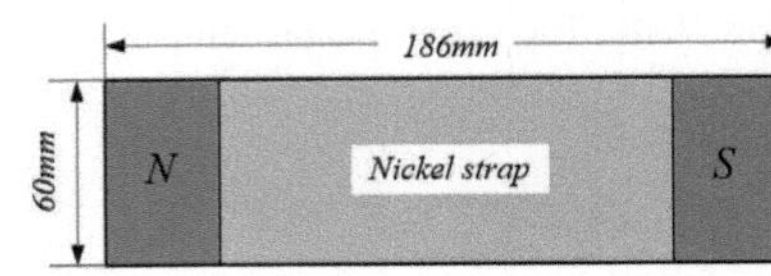

Fig. 7.20 The premagnetization of the nickel strap

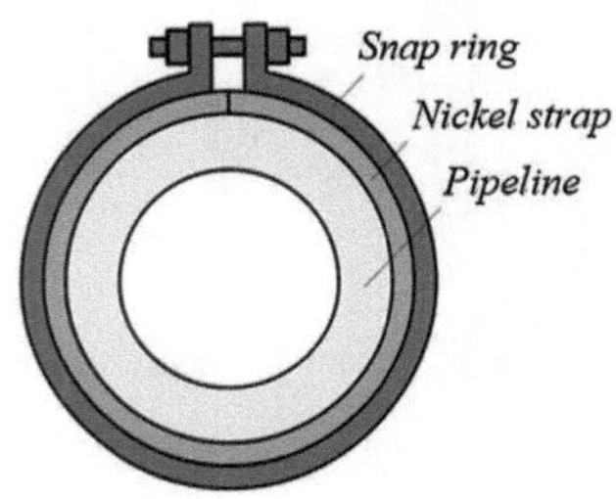

Fig. 7.21 The installation and fixation of the nickel strap

velocity is 0.3 %. It can therefore be verified that the excited longitudinal guided wave is in the L(0,2) mode.

In the circumstance of the torsional mode guided wave EMAT, the nickel strap should be premagnetized first to generate the static magnetic field, whose direction is along the length direction of the EMAT. The method is shown in Fig. 7.20. The length of the nickel strap is 186 mm, its width is 60 mm, and its thickness is 0.2 mm. N and S represent two permanent magnets with opposite polarity. After the premagnetization, the nickel strap is wrapped around the outside surface of the pipeline along the circumferential direction and fixed in place by a snap ring, so it has tight contact with the pipeline surface. The method is shown in Fig. 7.21.

The setup for the verification of the torsional mode guided wave by the excitation EMAT and receiving EMAT is shown in Fig. 7.22. The distance from the center position of the excitation EMAT to the left margin is 1.0 m. Copper varnished wire with a diameter of 1 mm is used to make the coils. The wire is divided into 4 layers, each layer having 35 turns, and the total width is 12.4 mm.

It is assumed that the excitation voltage amplitude of the excitation EMAT is 50 V, the excitation frequency is 100 kHz, and the cycle number is 6. A computer is used to collect and filter the voltage signal detected by the receiving EMAT, and the torsional mode signal is shown in Fig. 7.23.

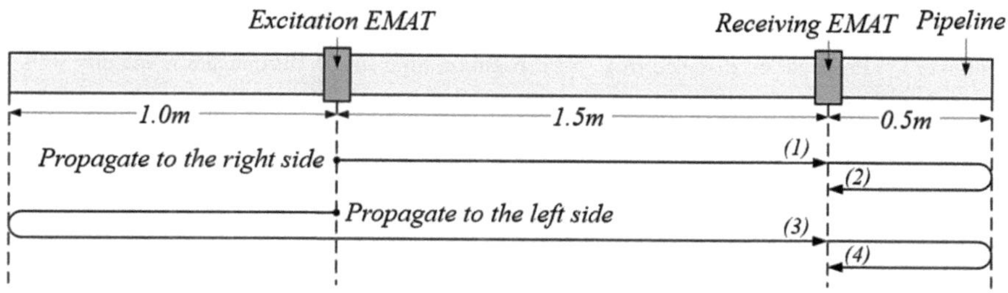

Fig. 7.22 The experimental setup for the verification of the torsional mode guided wave

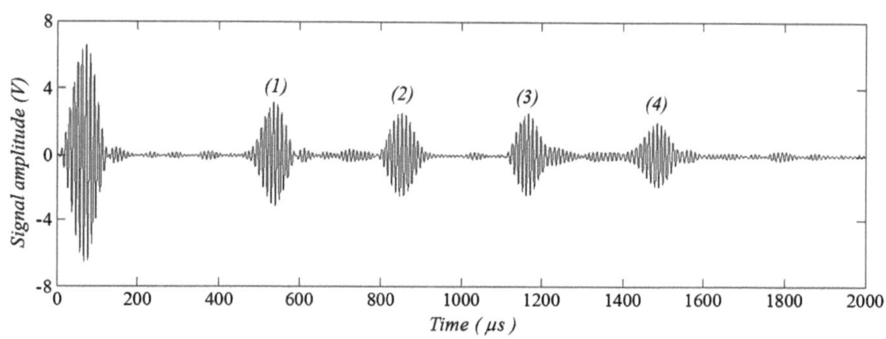

Fig. 7.23 The torsional mode guided wave signal

After the excitation signal is supplied to the excitation EMAT, the torsional mode guided waves excited by the EMAT are propagating to the left and right sides, respectively. The guided wave propagation path is shown in Fig. 7.22, and the precedence relationship of each signal is shown in Fig. 7.23. It is known that *signal* (1) and *signal* (2) are the direct signal propagating to the right side and the reflection signal from the right margin, respectively. *Signal* (3) and *signal* (4) are the reflection signal from the left margin propagating to the left side and the reflection signal from the right margin, respectively.

If *signal* (1) is taken as the reference, the difference in the propagation distance between each signal and *signal* (1) can be obtained from Fig. 7.22 and the retardation time of each signal relative to *signal* (1) can be obtained from Fig. 7.23. Then, the propagation velocity of the excited torsional mode guided wave can be calculated. The results are shown in Table 7.4.

Based on the group velocity dispersion curve of the torsional guided wave in the pipe at a frequency of 100 kHz, the theoretical value of the *T*(0,1) mode group velocity is 3200 m/s. As shown in Table 7.4, the average value of the propagation velocity of the torsional guided wave is 3181 m/s. The error of the *T*(0,1) mode guided wave group velocity is 0.6 %. Therefore, it can be verified that the mode of the excited torsional guided wave is the *T*(0,1) mode.

Table 7.4 Experimental results of the torsional guided wave propagation velocity

Number	Propagation distance (m)	Propagation time (μs)	Propagation velocity (m/s)
(2)	1.0	314	3185
(3)	2.0	627	3190
(4)	3.0	947	3168

When the electromagnetic ultrasonic technology is used to detect the pipeline, the excitation signals supplied to the EMAT are usually the high-frequency or pulse signals that contain more than one cycle. If the amplitude and frequency of the high-frequency signal remain unchanged, the energy from the excitation source to the EMAT is decided by its cycle number, and thus, the energy of the ultrasonic mechanical vibration inside the pipeline is also decided by the cycle number.

If the single cluster coil longitudinal guided wave EMAT is used, the voltage amplitude of the excitation EMAT is set at 20 V, the excitation frequency is 100 kHz, the magnetizing current of the electromagnet is 5 A, and the excitation signal cycle number N is adjusted to conduct the experiment. A multicluster coil longitudinal guided wave EMAT is used. Copper varnished wire with a diameter of 1 mm is used as the excitation EMAT coil, which contains 4 clusters, each cluster having 2 layers and each layer having 7 turns of wire. The space between the adjacent coil clusters is 27 mm. Copper varnished wire with a diameter of 0.31 mm is used as the receiving EMAT coil, which contains 4 clusters, each cluster having 2 layers and each layer having 15 *turns* of wire. The space between the adjacent coil clusters is 27 mm. The electric current amplitude of the excitation EMAT is set at 16 A, the excitation frequency is 10 kHz, the magnetizing current of the electromagnet is 7 A, and the excitation signal cycle number N is adjusted to conduct the experiment.

When the periodicity of the excitation signal is 4, the amplitude of the direct signal is taken as the denominator and the amplitudes of the direct signals under the excitation signals with other periodicities are taken as the numerator. Then, we obtain the curve of the relative value of the direct signal amplitude with the change of excitation signal cycle number, as shown in Fig. 7.24.

It can be seen that whether the single cluster coil EMAT or the multicluster coil EMAT is used, the amplitude of the guided wave will be improved with increase in the excitation signal cycle number, but there is a saturation effect. When the cycle number of the excitation signal increases from 4 to 8, it is found that the increasing effect of the amplitude will become weak through the increase in the excitation signal periodicity. Because the distance between the adjacent coil clusters in the multicluster coil has a matching relationship with the wavelength of the excited guided wave, which means the particle vibration of the adjacent clusters has constructive interference, and there is no corresponding relationship between the single cluster structure and the wavelength of the excited guided wave. The effect on the amplitude does not change significantly, as is the case for the multicluster structure coils.

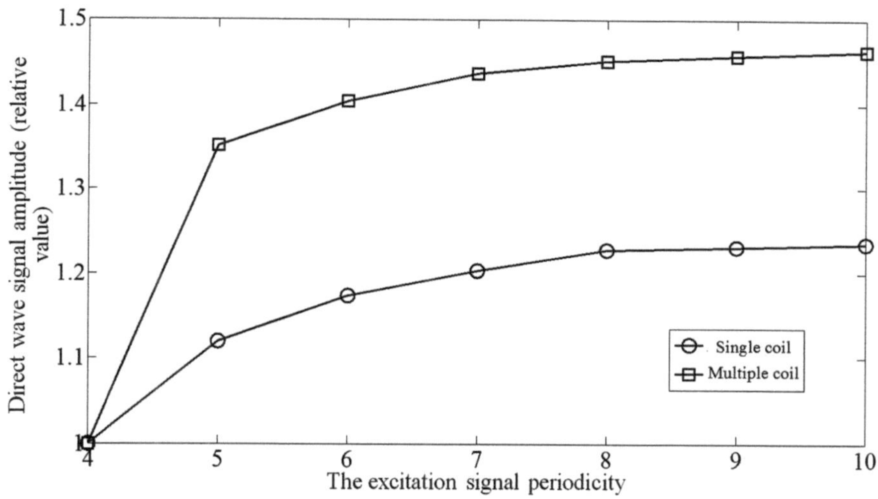

Fig. 7.24 The curves of the relative value of the direct signal amplitude with the change of excitation signal cycle number

In addition, as the excitation signal cycle number increases, the time width of the guided wave signal will also increase accordingly and the spatial resolution will decrease. So, when we select the excitation signal cycle number, its influence on the guided wave signal amplitude and spatial resolution should be considered comprehensively to achieve a good detection effect.

The intensity of the bias magnetic field can also affect the receiving signal. The peak excitation current of the excitation EMAT coil is set at 8 A and 14 A, respectively, adjusting the magnetizing current of the electromagnet. The data of the signal detected by the receiving EMAT are shown in Fig. 7.25.

It can be seen that before the magnetizing current of the electromagnet reaches 15 A, the direct signal amplitude will increase linearly with the increasing magnetizing current of the electromagnet. When the magnetizing current of the electromagnet is higher than 15 A, the direct signal amplitude will tend to be a stable value, and it does not increase with increase in the magnetizing current of the electromagnet. The finite element simulation shows that when the magnetizing current of the electromagnet is 15 A, the range of magnetic induction intensity is 1.6–2.0 T. At this time, the tested pipeline has already entered the saturation magnetization state. Thus, in practice, the EMAT DC bias magnetic field should be set at the state that can bring the pipeline close to the saturation magnetization state, so that a higher conversion efficiency between the electric energy and acoustic energy can be achieved.

In order to improve the detection ability of EMAT along the pipeline axial direction, an artificial defect is made on the outer wall of the tested pipeline, and its shape is shown in Fig. 7.26. The length of the artificial defect along the pipeline

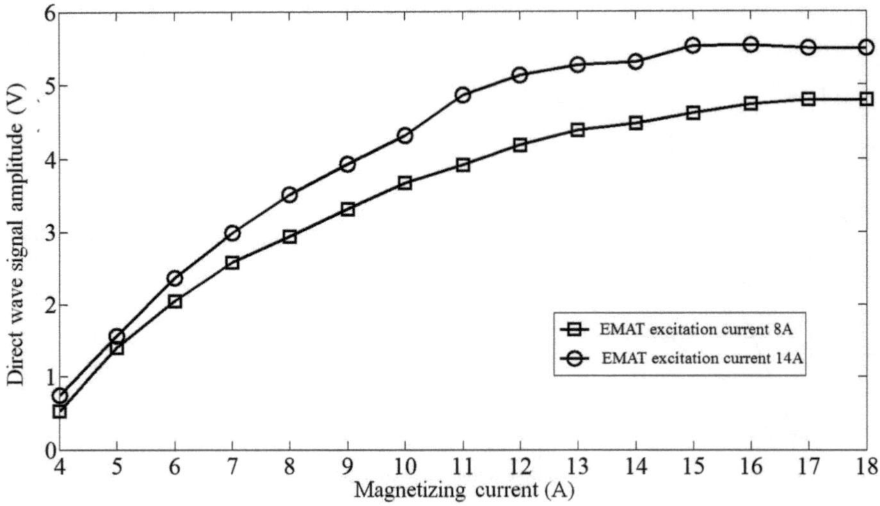

Fig. 7.25 The direct signal amplitude under different magnetizing current of the electromagnet

Fig. 7.26 A sketch of the artificial defect in the experimental pipeline

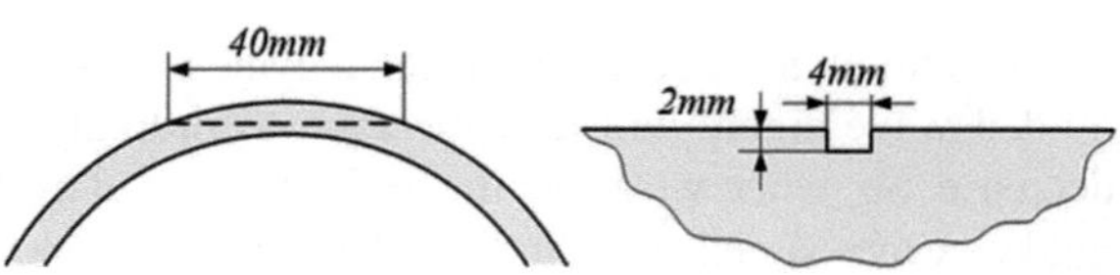

circumferential direction is 40 mm, and its width along the pipeline axial direction is 4 mm. Its maximum depth along the pipeline radial direction is 2 mm.

When the *L*(0,2) guided wave mode is used to detect the defect in the pipe, the experimental arrangement of the excitation EMAT and receiving EMAT is as shown in Fig. 7.27. The distance from the center position of the excitation EMAT to the left margin is 1.745 m. Copper varnished wire with a diameter of 1 mm is used to enwind the coil, which contains 8 clusters, each cluster having 2 layers and each layer having 3 turns of wire. The distance between the adjacent coil clusters is 10.8 mm. The distance from the center position of the receiving EMAT to the left margin is 2.4 m. Copper varnished wire with a diameter of 0.31 mm is used to enwind the coil, which contains 8 clusters, each cluster having 2 layers and each layer having 10 *turns* of wire. The distance between the adjacent coil clusters is 2.75 mm.

The excitation voltage of the excitation EMAT is set at 350 V, the excitation frequency is 250 kHz, the cycle number is 12, and the magnetizing current of the electromagnet is 15 A. The *L*(0,2) mode guided wave signal detected by the receiving EMAT is shown in Fig. 7.28.

After the excitation signal is supplied to the excitation EMAT, the *L*(0,2) guided wave modes excited by the EMAT are propagating to the left and right sides. The guided wave propagation path is shown in Fig. 7.27, and the precedence

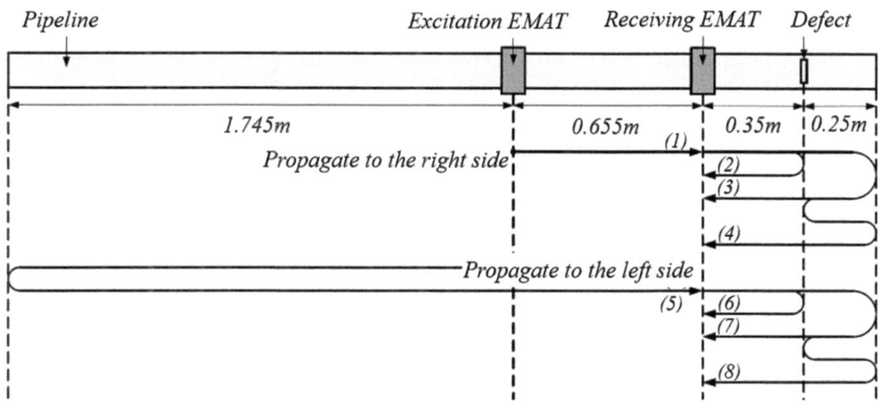

Fig. 7.27 The experimental setup of the defect detection in the pipe by the $L(0,2)$ guided wave mode

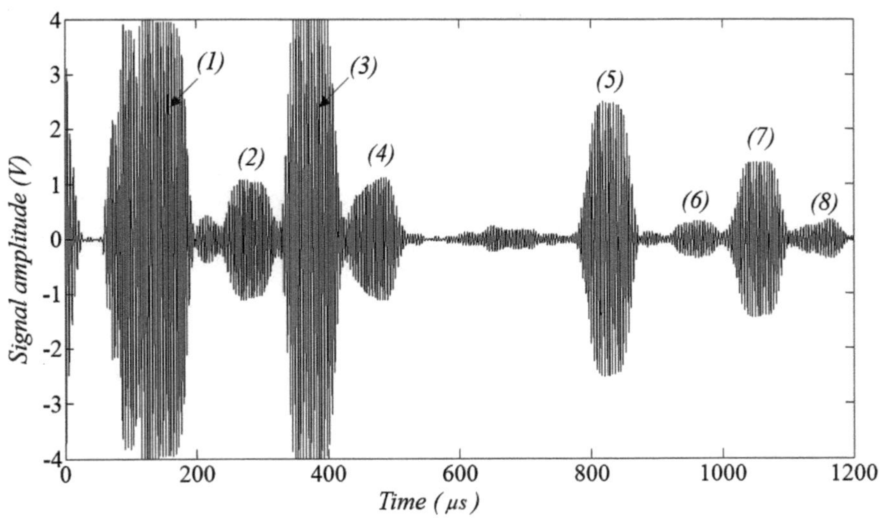

Fig. 7.28 The $L(0,2)$ mode guided wave signal in the pipeline with defect

relationship of each signal is shown in Fig. 7.28. This shows that *signal* (1), *signal* (2), *signal* (3), and *signal* (4) are the direct signal propagating to the right side, the reflection signal from the defect, and the first and second reflection signals from the right margin, respectively; *signal* (5), *signal* (6), *signal* (7), and *signal* (8) are the reflection signal from the left margin propagating to the left side, the reflection signal from the defect, and the first and second reflection signals from the right margin, respectively. It must be noted that since the artificial defect is an axisymmetric defect, when the reflection of the $L(0,2)$ mode guided wave meets the

Table 7.5 The results of the *L*(0,2) guided wave mode propagation velocity

Number	Propagation distance (m)	Propagation time (μs)	Propagation velocity (m/s)
(3)	1.20	230	5217
(5)	3.49	670	5209
(7)	4.69	900	5211

artificial defect, it can also generate the *L*(0,1) guided wave mode and torsion mode of the guided wave.

If *signal* (1) is taken as the reference, the differences in distance between each edge reflection wave and *signal* (1) can be obtained from Fig. 7.27 and the retardation time of each edge reflection wave signal relative to *signal* (1) can be obtained from Fig. 7.28, and then, the propagation velocity of the *L*(0,2) mode guided wave can be calculated. The results are listed as follows.

As shown in Table 7.5, the average value of the *L*(0,2) mode propagation velocity is 5212 m/s. It can be obtained from Fig. 7.28 that the retardation time of the reflection wave *signal* (2) relative to the direct *signal* (1) is 135 μs. The difference in propagation distance between the defect echo *signal* (2) and defect echo *signal* (1) is given in (7.5).

$$L = 5212 \times 135 \times 10^{-6} = 0.704 \text{ m} \tag{7.5}$$

So, the distance between the defect and the receiving EMAT is given in (7.6).

$$d = \frac{L}{2} = 0.352 \text{ m} \tag{7.6}$$

The actual distance between the artificial defect and the receiving EMAT is 0.35 m. Using the *L*(0,2) mode, the distance from the defect to the receiving EMAT is 0.352 m and the relative error is 0.6 %. It can be seen that the artificial defect on the outside surface of the pipeline can be detected effectively using the *L*(0,2) guided wave mode with a higher accuracy of defect location.

When the *T*(0,1) guided wave mode is used for defect detection in the pipeline, the setup of the excitation EMAT and receiving EMAT is as shown in Fig. 7.29. The distance between the center position of the excitation EMAT and the left margin of the pipeline is 1.415 m. Copper varnished wire with a diameter of 1 mm is used to enwind the coil, which contains 6 clusters. Each cluster has 2 layers, and each layer has 3 turns of wire. The distance between adjacent coil clusters is 8.4 mm, and the distance from the center position of the artificial defect to the left margin is 2.75 m.

The excitation voltage amplitude of the excitation EMAT is set at 350 V, the excitation frequency is 190 kHz, and the periodicity is 10. The *T*(0,1) mode guided wave detected by the receiving EMAT is shown in Fig. 7.30.

After the excitation signal is supplied to the excitation EMAT, the *T*(0,1) mode guided wave excited by the EMAT propagates to the left and right sides.

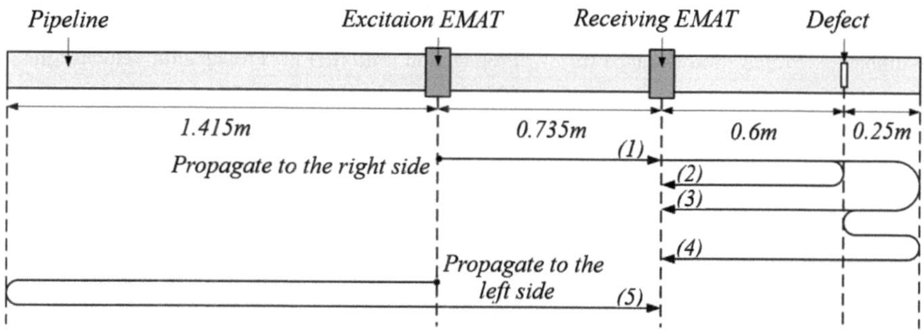

Fig. 7.29 The experimental setup for defect detection in the pipe by the *T*(0,1) mode

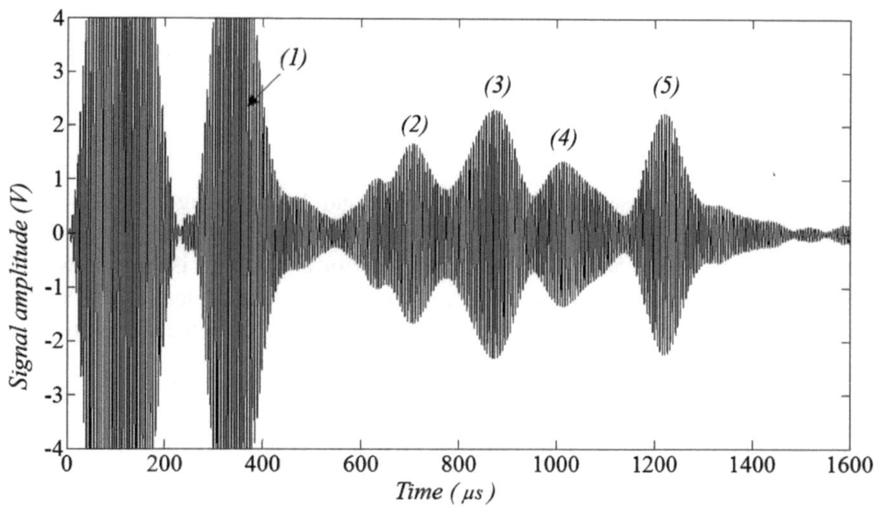

Fig. 7.30 The *T*(0,1) mode guided wave signal in the pipeline with defect

The guided wave propagation path is shown in Fig. 7.29. The precedence relationship of each signal shows that *signal* (1), *signal* (2), *signal* (3), and *signal* (4) are the direct signal propagating to the right side, the reflection signal from the defect, and the first and second reflection signals from the right margin, respectively. *Signal* (5) is the reflection signal from the left margin propagating to the left side. Similar to the *L*(0,2) guided wave mode, when the reflection of the *T*(0,1) mode guided wave is at the non-axisymmetry artificial defect, it can also generate the torsional mode.

If *signal* (1) is taken as the reference, the difference between the propagation distance of each edge reflection signal and that of *signal* (1) can be obtained from Fig. 7.29, and the retardation time of each edge reflection signal relative to *signal* (1) can be obtained from Fig. 7.30. The propagation velocity of the *T*(0,1) mode guided wave can then be calculated. The results are shown in Table 7.6.

Table 7.6 The results of the $T(0,1)$ guided wave mode propagation velocity

Number	Propagation distance (m)	Propagation time (μs)	Propagation velocity (m/s)
(3)	1.70	536	3172
(5)	2.83	885	3198

As shown in Table 7.6, the average value of the $T(0,1)$ mode propagation velocity is 3185 m/s. It can be seen from Fig. 7.30 that the retardation time of the defect reflection *signal* (2) relative to the direct *signal* (1) is 370 μs. The difference in propagation distance between the defect echo *signal* (2) and the direct *signal* (1) is given in (7.7).

$$L = 3185 \times 370 \times 10^{-6} = 1.178 \text{ m} \tag{7.7}$$

So, the distance between the defect and the receiving EMAT is given in (7.8).

$$d = \frac{L}{2} = 0.589 \text{ m} \tag{7.8}$$

The real distance between the artificial defect and the receiving EMAT is 0.6 m, the distance between the receiving EMAT and the defect detected by the $T(0,1)$ mode guided wave is 0.589 m, and the relative error is 1.8 %. It is thus shown that the artificial defect on the outside surface of the pipeline can be detected effectively by the $T(0,1)$ mode guided wave, with a higher accuracy of defect location.

7.3 Electromagnetic Ultrasonic Guided Wave Detection for Cracks in the Natural Gas Pipeline [5–7]

7.3.1 The Main Structure of the Detector

The detector includes three sections, each section containing three groups of detecting probes arranged evenly along the circumferential direction of the pipeline. The detecting probes of the adjacent sections are arranged with a staggered angle of 30°. Each group of detecting probes contains three electromagnetic ultrasonic transducers of different types: The circumferential guided wave electromagnetic ultrasonic probe is used to detect stress corrosion cracks; the surface wave electromagnetic ultrasonic probe is used to detect inner surface cracks and distinguish between cracks on the inside and outside walls of the pipeline; and the vertical incident wave electromagnetic ultrasonic probe is used to detect pipe thickness and identify the circumferential weld for the accurate location and quantification of the defect. The first section of the detector is equipped with a seal cup and a supporting wheel; the second section is equipped with a seal cup and an internal capsule equipped with a battery; and the third section is equipped with a seal cup and an

Fig. 7.31 A structure diagram of the electromagnetic ultrasonic detector

internal capsule equipped with an electronic information recorder. These three sections are connected by a cardan joint, as shown in Fig. 7.31.

In order to avoid cross talk between channels and improve the excitation power of the electromagnetic ultrasonic probe, an independent physical channel is used in the excitation circuit and signal detection circuit of the 27 probes of the detector, each probe having a dual role of excitation and receiving of the ultrasonic signals.

The output pulse signal is generated by the signal-generating unit and passed through the excitation coil of the excitation probe after the power amplification. Under the excitation of the pulse current, the ultrasonic wave is excited by the excitation coil. The electromagnetic ultrasonic signal is received by the receiving coil and then saved to the electronic hard disk after the amplification, smoothing, data collection, and data compression. After detection, the saved data are transferred onto a computer and the results analyzed using private data analysis software.

The pulsed power supply is composed of a signal-generating unit and a power amplification unit. The phase-shift full-bridge control chip UCC2895 is used in the signal-generating unit. Its characteristics are as follows: The output open time delay and self-adaption time delay can be set through programming; it can be used either in voltage mode or in current mode; it has a soft close function; its maximum operating frequency is 1 MHz; and the complementary square wave signal is output from two base pins OUTA and OUTB, the duty ratio being adjustable from 0 to 100 %. Because the working frequency range of the EMAT is usually from 100 kHz to 1 MHz, the UCC2895 can meet the requirements of the probe relative to the pulsed power supply.

The circuit connection diagram of the signal-generating unit is shown in Fig. 7.32.

The signal cycle is determined by the capacitance $C3$ and the resistance $R1$. The cycle satisfies (7.9).

$$t_{\mathrm{OSC}} = \frac{5 \times R1 \times C3}{48} + 120\ \mathrm{ns} \tag{7.9}$$

The duty ratio of the square signal is determined by the resistance $R2$, $R3$, and $R4$. In order to generate a group of pulse signals, the *UCC2895* should be controlled. The chip has the control pins SS/DISB. As long as the pins are set to a low level, the *OUTA* and *OUTB* will output low-level voltage. A single chip is used to output the high low-level current regularly to the SS/DISB so that the *UCC2895* chip can be controlled to output the square wave pulsed periodically. The pulsed square wave is amplified by the power amplification unit, and it becomes a large

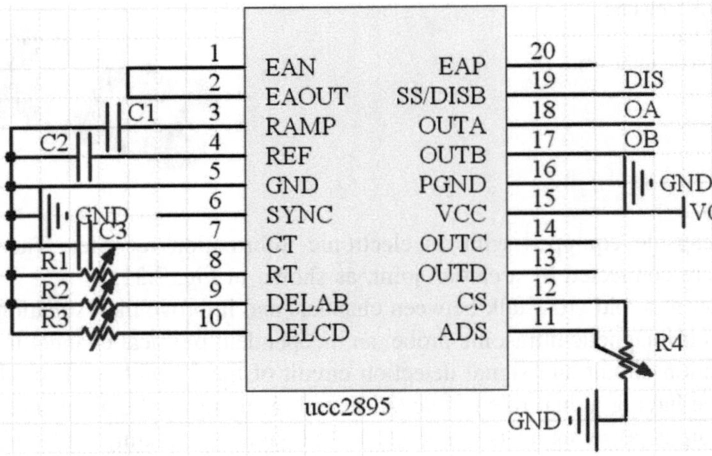

Fig. 7.32 The circuit connection diagram of *UCC2895*

current pulse. In the power amplification unit, the MOSFET is used as the power device, designed by a full-bridge circuit structure. MOSFET works in the switching state; the consumed power itself is very small, which causes the efficiency of the pulsed power supply to increase.

In order to improve the energy conversion efficiency of the electromagnetic ultrasonic probe, the capacitors in the parallel matching method are used for the receiving coil, so that the signal amplitude can be improved. Because the received signal is at a microvolt level, a perfect shielding and ground connection are needed, and a weak signal amplifier with high precision and low noise also needs to be designed.

AD797 has the characteristics of ultralow noise and low distortion. The bandwidth can reach up to 110 MHz and the response speed is 20 V/μs, which is suitable for preamplification. A *LT1568* chip is used in the design of the band-pass filter to clear the noise signal. Through the amplification of the main amplifier, the gain can reach up to 80 dB.

The 27-channel ADC is used to complete the parallel sampling of each channel. FPGA is used as the core of the acquisition system, in charge of the ADC parallel sampling, data flow organization, and data compression storage. Its functional block diagram is shown in Fig. 7.33.

The compression mainly contains of two parts: the wavelet transform and the self-adaptation arithmetic code. Its functional block diagram is shown in Fig. 7.34. Both the wavelet transform and self-adaptation arithmetic code can be realized through FPGA, with the main system established using the VHDL.

A hybrid programming method is adopted by using *Visual C* and *LabVIEW* to make the data analysis software. The software has a good human–computer interface and has a strong ability for data analysis and processing. The functions that can be achieved using this software are as follows: the automatic importing of

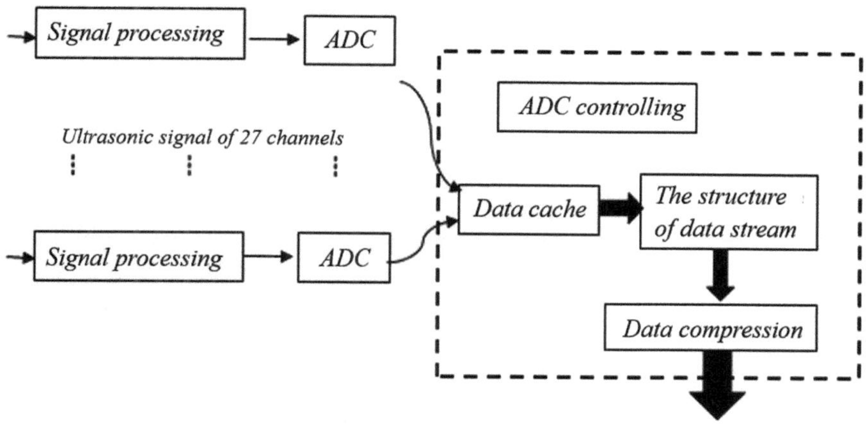

Fig. 7.33 The functional block diagram of the data acquisition and compression storage

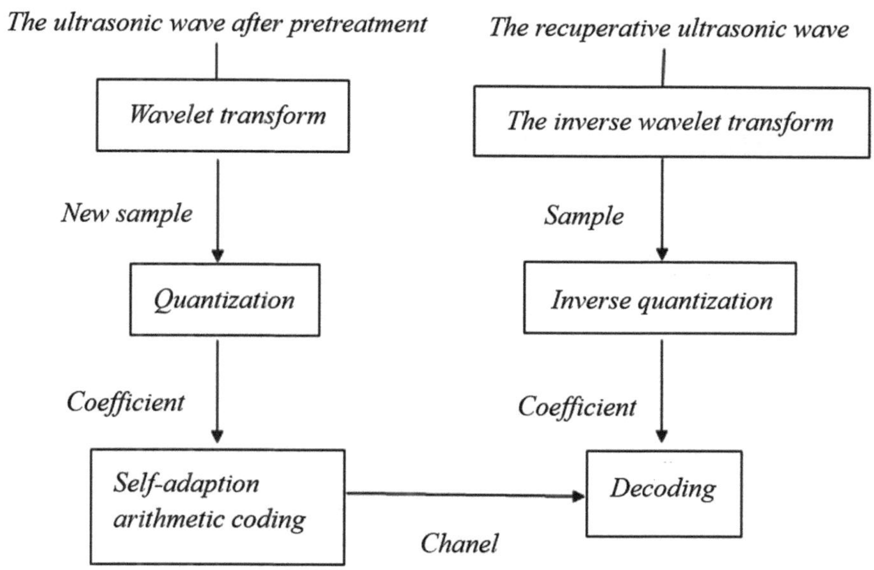

Fig. 7.34 The functional block diagram of the data compression

the data, displaying the image (*C-S* scan) of the planar scanning of the pipeline conditions and the waveform image of the crack defect, identifying the pipe welds automatically, quantifying and locating the crack defect, etc. For the test data, the alternate display between the spatial domain and time domain can be conducted conveniently. It also has functions such as smoothing, spectral analysis, automatic capture, and measure of the local peak value. A contrast display window is available in this software, which can enlarge the image locally. The signal can be

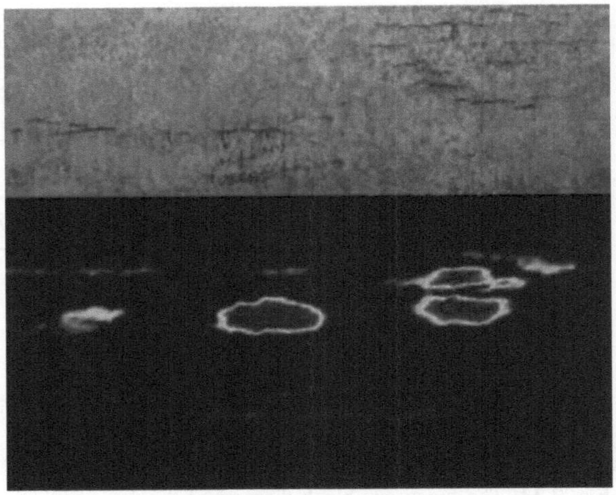

Fig. 7.35 Result comparison of magnetic particle inspection and electromagnetic ultrasonic detection of cracks in the pipeline

displayed according to the user needs: original signal, original signal envelope, and any kind, or all of, the filtered signals.

7.3.2 Relative Detection Experiment

The experiment of electromagnetic ultrasonic crack detection is conducted on the natural gas pipeline that has a stress corrosion crack. It shows that the crack with a minimum length of 20 mm and minimum depth of 1 mm can be detected effectively, and it has a resolving ability for crack clusters. The experimental result is shown in Fig. 7.35.

The hightest testing speed is 2 m/s and the longest detection distance is 100 km, and it can pass through the natural gas pipeline with 1.5D bend. Based on the development of the natural gas pipelines, higher pressure and higher throughput of the pipe will be used in the future. If we can improve the testing speed and single run distance of the inspector, the electromagnetic ultrasonic cracking detector will have better adaptability. With the development of electronic techniques and computer technology, online detection speed can be improved significantly if rapid acquisition and storage technology for high speeds and large volumes of data are introduced in electromagnetic ultrasonic inspectors for natural gas pipelines in the near future.

References

1. Xin, J.: Development of electromagnetic ultrasonic thickness measurement device. Tsinghua University, Beijing (2009). (In Chinese)
2. Li, P., Huang, S., Wang, S., et al.: Development of a pulsed exciting source for electromagnetic ultrasonic detection. Electr Measur & Instrum **49**(2), 76–79 (2012). (In Chinese)

3. Wang, S., Huang, S., Zhao, W.: Virtual instrument technology of electromagnetic ultrasonic data acquisition and analysis software design. Instrum Tech Sens **315**(5), 37–39 (2009). (In Chinese)
4. Zhang, Y., Huang, S., Zhao, W.: Development of the high frequency pulsed electrical source. Power Electron **42**(7), 49–50 (2008). (In Chinese)
5. Huang, S., Ye, C., Wang, S., et al.: Design of electromagnetic ultrasonic detector for the crack detection in natural gas pipe. Nondestr Test **31**(10), 827–829 (2009). (In Chinese)
6. Huang, S., Wang, S., Zhao, W., et al.: Study on oil and gas pipeline crack detection method based on EMAT-generated ultrasonic guided waves. In: The 3rd World Conference on Safety of Oil and Gas Industry (WCOGI 2010), pp. 491–493 (2010)
7. Huang, S., Wang, S., Zhao, W.: Study on oil and gas pipeline crack detection method based on EMAT-generated ultrasonic guided waves. Measur. Technol. **23**(S), 34–36 (2009) (In Chinese)

MIX
Papier aus verantwortungsvollen Quellen
Paper from responsible sources
FSC® C105338

If you have any concerns about our products,
you can contact us on
ProductSafety@springernature.com

In case Publisher is established outside the EU,
the EU authorized representative is:
Springer Nature Customer Service Center GmbH
Europaplatz 3, 69115 Heidelberg, Germany

Printed by Libri Plureos GmbH
in Hamburg, Germany